CHARACTERIZATION OF HIGH T_c MATERIALS AND DEVICES BY ELECTRON MICROSCOPY

This is is a clear and up-to-date account of the application of electron-based microscopies to the study of high T_c superconductors.

Written by leading experts, this compilation provides a comprehensive review of scanning electron microscopy, transmission electron microscopy and scanning transmission electron microscopy together with details of each technique and its applications. Introductory chapters cover the basics of high-resolution transmission electron microscopy, including a chapter devoted to specimen preparation techniques, and microanalysis by scanning transmission electron microscopy. Ensuing chapters examine identification of new superconducting compounds, imaging of superconducting properties by low-temperature scanning electron microscopy, imaging of vortices by electron holography and electronic structure determination by electron energy loss spectroscopy. The use of scanning tunneling microscopy for exploring surface morphology, growth processes and the mapping of superconducting carrier distributions is discussed. Final chapters consider applications of electron microscopy to the analysis of grain boundaries, thin films and device structures. Detailed references are included.

This text will be an indispensable reference for graduate students and researchers in materials science, physics and engineering.

CHARACTERIZATION OF HIGH T_c MATERIALS AND DEVICES BY ELECTRON MICROSCOPY

Edited by

NIGEL D. BROWNING
University of Illinois at Chicago

STEPHEN J. PENNYCOOK
Oak Ridge National Laboratory, Tennessee

CAMBRIDGE
UNIVERSITY PRESS

PUBLISHED BY THE PRESS SYNDICATE OF THE UNIVERSITY OF CAMBRIDGE
The Pitt Building, Trumpington Street, Cambridge, United Kingdom

CAMBRIDGE UNIVERSITY PRESS
The Edinburgh Building, Cambridge CB2 2RU, UK   http://www.cup.cam.ac.uk
40 West 20th Street, New York, NY 10011-4211, USA   http://www.cup.org
10 Stamford Road, Oakleigh, Melbourne 3166, Australia
Ruiz do Alarcón 13, 28014 Madrid, Spain

First published 2000

Printed in the United Kingdom at the University Press, Cambridge

Typeset in 11/14pt Times New Roman [KT]

A catalogue record for this book is available from the British Library

Library of Congress Cataloguing in Publication data
Characterization of high Tc materials and devices by electron
   microscopy / edited by Nigel D. Browning, Stephen J. Pennycook.
        p.   cm.
   ISBN 0 521 55490 X (hb)
      1. High temperature superconductors.   2. Electron microscopy–Technique.   I. Browning,
Nigel D.   II. Pennycook, Stephen J.
   QC611.98.H54C43   1999
   537.6′23′0284–dc21 99-18754 CIP

ISBN 0 521 55490 X hardback

Contents

Contributors

Lawrence M. Brown
University of Cambridge, Cavendish Laboratory, Madingley Road, Cambridge CB3 0HE, UK

Nigel D. Browning
Department of Physics, University of Illinois at Chicago, 845 West Taylor Street, Chicago, IL 60607-7059, USA

Matthew F. Chisholm
Solid State Division, Oak Ridge National Laboratory, 1 Bethel Valley Rd, Oak Ridge, TN 37831-6030, USA

Vinayak P. Dravid
Department of Materials Science & Engineering, Northwestern University, 2225 N. Campus Drive, Evanston, IL 60208, USA

Yufei Gao
Materials Science Division, Argonne National Laboratory, 9700 South Cass Avenue, Argonne, IL 60439, USA. Present address: Interfacial and Processing Sciences, Pacific Northwest National Laboratory, 3335 Q Ave, MS K8-93, Richland, WA 99352, USA

Marilyn E. Hawley
Center for Materials Science, Los Alamos National Laboratory, MS K-765, Los Alamos, NM 87545, USA

Lianlong He
National Institute for Research on Inorganic Materials, Namiki 1-1, Tsukuba, Ibaraki 305, Japan

Shigeo Horiuchi
National Institute for Research on Inorganic Materials, Namiki 1-1, Tsukuba, Ibaraki 305, Japan

Rudolph P. Huebener
Physikalisches Institut, Lehrstruhl Experimentalphysik II, Universität Tübingen, W-7400 Tübingen, Germany

Chunlin L. Jia
Institut für Festkörperforschung, Forschungszentrum Jülich GmbH, D-52425 Jülich, Germany

Thomas Krekels
University of Antwerp, RUCA, Groenenborgerlaan 171, B-2020 Antwerp, Belgium

Ann F. Marshall
Center for Materials Research, McCullough Building, Stanford University, Stanford, CA 94305-4045, USA

Karl L. Merkle
Materials Science Division, Argonne National Laboratory, 9700 South Cass Avenue, Argonne, IL 60439, USA

Eva Olsson
Department of Physics, Chalmers University of Technology, S-412 96 Göteborg, Sweden. Present address: The Ångström Laboratory, Uppsala University, Box 534, Uppsala, SE-751 21, Sweden

Stephen J. Pennycook
Solid State Division, Oak Ridge National Laboratory, 1 Bethel Valley Rd, Oak Ridge TN 37831-6030, USA

Akira Tonomura
Advanced Research Laboratory, Hitachi Ltd, Hatoyama, Saitama 350-03, Japan

Knut Urban
Institut für Festkörperforschung, Forschungszentrum Jülich GmbH, D-52425, Jülich, Germany

Gustaaf Van Tendeloo
University of Antwerp, RUCA, Groenenborgerlaan 171, B-2020 Antwerp, Belgium

Boris V. Vuchic
Materials Science Division, Argonne National Laboratory, 9700 South Cass Avenue, Argonne, IL 60439, USA

Yun-Yu Wang
Northwestern University, Department of MS & E, 2225 Sheridan Road, 3013A MLSF, Evanston, IL 60208, USA

Jian-Guo Wen

Superconductivity Research Laboratory, International Superconductivity Technology Center, 1-10-13 Shinonome, Koto-ku, Tokyo 135-0062, Japan

Jun Yuan

University of Cambridge, Cavendish Laboratory, Madingley Road, Cambridge CB3 0HE, UK

Yimei Zhu

Department of Applied Science, Brookhaven National Laboratory, Upton, Long Island, NY 11973, USA

Preface

Discovered just over a hundred years ago, the ubiquitous electron now forms the basis for a remarkably large range of characterization tools. Surface roughness and morphology, local atomic and electronic structure, vortex motion and superconducting properties can all be imaged thanks to the electron. Being light in mass, samples withstand appreciable irradiation without destruction. Carrying a charge, electrons can be accelerated to high energies and focussed to form transmission images or fine probes, which enables the interior of bulk samples or thin films to be investigated. Electrons may be scattered elastically to provide images of defects and interfaces at atomic resolution, or inelastically, facilitating spectroscopic studies of electronic structure in the vicinity of individual defects or interfaces. Low energy electrons, guided by a metal probe, form the basis for scanning tunneling microscopy, revealing insights into the atomic and electronic structure of surfaces.

This book presents the entire range of electron-based microscopies as applied to high T_c superconductors, scanning electron microscopy, transmission electron microscopy and scanning tunneling microscopy. Introductory chapters cover the basics of high-resolution transmission electron microscopy and microanalysis by scanning transmission electron microscopy. One chapter deals in detail with the difficult procedures of specimen preparation. Other chapters deal with imaging techniques specific to superconductors, the imaging of vortices by electron holography and the mapping of weak links by low temperature scanning electron microscopy. Several chapters deal with specific applications to subjects such as grain boundaries, thin films and device structures. We hope that by covering the techniques from an introductory level to a detailed description of specific methods, from an applications perspective as well as fundamental research interests, that this book will be of value to all groups involved in high T_c superconductivity.

Grateful thanks are due to all contributors for their patience during the production of this book, and especially to Sharon Jesson for final editing.

NDB and SJP

1

High-resolution transmission electron microscopy

S. HORIUCHI and L. HE

1.1 Introduction

High-resolution transmission electron microscopy (HRTEM) has been widely and effectively used for analyzing crystal structures and lattice imperfections in various kinds of advanced materials on an atomic scale. This is especially the case for high T_c superconductors (HTSCs). The most characteristic feature in crystal structures of HTSCs is that there is a common structural element, a CuO_2 plane, in which superconductive carriers (positive holes or electrons) are transported. The remaining part, sandwiching the CuO_2 planes, accommodates additional oxygen atoms or lattice defects to provide carriers to the CuO_2 planes. This is known as the charge reservoir. The transition temperature between superconductive and non-superconductive states, T_c, strongly depends on the concentration of carriers in CuO_2 planes and the number of CuO_2 planes. Any charge reservoir is composed of some structural elements, including lattice defects. An aim of HRTEM is to clarify the structure of the charge reservoirs. Additionally, a variety of microstructures strongly affect the critical current density, J_c, since they closely relate to the weak link at boundaries between superconductive grains as well as to the pinning of magnetic fluxoids. The characterization of point defects, dislocations, stacking faults, precipitates, grain boundaries, interfaces and surface structures is another important aim of HRTEM. In this chapter, we describe some fundamental issues in analyzing crystal structures and microstructures in HTSCs by HRTEM.

1.2 Theoretical background for HRTEM

HRTEM images closely depend not only on some optical factors in the imaging process by the electron lens, but also on a scattering process of the electrons incident on the crystal specimen [1.1]. This section describes the electron-optical background for HRTEM.

1.2.1 Phase contrast

Let us begin with a simple case where a central beam and one diffracted beam pass through the objective aperture of an electron microscope. Both beams starting from a site in the bottom surface of a thin specimen meet again at the image plane to form an image. With their contributions at the site x_i on the image plane, $\Psi_o(x_i)$ and $\Psi_g(x_i)$, the amplitude and intensity of the resultant wave, $\Psi(x_i)$ and $I(x_i)$, can be expressed as

$$\Psi(x_i) = \Psi_o(x_i) + \Psi_g(x_i)$$

$$I(x_i) = \Psi(x_i)\Psi^*(x_i)$$

$$= I_o(x_i) + I_g(x_i) + 2\text{Re}[\Psi_o(x_i)\Psi_g^*(x_i)], \quad (1.1)$$

where $I_o(x_i) = |\Psi_o(x_i)|^2$ and $I_g(x_i) = |\Psi_g(x_i)|^2$. Re means that only the real part in the bracket should be considered. Here it is assumed that both waves can interfere coherently.

A phase difference arises between $\Psi_o(x_i)$ and $\Psi_g(x_i)$ mainly because of the difference in the path length. As a result, interference fringes (a lattice image) appear in the image. The image contrast is called phase contrast since it owes its origin to the phase difference. For a very thin specimen with an incident beam of unit amplitude, $I_0 = 1 \gg I_g$,

$$I(x_i) = 1 + 2\,\text{Re}[\Psi_g^*(x_i)], \quad (1.2)$$

where the magnification is assumed to be 1. The image contrast is proportional to the diffraction amplitude. This means that appreciable contrast can be obtained even from very small subjects like a single atom, a fine particle or a thin film, whose scattering power is very small. For comparison, the so-called diffraction contrast, another image contrast mechanism widely used for characterizing large scale defects like dislocations, is proportional to the diffracted intensity.

1.2.2 Lattice image and structure image

Let us consider the imaging mechanism of a lattice image using Fig. 1.1(a) [1.1]. First, diffraction waves are excited by a substance. After passing through the electron lens the diffraction pattern is formed on the back focal plane. The intensity maximum of each reflection is designated as 0, ± 1, ..., $\pm h$, ... Secondly, these spots become new sources so that the electrons starting here

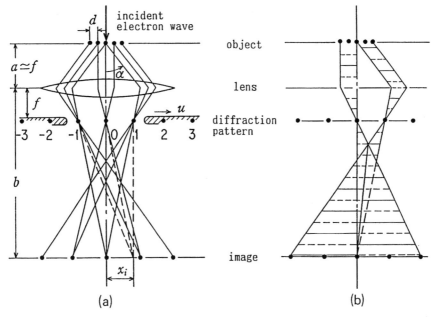

Fig. 1.1. (a) Optical geometry for HRTEM; (b) phase of the central and scattered waves under the imaging conditions for forming a crystal structure image. The phase difference of these two waves becomes π at the sites of atoms in the image plane.

meet again at the image plane to form a lattice image. Since the distance between the back focal plane and the image plane ($= b-f$) is actually very large as compared to the extension of the diffraction spots, the interference pattern on the image plane can be taken to be due to Fraunhofer diffraction. The amplitude of a lattice image $\Psi(x_i)$ is then described by

$$\Psi(x_i) = \mathscr{F}[Q(h)A(h)\exp\{2\pi i\chi(h)\}], \tag{1.3}$$

where Q is the amplitude of diffracted waves at site h, A the effect of the objective aperture, and $2\pi\chi$ the amount of phase change on passing through the electron lens. $\mathscr{F}$ denotes Fourier transformation.

In one of the through-focal series of lattice images, which are taken using many diffracted beams, the sites of dark spots coincide with those of atom columns parallel to the incident electron beam. This image is called the crystal structure image or simply structure image. Examples of structure images will be shown later (cf. Figs. 1.4, 1.8 and 1.11). In a structure image, the arrangement of dark spots corresponds to that of atom columns uniquely. In many cases, we can read out only the sites of heavy atoms. However, if we can speculate on the sites of all the atoms, including the light atoms, with help from crystal symmetry and crystal chemistry, we may call this a structure image as well.

1.2.3 Phase contrast transfer function

Electrons change the phase on passing through an electron lens due to the spherical aberration and defocus. The aberration function $\chi(u)$, proportional to the change in phase of the electron wave, is described by

$$\chi = \varepsilon\lambda u^2/2 - C_s\lambda^3 u^4/4, \tag{1.4}$$

where ε is the amount of defocus, λ the wavelength of electrons, u the spatial frequency and C_s the spherical aberration constant [1.2].

The phase factor $\exp\{2\pi i\chi(u)\}$ in eq. (1.3) strongly affects the intensity of lattice images. The function $\sin(2\pi\chi)$ is very important to determine, and is called the phase contrast transfer function. Fig. 1.2 shows some calculated results of the function with parameters, E (accelerating voltage) $= 200$ kV and $C_s = 1.2$ mm, for the range between $\varepsilon = 900$ and -500 Å [1.1]. The horizontal axis is scaled by the value of u ($= 1/d$ where d is the interplanar spacing). The value of the function fluctuates between 1 and -1. The fluctuation is more prominent at higher ranges of u. It is noted that $\sin(2\pi\chi) = 1$ in the range between $u = 1/2.5$ and $1/6$ Å^{-1} at $\varepsilon = 650$ Å.

1.2.4 Weak phase object approximation

Electrons entering into any material are affected by the electrostatic potential field V and, as a result, change their phase. For electrons running in the z direction, the phase change is described by

$$q(r_0) = \exp\left(i\sigma\int V\,dz\right), \tag{1.5}$$

where σ is the interaction parameter. q is called the transmission function. When the crystal is so thin that the relation

$$\sigma\int V\,dz(= \sigma V_p) \ll 1 \tag{1.6}$$

holds, eq. (1.5) can be expanded as follows:

$$q(r_0) = 1 + i\sigma V_p(r_0), \tag{1.7}$$

where V_p is the projected potential of crystal, $r_0 = (x_0, y_0)$ the two-dimensional positional vector in the objective plane. This is the approximation of a weak phase object. Since i represents the phase change of $\pi/2$, eq. (1.7) means the sum of the central beam with amplitude 1 and the scattered electron waves with amplitude $i\sigma V_p(r_0)$.

On the Fourier transformation of eq. (1.7) we get

$$\mathscr{F}[q(r_0)] = Q(u) = \delta(u) + i\sigma\mathscr{F}[V_p(r_0)] \tag{1.8}$$

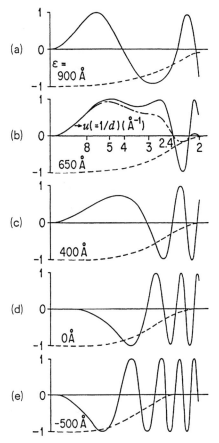

Fig. 1.2. Phase contrast transfer function $\sin(2\pi\chi)$ (solid line) and attenuation function $-E_D E_j$ (broken line) against u, for $E = 200$ kV and $C_s = 1.2$ mm.

where $\delta(u)$ is the delta function. The Fourier transform of the second term has the following value

$$\mathscr{F}[V_p(r_0)] = V_u \Delta z = (48.0/\Omega)F(u)\Delta z, \tag{1.9}$$

where V_u is the Fourier coefficient of potential, Ω the volume of unit-cell and $F(u)$ the crystal structure factor, the atomic structure factor f_j summed over sites j,

$$F(u) = F_{hk0} + \sum_j f_j \exp[-2\pi i\{h(x_{0j}/a) + k(y_{0j}/b)\}]. \tag{1.10}$$

1.2.5 Scherzer imaging condition

On substituting eqs. (1.8) and (1.9) into eq. (1.3) we get

$$\Psi(r_i) = \mathscr{F}[\{\delta(u) + i\sigma V_u \Delta z\}\{\cos 2\pi\chi(u) + i\sin 2\pi\chi(u)\}]$$
$$= 1 - \sigma \Delta z \mathscr{F}[V_u\{\sin 2\pi\chi(u) - i\cos 2\pi\chi(u)\}] \tag{1.11}$$

where the size of objective aperture is assumed to be infinite. The image intensity is

$$I(r_i) = 1 - 2\sigma \Delta z \mathscr{F}[V_u\{\sin 2\pi\chi(u)]. \tag{1.12}$$

We note from this formula that the image contrast closely depends on the phase contrast transfer function $\sin 2\pi\chi(u)$. As shown in Fig. 1.2, the value of $\sin 2\pi\chi(u)$ changes strongly depending on the defocus ε. However, if such a condition as

$$\sin 2\pi\chi(u) = 1 \tag{1.13}$$

holds, the image intensity will be [1.3]

$$I(r_i) = 1 - 2\sigma V_p(r_0). \tag{1.14}$$

Therefore, we can expect that the projected potential $V_p(r_0)$ is reflected in the image; the site with large value of V_p shows a dark contrast, while that with low V_p shows a bright one. This is a general principle for the formation of the structure image; sites of heavy metal atoms having high potential are imaged as dark dots. In order to realize the relation of eq. (1.13) as widely as possible we assume

$$2\pi\chi_{max} = 0.7\pi$$
$$\therefore \sin 2\pi\chi_{max} = 0.81.$$

The defocus value corresponding to this, ε_s, is

$$\varepsilon_s = 1.2 C_s^{1/2} \lambda^{1/2}. \tag{1.15a}$$

At this defocus the value of spatial frequency, u_s, for which the value of χ first becomes zero, is

$$u_s = 1.5 C_s^{-1/4} \lambda^{-3/4}. \tag{1.15b}$$

Therefore, a structure image can be obtained from a weak phase object first by setting the defocus value at ε_s and then by cutting out those diffracted waves whose spatial frequencies are larger than u_s. These observation conditions derived from eq. (1.13) are called the Scherzer imaging conditions and the defocus value ε_s is called the Scherzer focus.

Let us consider the phase change at the Scherzer focus geometrically. In Fig. 1.1(b) solid and broken lines represent the peaks and valleys of the phase, respectively, for the central and diffracted waves [1.1]. For simplicity, only one diffracted wave is shown. The phase of the diffracted wave is advanced first by $\pi/2$ on scattering and secondly also by $\pi/2$ on passing through an electron lens (since $\sin 2\pi\chi(u) = 1$). As a result, the phase difference between the central wave and the diffracted wave becomes π. On the image plane, therefore, destructive interference occurs between them to cause the atom sites to appear dark.

1.2.6 Resolution limit for HRTEM

The maximum scattering angle under the Scherzer condition, α_{max}, and the corresponding lattice spacing, d_s, are

$$\alpha_{max} = \lambda u_s = 1.5 C_s^{-1/4} \lambda^{1/4} \qquad (1.16a)$$

$$d_s = 1/u_s = 0.65 C_s^{1/4} \lambda^{3/4} \qquad (1.16b)$$

d_s is the minimum spacing in the information contributing to imaging and is called the resolution limit due to spherical aberration, or the Scherzer resolution limit. It is essentially the resolving power of HRTEM. We note from this that C_s and λ must be made smaller to obtain a higher resolving power.

1.2.7 Extension of weak phase object approximation

The weak phase object approximation (eq. (1.6)) is satisfied when $V_p(= V_0 \Delta z) \ll 1.2 \times 10^3$ VÅ for $E = 200$ kV. This means $\Delta z \ll 120 \sim 40$ Å since V_0 (mean inner potential) $= 10 \sim 30$ V for most inorganic crystals. In reality, however, structure images are obtained mostly for thickness between 15 and 50 Å, depending on the material, crystal structure and orientation, and accelerating voltage. That is to say, the condition for the weak phase object is not satisfied in most actual cases.

In order to overcome this contradiction we extend the theory of weak phase object approximation to slightly thicker crystals as follows; from dynamical calculations on the amplitude and phase of many waves we note the existence of such a relation as

$$Q(u) = Q'(u)\exp(iB\Delta z) \qquad (1.17)$$

where B is constant for each of the scattered waves and 0 for the central wave as long as the crystal thickness is less than that for the first extinction. Besides, the amplitude of the scattered waves is still considerably less than that of the central wave, and the relative intensities of the scattered waves are almost the same as for kinematical scattering [1.4]. It is therefore reasonable to assume that $\mathscr{F}[Q'(u)]$ essentially resembles $\exp(i\sigma V_p)$. Then we obtain

$$\Psi(r_i) \propto \exp\{i\sigma V_p(r_0)\} * \mathscr{F}[\exp\{2\pi i(\chi + B\Delta z)\}] \tag{1.18}$$

where $*$ means the convolution integral and the size of the objective aperture is assumed to be infinite. A similar formula can be derived from wave mechanical considerations [1.1]. Eq. (1.18) means that we have only to consider the effect of the total phase change. The wave aberration $2\pi\chi$ and the phase change $2\pi B\Delta z$ due to the dynamical effect of electron diffraction are therefore equivalent from the viewpoint of the transfer of phase contrast. The Scherzer imaging condition (eq. (1.13)) should then be modified to

$$\sin\{2\pi(\chi + B\Delta z)\} = 1. \tag{1.19}$$

In fact, the optimum focus at which the structure image is obtained is in most cases not at the Scherzer focus ε_s (eq. (15a)) but slightly shifted towards the Gaussian focus, depending on thickness. This is due to the effect of the phase shift by the dynamical scattering. According to a numerical calculation, for example [1.1], the main part of the modified transfer function already deviates from the condition of $\sin 2\pi\chi = 1$ at a thickness of only 16 Å for a crystal with heavy elements such as $2Nb_2O_5.7WO_3$, and the optimum defocus shifts from 1000 to 500 Å for a thickness near 50 Å.

1.2.8 Effect of the coherence among electron waves

On the formation of lattice images the amplitudes of waves are integrated. This means that we must consider the interference among them under coherent conditions. In fact, the chromatic aberration and the beam convergence effects seriously deteriorate the coherence and, as a result, the image intensity is decreased. The degree of coherence among electron waves can be described by a function which is called the transmission cross coefficient. Using this function we can estimate the effect of the coherence degradation for the weak phase object as follows

$$I(r_i) = 1 - 2\sigma\Delta z\mathscr{F}[V_u \sin\{2\pi\chi(u)\}E_D(u, \varepsilon)E_j(u, \varepsilon)] \tag{1.20a}$$

$$E_D(u, \varepsilon) = \exp(-0.5\pi^2\lambda^2\Delta^2 u^4) \tag{1.20b}$$

$$E_j(u, \varepsilon) = \exp\{-(\pi u_0)^2[(\varepsilon - C_s\lambda^2 u^2)\lambda u]^2\} \tag{1.20c}$$

where Δ is the mean fluctuation of focus due to the chromatic aberration and u_0 the effective size of the electron source ($u_0 = \beta/\lambda$, where β is the semi-angle of the illumination convergence). On comparing with eq. (1.12) we note that the phase contrast transfer function $\sin\{2\pi\chi(u)\}$ is attenuated by the modulation function $E_j(u, \varepsilon)E_D(u, \varepsilon)$. In Fig. 1.2 the values of $-E_jE_D$ are plotted by broken lines. An example of the effective transfer function, obtained by the product between this and $\sin\{2\pi\chi(u)\}$, is shown by a chain line in Fig.

1.2(b). In the region of large spatial frequency the attenuation of the transfer function is prominent.

1.3 Techniques relevant to HRTEM

1.3.1 Electron-optical conditions to obtain an HRTEM image

Many-beam lattice images are usually obtained under the following electron-optical conditions. (1) The illuminating electron beam is axial. (2) It is incident along a low-order zone axis of the crystal. (3) The objective aperture has the size given by eq. (1.15b) or (1.16). (4) The specimen crystal is very thin. (5) The image is observed at the defocus given by eq. (1.15a), or, it is slightly shifted toward the Gaussian focus, depending on the thickness, as mentioned in relation to eq. (1.19).

A structure image appears in a through-focal series of many-beam lattice images. For getting a structure image, the specimen thickness must be thinner than a few tens of ångströms. Except for the case when a specimen is originally prepared as a thin film, this is achieved only by crushing a bulk material. The crushing method is applicable to crystals which break into fragments by cleavage.

On the other hand, for the observation of cross-sections of materials, for example prepared by CVD methods, specimens must be thinned by an ion-milling method. The resulting thickness is generally more than 100 Å. This means an optical artifact inevitably arises in the image, i.e. the correspondence between the image contrast and the crystal structure is no longer unique. The interpretation of such an image must be done even more carefully, using computer simulations of image intensities.

1.3.2 Procedure for observing an HRTEM image

A practical procedure for observing a many-beam lattice image is as follows. For simplicity, we assume a specimen fragment in the form of a sharp wedge is used, which has been prepared by the crushing method and supported on a carbon microgrid. (1) The specimen is observed with a magnification of about 1×10^4 times to look for an area which is very thin, clean and not distorted, at the edge of a fragment. (2) The orientation of a small crystal area is examined by means of a diffraction pattern. When Laue zones are found suggesting that the orientation is near to that intended, the tilting stage is operated so that the zone axis becomes parallel to the optical axis. (3) An objective aperture is inserted at the center of the diffraction pattern. (4) After confirmation of voltage center, astigmatism is corrected by use of a stigmator. This is done by observing the granular structure of a microgrid film enlarged on a TV screen. (5) On changing slowly the amount of defocus, photographs are taken when the images anticipated in advance by means of the computer simulation of image intensity appear. For an unknown structure an image with high contrast appearing nearly at the Scherzer focus is recorded, and then several images are taken changing the defocus value for each in the direction to the Gaussian focus (a through-focal series of images). (6) A diffraction pattern is recorded to check how far the orientation has deviated from the zone axis during this procedure.

1.3.3 Computer simulation of an HRTEM image

Whether a structure model of a crystal or a defect obtained from an HRTEM image is correct or not must be examined. The method most widely practiced at present is to compare experimental images to calculated images obtained by computer simulation based on the structure model. The image intensity can be computed using commercially available or home-made software [1.1]. The calculation consists of two stages, i.e. the scattering stage and the imaging stage. In the former, dynamical diffraction amplitudes are computed usually by the multi-slice method [1.5]. During the course of this stage the projected potential of the crystal is also calculated.

1.3.4 Use of information from electron diffraction

Electron diffraction patterns include a great deal of information on crystal structure. They give valuable information that is complementary to that obtained from HTREM images, because they give more average and statistically significant information than images. For an example, lattice parameters are obtained more precisely from diffraction patterns than from images. Another example is the crystal symmetry and the related space group, which is derived from the extinctions of diffraction spots [1.1]. These are demonstrated by some concrete examples later.

1.4 HRTEM analysis of high T_c superconductors

In this section, some recent results on the structure analysis of HTSCs are shown in order to demonstrate the usefulness of HRTEM. In all cases mentioned here small blocks of specimens were lightly crushed in an agate mortar and the fragments obtained were observed in a high-resolution, high-voltage electron microscope (model H-1500) at an accelerating voltage of 800 kV. The point-to-point resolving power has been measured to be 1.4 Å.

1.4.1 Structure analysis of $Ga_2(Sr, Nd)_4Nd_3Cu_4O_z$

Powders of Ga_2O_3, PbO, $SrCO_3$, Nd_2O_3 and CuO were mixed with a nominal composition of $(Ga_{1.78}Cu_{0.22})(Sr_{3.16}Nd_{0.84})Nd_3Cu_4O_{16-y}$. The mixture was heated in air at 1000 °C for 20 h with intermediate grindings. It was then heated at 1070 °C for 2 h under the oxygen pressure of 20 MPa, using a hot isostatic pressing (HIP) apparatus [1.6]. According to a powder X-ray diffraction the unit cell of the product crystal is $a = 5.458$, $b = 5.535$ and $c = 51.302$ Å (orthorhombic). Using energy dispersive X-ray spectroscopy (EDS) the composition of the product was measured to be $(Ga_{1.8}Cu_{0.2})(Sr_3Nd)Nd_3Cu_4O_z$. We call this compound Ga-2434 hereafter for simplicity. Although it did not show any d.c. susceptibility, we have analyzed the crystal structure by HRTEM, since it appeared to be a new type of structure with interesting possibilities for superconductivity.

Electron diffraction patterns were taken from many crystal fragments of Ga-2434. Some typical patterns are shown in Fig. 1.3. From the reflection conditions ($h + k = 2n$ for $hk0$, $l = 2n$ for $0kl$, $h0l$ and $00l$, $h = 2n$ for $h00$ and $k = 2n$ for $0k0$) the space group is uniquely determined to be $Pccn$ (56). The lattice parameters are $a = 5.46$, $b = 5.54$ and $c = 51.3$ Å, being in agreement with those obtained from X-ray diffraction. Weak diffuse streaks along [010] in Fig. 1.3(a) must be due to stacking faults, which have been observed locally as intergrowth defects.

Fig. 1.4 is an HRTEM image corresponding to the diffraction pattern in Fig. 1.3(c), in which the electron beam is incident along the [110] direction. We see two adjoining lines and a single line of the darkest spots, as marked by large and small, leftward arrows, respectively. The image was taken near to Scherzer focus (the calculated Scherzer focus ε_s is 570 Å underfocus, while the real defocus is about 450 Å underfocus) so that the darker spots mark the sites of the heavier atom columns. Therefore, we may correlate the darkest spots to Nd atom columns, the second darkest spots to Sr columns and the least dark ones to Ga or Cu atoms. As a result, we can depict the structure of the present crystal like that in Fig. 1.5, i.e. the crystal is composed of a stacking of different atomic sheets, having the sequence $GaO/SrO^{*1}/CuO_2/(NdO)_2/CuO_2/SrO^{*1}/GaO/SrO^{*2}/CuO_2/Nd/CuO_2/SrO^{*2}/GaO$.

When we examine the image contrast of Fig. 1.4 in more detail, we note that there are two kinds of darkness for the SrO planes, i.e. the atom column sites of SrO^{*1} plane, marked by arrow 1, are slightly darker than those of the SrO^{*2}

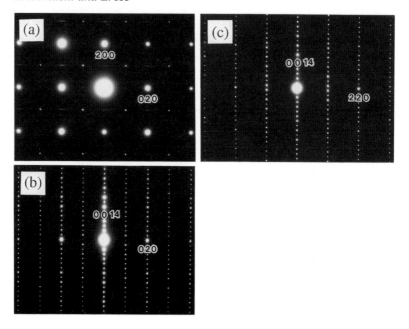

Fig. 1.3. Electron diffraction patterns taken from a Ga-2434 crystal. The electron beam is incident along [001] (a), [100] (b) and [110] (c). From the reflection conditions the space group of the crystal is determined to be *Pccn*.

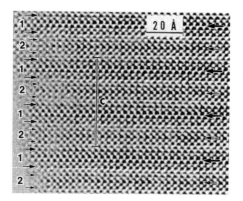

Fig. 1.4. HRTEM image of Ga-2434 corresponding to the diffraction pattern of Fig. 1.3(c). The darkest spots on the lines marked by large and small leftward arrows indicate the columns of Nd atoms in $(NdO)_2$ and Nd planes, respectively. Rightward arrows 1 and 2 show the SrO^{*1} and SrO^{*2} planes in Fig. 1.5, respectively. $c = 51.3$ Å.

plane, marked by arrow 2. This suggests that the planes labeled 1 may include Nd atoms to some extent. In order to solve the problem, we have computed the image intensity on changing the occupational probability of Nd atoms in the SrO^{*1} plane, which is now denoted by $(Sr_{1-x}Nd_x)O$. The computed images for $x = 0.3$, 0.5 and 0.7 are shown in Fig. 1.6. On comparing them to the real image of Fig. 1.4, the best fit is obtained for $x = 0.5$. The composition corresponding to $x = 0.5$ is identical with what is expected from the composition measured by EDX mentioned above. If the composition of $x > 0.5$ could be realized in an oxidizing atmosphere, the crystal could be superconducting, although this has not been the case so far.

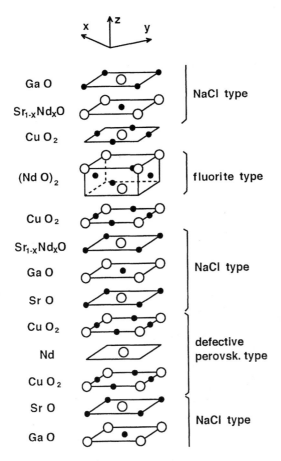

Fig. 1.5. Stacking sequence of atomic planes in a Ga-2434 crystal, constructed directly from an HRTEM image in Fig. 1.4. That corresponding to a half unit-cell length is shown ($c/2 = 25.65$ Å). Open circles stand for cations, closed ones for oxygens.

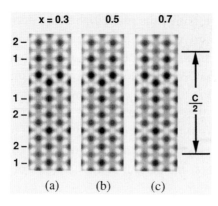

Fig. 1.6. Computer-simulated images of a Ga-2434 crystal, calculated for different occupational probabilities of Nd atoms in the SrO^{*1} plane, whose composition is denoted by $(Sr_{1-x}Nd_x)O$; $x = 0.3$ (a), $x = 0.5$ (b) and $x = 0.7$ (c). 1 and 2 indicate the SrO^{*1} and SrO^{*2} planes in Fig. 1.5, respectively.

1.4.2 Crystal structure and projected potential in YBa₂Cu₄Oₓ

Figure 1.7(a) shows a crystal structure of $YBa_2Cu_4O_x$ (124 phase, $a = 3.84$, $b = 3.87$ and $c = 27.24$ Å, orthorhombic). It contains the —Cu—O—Cu— single and double chains [1.7, 1.8] along [010], marked by S and D respectively. T_c of the 124 phase is known to be almost fixed (80 ∼ 82 K), since the range of variable oxygen content is very small. Fig. 1.8 is an HRTEM image of the 124 crystal, prepared under the pressure of 1.3 GPa [1.9]. The electron beam is incident along the [100] direction. Each site of cation columns is imaged as a dark spot. It is moreover noted that a Cu site in the double chains, which consists of a column of only Cu atoms, show slightly less darkness than another Cu site in the single chain which consists of a —Cu—O—Cu— column. This is due to the difference in the electrostatic potential at the two sites which can be checked by calculation. Fig. 1.7(b) shows the calculated image of the 124 phase. It is clear that the column sites of Cu atoms in the double chains show slightly less darkness than those of —Cu—O—Cu— in the single chain, in agreement with the observation. A computed projected potential (cf. eq. (1.6)) is shown in Fig. 1.7(c). It is clear that the projected potential of the column sites of Cu ions in the double chains is slightly lower than that of —Cu—O—Cu— ions in the single chain. This result indicates that the computer simulation of the projected potential is sometimes useful for the correct interpretation of the image contrast, in addition to simulation of the image intensity.

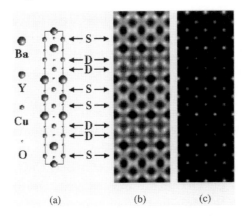

(a)   (b)   (c)

Fig. 1.7. (a) A structure model of YBa$_2$Cu$_4$O$_x$ (124 phase) projected along [100]. S and D mean the —Cu—O—Cu— single and double chains, respectively. (b) is the computer-simulated image and (c) the projected potential. For (b) $\varepsilon = 450$ Å (underfocus), $\Delta z = 30$ Å, $E = 800$ kV, $C_s = 2.2$ mm, $\Delta = 100$ Å and $\beta = 0.5$ mrad.

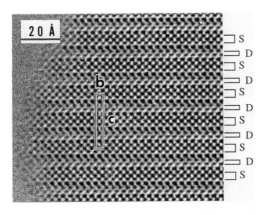

Fig. 1.8. HRTEM image of the 124 crystal taken with the incident beam along the [100] direction. The Cu column sites in the double chains (marked by D) show weak contrast as compared to —Cu—O—Cu— column sites in the single chain (S). $c = 27.24$ Å.

1.4.3 Visualization of oxygen atoms in YBa$_2$Cu$_3$O$_{+x}$

HRTEM would be more useful if all species of constituent atoms including light elements like oxygen could be visualized. This has become possible by a new ultra-high-resolution high-voltage electron microscope (UHR-HVEM, model H-1500) [1.10, 1.11]. Using the UHR-HVEM we have observed oxygen atoms in high T_c superconductors, $YBa_2Cu_3O_{7.7}$ [1.12] and $YBa_2Cu_3O_{6.4}$ [1.13]. Here we clarify the conditions for observing oxygen atoms, with the help of a map showing the relationship between crystal thickness and defocus value.

It is known [1.14] that $YBa_2Cu_3O_{6+x}$ has an orthorhombic phase for $0.35 < x < 1$ (e.g. $a = 3.82$, $b = 3.89$ and $c = 11.68$ Å, for $x = 0.93$) and a tetragonal phase for $0 < x < 0.35$ (e.g. $a = b = 3.86$ and $c = 11.78$ Å, for $x = 0.34$). They are essentially similar to each other, i.e. the positions of the metal atoms are almost the same for both structures. The structure can simply be denoted by the stacking sequence of $BaO/CuO_2/Y/CuO_2/BaO/CuO_y$ planes. The total content $6 + x$ depends on how many oxygens exist in the CuO_y planes. In the present calculation, a tetragonal structure model, as shown in Fig. 1.9(a), is used for simplicity. The metal atoms are in the positions mentioned above. Excess oxygens are assumed to occupy all possible sites in the CuO_y planes. In the actual process of calculation the structure was simplified; there are three subcells in the model of Fig. 1.9(a). They are averaged so that a simplified unit-cell as shown in Fig. 1.9(b) can be assumed. The unit-cell is substantially the same as that for the perovskite-type structure. The site at the center (A site) is occupied by $(2/3)Ba + (1/3)Y$, sites at the corners by Cu, and sites at the edge centers by oxygen. The sites O1, O2 and O3 are fully occupied in the first stage of the calculation.

The image pattern and the image contrast strongly depend not only on the amount of defocus but also on the crystal thickness [1.15]. Figure 1.10 shows a defocus vs. thickness map for simulated images for [001] incidence. Some characteristic image patterns, which are apparently related to the real structure, appear in the definite areas outlined in the diagram. In an area marked by D1, column sites of Ba(Y) atoms are imaged as strong dark spots. Columns of —Cu—O—Cu— are imaged as medium dark spots. Columns containing only oxygen atoms are imaged as weak dark spots. The defocus is near the Scherzer condition. Since the scattering power of the —Cu—O—Cu— column is determined mainly by the Cu, this column is hereafter referred to as a 'column of Cu'.

In area D2 an image with dark spots is also obtained. The sites of metal atom columns are imaged but those of oxygens are not. In areas D3 and D4 the oxygen column sites are seen dark and the metal atom columns are less dark. In areas B1 and B2, on the other hand, bright-spot images appear; the image patterns are similar to those mentioned above but the contrast is reversed. In the area B1, sites of both Ba(Y) and Cu columns are imaged as strong bright spots, while the oxygen columns are seen less bright. In B2, the sites of oxygen columns are the brightest, while those of metal atom columns are less bright. An arrow on the lateral axis shows the center position of the first thickness

(a)

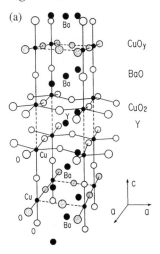

CuO$_y$

BaO

CuO$_2$

Y

(b)

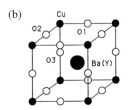

Fig. 1.9. (a) A structure model of YBa$_2$Cu$_3$O$_{8.0}$; (b) a simplified structure model for computer simulation. $a = 3.86$ Å and $c = 3.91$ Å (tetragonal).

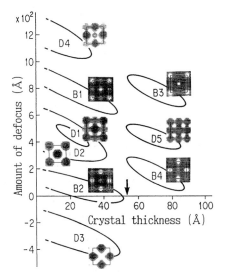

Fig. 1.10. Defocus vs. thickness map, showing computer simulated image contrast. D denotes a dark-spot image, B a bright-spot image. For each defocus value, images are calculated for crystal thicknesses of 23.5 Å and 74.3 Å, with $E = 800$ kV, $C_s = 2.2$ mm, $\Delta = 100$ Å and $\beta = 0.5$ mrad.

contour. In a crystal thicker than this there are also regions of bright- as well as dark-spot images; areas B3, B4 and D5, but no structure images are observed in these areas. In B4, for example, the oxygen column sites are very bright, while the metal atom columns are less bright. In Fig. 1.10 we note that each area extends in a downward direction with increasing thickness. This is because each of the scattered waves slowly increases its phase with increasing thickness due to their dynamical interaction [1.4], as mentioned in relation to eqs. (1.17)–(1.19).

A small sintered block of $YBa_2Cu_3O_{6.4}$ crystal ($T_c = 27$ K, orthorhombic, $a = 3.84$, $b = 3.87$ and $c = 11.73$ Å) [1.13] was lightly crushed, and the fragments obtained were observed in the UHR-HVEM. The accelerating voltage was selected at 800 kV in order to suppress the irradiation damage as far as possible [1.12, 1.13]. Figure 1.11 shows a through-focal series of images from a very thin part of the crystal, taken with incident electrons parallel to the [001] direction. The direct magnification was 3.5×10^5 times, and the exposure time was 2 s.

Figures 1.11(a), (b), (c), (d) and (e) are electron micrographs taken at a defocus of about 800 Å (underfocus), 450 Å, 300 Å, 50 Å and -300 Å (overfocus), respectively. Figures 1.11(a) and (d) are bright-spot images, while Figs. 1.11(b), (c) and (e) are dark-spot images. The image contrast is apparently reversed between (a) and (b), (c) and (d), and (d) and (e). Images calculated for the areas B1, D1, D2, B2 and D3 in Fig. 1.10 are inserted in Fig. 1.11 with the same magnification. It is noted that they almost fit to each other. In (b) small dark spots are clear at the sites of the arrowheads; the columns of oxygen atoms can clearly be discriminated beside those of the metal atoms.

It has been shown in a previous paper on ZrO_2 [1.15] that 'dark-spot images' correctly reflect the real structures of defects, while 'bright-spot images' do not. In order to confirm this also in the present case we have carried out further computer simulations of the image contrast. In the structure model of Fig. 1.9(a) all the oxygen atoms in the CuO_y planes were deleted; the total composition is then $YBa_2Cu_3O_{6.0}$. The occupational probability of the oxygen sites O1 and O2 in Fig. 1.9(b) becomes $2/3$. This means that the oxygen columns in the [001] projection of Fig. 1.12(a) have an occupational probability of $2/3$.

Figures 1.12(b) and (c) are the results of image calculations for the underfocus of $\varepsilon = 800$ and 450 Å, respectively. The crystal thickness is assumed to be 23.5 Å. For comparison, Figs. 1.12(d) and (e) show simulations using the same defocus values as the crystal with fully occupied oxygen columns. These are the values used for B1 and D1 in Fig. 1.10, respectively. The difference in the oxygen occupation is clearly visible in the contrast of the dark-spot images

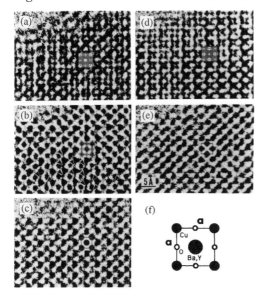

Fig. 1.11. Through-focal series of UHR-HVEM images for a very thin part of YBa$_2$Cu$_3$O$_{6.4}$, taken with the incident beam parallel to [001] at an accelerating voltage of 800 kV. The defocus value is 800 Å (underfocus) (a), 450 Å (b), 300 Å (c), 50 Å (d) and −300 Å (e). (f) is the projected structure of the unit-cell. At the sites arrowed in (b) small but clear dark spots are visible, showing the sites of oxygen atom columns.

(Figs. 1.12(c) and (e)). This is not the case in the bright-spot images ((b) and (d)); bright spots always appear at the oxygen column sites regardless of the oxygen occupation.

These results show that the fluctuation in intensity at the oxygen column sites of Fig. 1.11(b) is intuitively interpretable, while that of Fig. 1.11(a) is not; the dark spots of oxygen columns are clearly visible only at the sites marked by small arrows in Fig. 1.11(b). Similar investigations to these can be carried out for other regions in Fig. 1.10. In general, we conclude that the oxygen defects are visible and intuitively interpretable only in the area B1 of Fig. 1.10. The observation that the oxygen column sites become unclear near the edge of the fragment in Fig. 1.11(b) is because the contrast becomes very weak there. Local fluctuation of oxygen occupancy may be another possible reason.

Figure 1.13 shows the calculated image contrast vs. crystal thickness. The image contrast defined as 100 × (background − peak intensity)/background, increases with increasing thickness, and at the thickness of about 25 Å, reaches about 80% for Ba(Y) columns and about 20% for oxygen columns.

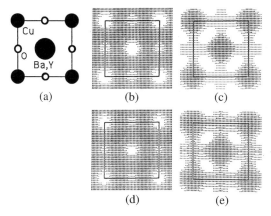

(a)   (b)   (c)

(d)   (e)

Fig. 1.12. Calculated images for $YBa_2Cu_3O_{8.0}$. (a) Projected structure in the [001] direction, $a = b = 3.86$ Å. (b) and (c) are computer simulated images of a crystal with oxygen defects at a thickness of 23.5 Å, with $\varepsilon = 800$ Å for (b) and $\varepsilon = 450$ Å for (c). (d) and (e) are calculated images for a defect-free crystal at a thickness of 23.5 Å, with $\varepsilon = 800$ Å and 450 Å, respectively.

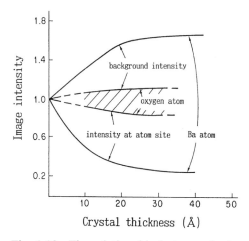

Fig. 1.13. The relationship between the image intensity and the thickness for Ba and oxygen columns in a $YBa_2Cu_3O_{8.0}$ crystal. The image contrast at Ba and oxygen columns sites are 80% and 20%, respectively, at a thickness of about 25 Å.

References

[1.1] S. Horiuchi, *Fundamentals of high-resolution transmission electron microscopy*, chs. 6–8 (North-Holland, 1994).
[1.2] O. Scherzer, *J. Appl. Phys.* **20**, 20 (1949).
[1.3] J. M. Cowley & S. Iijima, *Z. Naturforsch.* **27**, 445 (1972).

[1.4] S. Horiuchi, *Ultramicrosc.* **10**, 229 (1982).
[1.5] J. M. Cowley & A. F. Moodie, *Acta Crystallog. A* **30**, 280 (1974).
[1.6] A. Ono, L. L. Hu & S. Horiuchi, *Physica C* **247**, 91 (1995).
[1.7] J. Karpinski *et al.*, *Nature* **336**, 660 (1988).
[1.8] P. Marsh *et al.*, *Nature* **334**, 141 (1988).
[1.9] A. Ono & S.Horiuchi, *Physica C* **247**, 319 (1995).
[1.10] S. Horiuchi *et al.*, *Ultramicrosc.* **39**, 231 (1991).
[1.11] Y. Matsui *et al.*, *Ultramicrosc.* **39**, 8 (1991).
[1.12] S. Horiuchi, Y. Matsui & B. Okai, *Jpn. J. Appl. Phys.* **31**, L59 (1992).
[1.13] S. Horiuchi, *Jpn. J. Appl. Phys.* **31**, L1335 (1992).
[1.14] J. D. Jorgensen *et al.*, *Phys. Rev. B* **41**, 1863 (1990).
[1.15] S. Horiuchi & Y. Matsui, *Jpn. J. Appl. Phys.* **31**, L283 (1992).

2

Holography in the transmission electron microscope

A. TONOMURA

2.1 Introduction

In conventional electron microscopy, specimens are observed using the intensity of an electron beam. However, in electron holography [2.1], the phase as well as the intensity of the electron beam transmitted through a specimen is first recorded on film as an interference pattern, which is called a 'hologram'. The illumination of a laser beam onto this hologram then produces an optical image of a specimen in three dimensions. To be more exact, the wavefronts of the scattered electron beam are reproduced as wavefronts of a laser beam. Although the optical wavelength is 10^5 times larger than the electron wavelength, the two wavefronts are otherwise alike.

Once the image is completely transferred from inside the electron microscope onto an optical bench, versatile optical techniques can be used for electron optics. For example, the effect of the spherical aberration in the objective lens of the electron microscope can be compensated for to improve the resolution which was the original objective for which Gabor invented holography [2.1]. This is done by optically adding aberration with an opposite sign. The phase distribution of the electron beam can also be drawn on an electron micrograph by using an optical interferometer in the optical reconstruction stage of electron holography [2.2]. An electron microscope with an electron biprism [2.3] can provide an interferogram, but not a contour map nor a phase-amplified interference micrograph.

These possibilities were opened up by the development of a 'coherent' field-emission electron beam [2.4] which is indispensable for forming high-quality electron holograms. In fact, an electron hologram formed with this type of electron beam can have such high contrast and precision that a phase distribution can be reconstructed within an accuracy of $1/100$ of its electron wavelength [2.5]. Furthermore, since the field-emission beam has a high current density even under collimated illumination, a necessary condition for forming an electron hologram, a hologram on a fluorescent screen can be seen with the naked eye, thus making dynamic observation using computers and optical devices possible. Thanks to these technical developments, phase information from an electron beam can now be used to observe and measure objects hitherto inaccessible by electron microscopy.

2.2 Electron holography

Electron holography is a two-step imaging method. In the first step, an interference pattern is formed between an object wave and a reference wave in a field-emission transmission electron microscope, and this pattern is recorded on film as a hologram. A field-emission electron beam is needed to make high-quality electron holograms within a reasonable exposure time. The brightness (the current density per unit solid angle) of this beam is much greater than that of conventional thermionic electron beams, so the current density is higher even when an electron beam is used with a narrow illumination angle to make holograms.

In general, off-axis holograms are formed using an electron biprism [2.3] (see Fig. 2.1). The biprism, composed of a thin central wire less than 1 μm in diameter and a ground potential electrode on each side, is installed below the objective lens. Applying a positive potential of up to 300 V to the central wire causes the beams passing through on each side of the central wire to overlap. A specimen is placed in one of the two beams to form an object wave, and the other beam acts as a reference beam. The interference pattern made by the

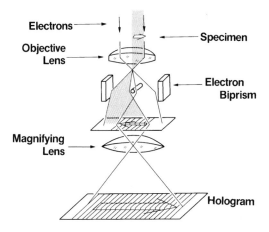

Fig. 2.1. Schematic for electron-hologram formation.

object and reference beams is magnified by the magnifying lenses and recorded on film as a hologram.

The hologram is subsequently illuminated by a collimated laser beam and an exact three-dimensional image is produced in a diffracted beam. An additional image, called a 'conjugate image', is also produced with the same amplitude, but the opposite phase.

An interference micrograph, or a contour map of the wavefronts, can be obtained by simply overlapping an optical plane wave with this reconstructed image. An example of such an optical system is shown in Fig. 2.2. A collimated laser beam is split by a Mach–Zehnder interferometer into two beams (A and B, for convenience) traveling in different directions. Each beam illuminates the hologram and produces a reconstructed image and its conjugate. An interference micrograph is observed when the reconstructed image of beam A, and the transmitted beam of beam B, or vice versa, are made to pass through a slit and overlap in the observation plane. The precision of phase measurement in a micrograph obtained in this way is $2\pi/4$.

In electron interferometry, there are often cases where great precision is required, for example, to measure the thickness distribution in atomic dimensions or to observe microscopic electromagnetic fields. To achieve such precision, phase amplification techniques peculiar to holography have been developed and used. Using these techniques, phase shifts as small as $1/100$ of the wavelength can be detected [2.5].

The interference microscopy described up to now uses a single electron hologram. The problem is how to form a high-quality electron hologram and how to precisely detect the peak positions of interference fringes from the hologram. Another method using many electron holograms has been developed so that the phase distribution can be determined more precisely [2.6]. In these multiple holograms, the image of an object is fixed and the positions of the

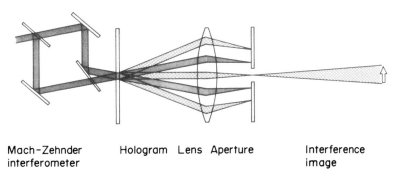

Mach–Zehnder   Hologram Lens Aperture   Interference
interferometer   image

Fig. 2.2. Optical reconstruction system for interference microscopy.

interference fringes are displaced relative to the image, by changing the initial phases of a reference wave. From these holograms, the phase distribution can be precisely calculated. Therefore, this is called a phase-shifting (or fringe-scanning) method [2.7]. It was originally developed for optical interferometry and is now applied to electron holography [2.8, 2.9].

Optical reconstruction with laser light is simple, but is off-line as a result of the time involved in developing the film. Therefore, on-line reconstruction techniques that use computers and optical devices are being developed. An image can be numerically reconstructed from a hologram recorded on film or on a charge-coupled device (CCD) attached to an electron microscope. In this way, an amplitude image, a phase image, and an interference image can be displayed. These images can be obtained fairly quickly, depending on the performance of the computer used, but not yet in real-time.

A real-time method in which a liquid crystal panel is used as a phase hologram [2.10] has been developed. The image signal of a hologram is detected with a TV camera attached to the electron microscope and transferred to the liquid-crystal panel, where the intensity distribution is transformed into a phase shifting function for an illuminating light beam. The time resolution of $1/30$ s is limited only by the scanning rate of the TV system. Dynamic phenomena can therefore be observed in real time with this method [2.11]. Due to the development of both a 'coherent' field-emission electron beam and electron holography, electron–optical techniques using phase information have made remarkable progress, thus opening up various fields of application [2.12].

2.3 Applications

When a parallel electron beam is incident to an electromagnetic field, the electron beam is deflected, or phase shifted. The phase shift $\Delta S/\hbar$ is calculated from the Schrödinger equation as,

$$\Delta S/\hbar = (1/\hbar)\oint(m\upsilon - e\boldsymbol{A})\cdot\mathrm{d}\boldsymbol{s}$$

$$= (1/\hbar)\oint(\sqrt{2meV} - e\boldsymbol{t}\cdot\boldsymbol{A})\,\mathrm{d}\boldsymbol{s}, \qquad (2.1)$$

where integration is carried out along a route connecting two electron trajectories and t is the unit tangent vector of the electron trajectories. This equation shows that electromagnetic potentials $(\boldsymbol{A}, V)$ can be detected by measuring the phase shift in an electron beam, although what we detect is not the electromagnetic potentials themselves but their integrals along the electron beam trajectory.

2.3.1 Specimen thickness distribution

An electron is accelerated by the inner potential V_0 when it enters a specimen. The specimen is regarded as a space within which the electrostatic potential is different from that in a vacuum, or as a material whose refractive index n is given by $n = 1 + V_0/2E$, where E is the accelerating voltage of an electron beam. An electron beam transmitted through the specimen thus receives a phase shift depending on the specimen material and thickness. When a specimen is made of a uniform material, a contour map of the electron phase distribution can indicate its thickness contours. Contour lines appear for every increase in thickness equal to $(2E/V)\lambda$, where λ is the electron wavelength. When $E = 100$ kV and $V_0 = 10$ to 30 V, the thickness change for a 2π phase shift is 200 to 700 Å. An example of the measurement of a thickness distribution is shown in Fig. 2.3. The specimen is a fine beryllium particle. A conventional interference micrograph (b) yields thickness contour lines in 440 Å steps, whereas the contour lines in the 32-times amplified interference micrograph (c) indicate steps of only 13 Å. One might wonder, though, whether these fringes precisely represent the thickness distribution or are only rough interpolations.

This problem was examined through the observation of surface steps by Tonomura *et al.* [2.4]. Figure 2.4 shows a cleaved molybdenite thin film phase-amplified 24 times. The phase distribution is displayed here as a deviation from regular fringes (i.e., as an interferogram). Steps A, B, and C in the micrograph correspond to steps of one, three, and five layers of the atomic surface. The thickness change at step A is only 6.2 Å (one-half of the *c*-axis spacing) and produces a phase shift of $2\pi/50$. This figure shows that a phase shift on the order of $2\pi/100$ can be detected.

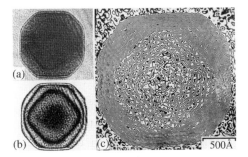

Fig. 2.3. Interference micrograph of a fine beryllium particle: (a) electron micrograph; (b) interference micrograph; (c) 32-times phase-amplified interference micrograph.

When many holograms are formed at different incident angles of the electron beam transmitted onto a specimen, a three-dimensional view of an object can be obtained numerically using methods similar to those of X-ray computer tomography [2.12]. An arbitrary view seen from any direction, or any cross-section can be obtained. An example of two views of latex spheres reconstructed from 24 different holograms is shown in Fig. 2.5.

Here, the physical quantity detected at each point is the refractive index for electrons which is a scalar. Computer tomography for a vector quantity has also been developed and applied to the three-dimensional distribution of magnetic fields [2.13] which are discussed in the following section. In vector computer tomography, holograms have to be formed by rotating around two axes rather than just one.

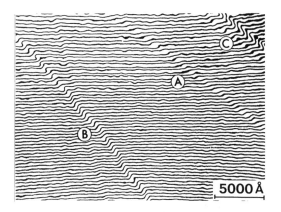

Fig. 2.4. Interferogram of a MoS_2 film (phase amplification: $\times 24$).

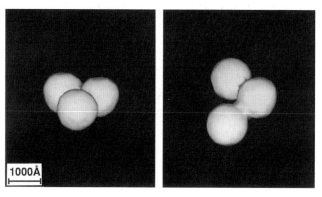

Fig. 2.5. Two views of latex spheres reconstructed from 24 holograms formed by an incident electron beam with different illumination angles.

In the transmission mode described so far, surface topography can only be investigated by measuring thickness. In the reflection mode, thickness can be measured with a high degree of sensitivity because surface height differences are directly measured in units of extremely short electron wavelengths as geometrical path differences [2.14]. An example is shown in Fig. 2.6. The specimen here is single-crystal GaAs with an atomically flat surface that has a single screw dislocation. One fringe displacement in the interferogram corresponds to a height difference of only 0.5 Å. The monatomic step of height 2 Å is terminated by a dislocation core. One can see at a glance how the 2 Å difference in height is relaxed in the region surrounding the core.

Inside the crystal, this relaxation was supposed to be just like a spiral staircase around the dislocation line. This micrograph however reveals that the slope of the staircase is not uniform but is steep in only one direction. The relaxation extends even to a region a few micrometers away. In this mode of holography, surface topography can be measured to a precision of 0.1 Å even in an unamplified interference micrograph.

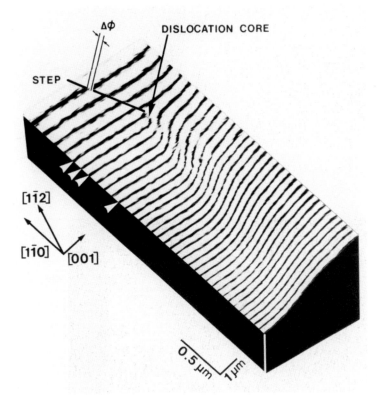

Fig. 2.6. Interferogram of a GaAs surface with a single screw dislocation. One fringe displacement corresponds to a height difference of 0.5 Å.

2.3.2 *Magnetic field observation*

For a pure magnetic object, the phase difference between two electron beams passing through the object is given by

$$\Delta S/\hbar = -(e/\hbar)\oint \boldsymbol{A}\cdot\mathrm{d}\boldsymbol{s} = -(e/\hbar)\int \boldsymbol{B}\cdot\mathrm{d}\boldsymbol{S}, \tag{2.2}$$

where the first integral is carried out along a closed path along two electron trajectories, and the second integral is carried out over the surface determined by the two paths.

The following conclusions can be drawn from this equation; see reference [2.15].

(1)   Contour fringes in the interference micrograph indicate magnetic lines of force, because the phase difference $\Delta S/\hbar$ vanishes between two beams passing through arbitrary points along a magnetic line.

(2)   A magnetic flux of h/e flows between two adjacent contour fringes.

An example of this is shown in Fig. 2.7. The object here is a smoke cobalt particle prepared by gas evaporation in an inert-gas atmosphere. No contrast can be seen in electron micrograph (a) which represents the electron intensity distribution. However, circular contour fringes appear in contour map (b). Since the phase distribution is amplified twice, these contour lines indicate magnetic lines of force in $h/2e$ units. It can be seen at a glance how magnetic lines of force rotate in such a fine particle. The principle behind this technique is the same as that of a flux-meter SQUID except that its sensitivity is h/e rather than $h/2e$ as in a SQUID. Therefore, we may call an electron interferometer a 'SQUID in a microscopic world'.

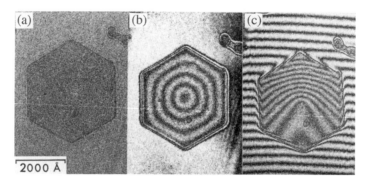

Fig. 2.7. Example of an interference micrograph of a fine cobalt particle; (a) electron micrograph; (b) contour map (phase amplification: ×2); (c) interferogram (phase amplification: ×2).

We cannot determined from this contour map alone whether the magnetization rotates clockwise or counterclockwise. The direction can be determined, however, from interferogram (c). Interference fringes are displaced at the particle's edges and go upward inside the particle. This can be interpreted as follows: an electron travels faster inside the particle than in a vacuum and consequently has a shorter wavelength. Therefore, the wavefront of the transmitted electron wave is retarded. In addition, the wavefront is either advanced if the magnetization is clockwise or retarded if the magnetization is counterclockwise. Therefore, in this case, the rotation direction proves to be clockwise.

The diameter of this particle is around 3000 Å. For smaller or anisotropic particles, magnetization cannot be kept inside, and the particle is uniformly magnetized. An example of a barium-ferrite particle [2.16] is shown in Fig. 2.8. Here, magnetic fields leak from the north pole at the top of the particle and are then sucked up at the south pole. The particle can therefore be seen to have a single magnetic domain.

Because the electron phase shift is produced not only due to magnetic fields but also due to changes in the specimen thickness, magnetic lines of force can be directly observed through an interference micrograph only for specimens whose thickness is uniform. A three-dimensional magnetic domain structure therefore cannot be determined from an interference micrograph.

In that case, electric and magnetic contributions to the phase can be separated by using two holograms of an object [2.17]. This method is based on the different behavior of the electric and magnetic fields when the electron incidence is reversed. Two holograms are formed, one in the standard specimen position and one with the face of the specimen turned down. In these holograms, the electric contributions to the electron phase are the same but the magnetic contributions have opposite signs. A holographic technique then displays the electric and magnetic images separately.

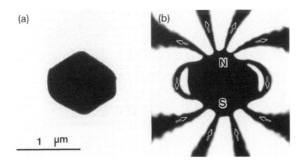

Fig. 2.8. Interference micrograph of a barium-ferrite particle.

This can be easily understood from eq. (2.1):

$$\Delta S/\hbar = (1/\hbar)\oint(mv - eA)\cdot ds,\qquad(2.3)$$

where the line integral is performed along the electron path. If an electron beam is incident to the specimen from the opposite direction ($t \to -t$, $v \to -v$, $s \to -s$), then the electric contribution to the phase shift (the first term in eq. (2.1)) is the same but the magnetic contribution reverses the sign of the phase.

If we consider the phase distribution in the micrograph seen from above to consist of an electric phase ($S_e/\hbar$) plus a magnetic phase ($S_m/\hbar$), then the phase distribution in the micrograph seen from below consists of the electric phase minus the magnetic phase ($S_e - S_m$)/$\hbar$. If these optically reconstructed images are made to overlap to display the phase difference between them, the resultant phase distribution is double the magnetic phase, $2S_m/\hbar$. In this way, a purely magnetic image can be obtained.

The thickness image, on the other hand, can be obtained when we make full use of the conjugate image, which has a phase distribution with a sign opposite to the original sign. The thickness-only image ($2S_e/\hbar$) can be obtained by overlapping the reconstructed image of the top view, which has a phase ($S_e + S_m$)/$\hbar$, and the conjugate image of the bottom view, which has a phase $-(S_e - S_m)/\hbar$.

The results are shown in Fig. 2.9, where micrographs (a) and (b) respectively show thickness contours and in-plane magnetic lines of force.

Although many choices are available in the domain structure of a three-dimensional particle, the magnetization in this sample can be determined to

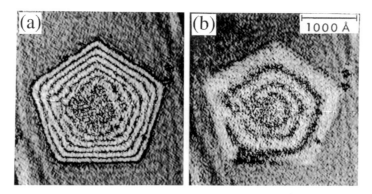

Fig. 2.9. Interference micrographs of a dodecahedron cobalt particle (phase amplification: ×2); (a) thickness contour map; (b) magnetic lines of force.

rotate along the common side of the five regular tetrahedrons forming a particle like that shown in Fig. 2.10.

2.3.3 Vortices in superconductors

Vortices in superconductors can be observed quantitatively by interference microscopy [2.18, 2.19] and Lorentz microscopy [2.20] with our 350 kV holography electron microscope [2.21]. In the experiments we conducted, a superconductive thin film was tilted with respect to both the electron beam and the magnetic field.

The experimental arrangement is shown in Fig. 2.11. A Nb thin film, set on a low-temperature stage, was tilted 45° to an incident beam of 300 kV electrons so that the electrons could be influenced by vortex magnetic fields. An external magnetic field of up to 150 G was applied horizontally. An example of a vortex array in a single-crystalline Nb thin film [2.18] is shown in Fig. 2.12. In this interference micrograph, projected magnetic lines of force can be observed.

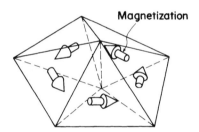

Fig. 2.10. Schematic diagram of the magnetic domain structure in a cobalt particle.

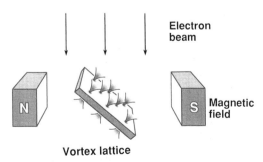

Fig. 2.11. Schematic diagram for vortex lattice observation.

Fig. 2.12. Interference micrograph of a superconducting Nb film at $B = 100$ G (phase amplification: $\times 16$).

These lines become dense in the localized regions indicated by circles in the photograph, which correspond to individual vortices.

Although interference microscopy has high resolution and produces quantitative results, Lorentz microscopy is more convenient for observing the dynamic behavior of vortices. In this experiment, the sample was first cooled to 4.5 K and the magnetic field B was gradually increased. Vortices suddenly began to penetrate the film at $B = 32$ G, and the number of vortices increased as B was increased further. Their dynamic behavior was quite interesting. At first, only a few vortices appeared here and there in a 15×10 μm field of view. They oscillated around their own pinning centers and occasionally hopped from one center to another. These movements continued as long as the vortices were not closely packed ($B \leqslant 100$ G).

An equilibrium Lorentz micrograph at $B = 100$ G [2.20] is shown in Fig. 2.13. The film has a fairly uniform thickness in the region shown, but is bent along the black curves, called bend contours, caused by Bragg reflections at the atomic plane brought to a favorable angle by bending. Each spot showing a black and white contrast is an image of a single vortex. This contrast reversed, as expected, when the applied magnetic field was reversed. The tilt direction of the sample can be seen from the line dividing the black and white parts of the spots. Because the black part is on the same side of all the spots, the polarities of all the vortices seen in the region are the same. At low B, i.e., up to 30 to 50 G, the vortices are too sparsely distributed to form a lattice, even at equilibrium. At $B = 100$ G, the vortex density is so high that they cannot form anything but a hexagonal lattice.

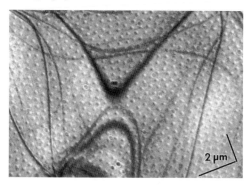

Fig. 2.13. Lorentz micrograph of a two-dimensional array of vortices in a superconducting Nb film.

A high T_c superconductor has been investigated by Lorentz microscopy [2.22]. High T_c superconductors are difficult to use practically, because the critical current vanishes at high temperatures and in high magnetic fields, even when the temperature is well below the critical temperature T_c. This phenomenon probably arises from the behavior of vortices but this has not yet been proven conclusively. Some researchers believe that these vortices melt like molecules in a liquid, and as a result it is difficult to fix vortices at some pinning sites [2.23]. Evidence for vortex-lattice melting was provided by a Bitter BSCCO figure in which the vortex image was blurred even at 15 K and 20 G ($T_c = 85$ K) [2.24]. Accordingly, the temperature for practical use would not be T_c but rather the melting temperature T_m [2.25]. Others however disagree with this, and attribute this phenomenon to weak pinning effects.

The vortices were dynamically observed to test whether vortices begin to move under such conditions. The observation was made under a fixed magnetic field B increasing the sample temperature from 4.5 K to above T_c. A Lorentz micrograph at $T = 4.5$ K and $B = 20$ G is shown in Fig. 2.14(a). Vortices are distributed at random. When the temperature was raised stepwise by a few Kelvin, vortices moved. After a few minutes, they arrived at an equilibrium state and became still. They did not melt even at 20 K. The vortex configuration changed between 40 K and 50 K.

Vortices form a regular lattice (c) above this transition region. The vortex lattice persisted at higher temperatures (see Fig. 2.4(d) at 68 K) though the image contrast gradually decreased and then disappeared above 77 K.

2.4 Conclusions

The performance of electron phase microscopy has been improved thanks to the development of a coherent electron-beam source and electron holography.

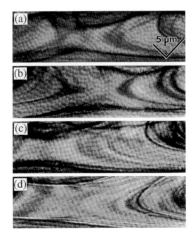

Fig. 2.14. Lorentz micrograph of a BSCCO (2212) film; (a) $T = 4.5$ K, (b) $T = 20$ K, (c) $T = 56$ K, and (d) $T = 68$ K.

Phase distributions in electron beams can be measured to within $2\pi/100$, which has opened the way to measuring microscopic objects and fields with ultra-high precision. These developments allow the direct observation of individual vortices in a superconductor. Electron phase microscopy can be used to clarify the fundamental and practical applications of superconductivity, especially in the field of high T_c superconductors.

References

[2.1] D. Gabor, *Proc. R. Soc. A* **197**, 454 (1949).
[2.2] For example, see A. Tonomura, *Rev. Mod. Phys., Electron Holography*, (Springer Verlag, Heidelberg, 1993).
[2.3] G. Möllenstedt & H. Düker, *Z. Phys* **145**, 375 (1956).
[2.4] A. Tonomura *et al.*, *J. Electron Microsc.* **28**, 1 (1979).
[2.5] A. Tonomura *et al.*, *Phys. Rev. Lett.* **54**, 60 (1985).
[2.6] Q. Ru *et al.*, *Ultramicroscopy* **53**, 1 (1994).
[2.7] J. H. Bruning, *Optical Shop Testing*, ed. D. Malacara, p. 409 (Wiley, New York 1978).
[2.8] T. Yatagai *et al.*, *Appl. Opt.* **26**, 377 (1987).
[2.9] S. Hasegawa *et al.*, *Appl. Phys.* **65**, 2000 (1989).
[2.10] J. Chen *et al.*, *Optics Communications* **110**, 33 (1994).
[2.11] T. Hirayama *et al.*, *Ultramicroscopy* **54**, 9 (1994).
[2.12] For most recent technical developments and applications, see *Proc. Intern. Workshop on Electron Holography*, Knoxville, August 29–31, 1994, eds. A. Tonomura, L. F. Allard, G. Pozzi, D. C. Joy & Y. A. Ono, (Elsevier Science B. V., Amsterdam, 1995).
[2.13] G. Lai *et al.*, *Appl. Opt.* **33**, 829 (1994) .
[2.14] G. Lai *et al.*, *J. Appl. Phys.* **75**, 4593 (1994).

[2.15] N. Osakabe *et al.*, *Phys. Rev. Lett.* **62**, 2969 (1989).
[2.16] A. Tonomura *et al.*, *Phys. Rev. Lett.* **44**, 1430 (1980).
[2.17] T. Hirayama *et al.*, *Appl. Phys. Lett.* **63**, 418 (1993).
[2.18] A. Tonomura *et al.*, *Phys. Rev. B* **34**, 5 (1986).
[2.19] J. Bonevich *et al.*, *Phys. Rev. Lett.* **70**, 2952 (1993).
[2.20] J. Bonevich *et al.*, *Phys Rev. B* **49**, 6800 (1994).
[2.21] K. Harada *et al.*, *Nature* **360**, 51 (1993).
[2.22] T. Kawasaki *et al.*, *Jpn. J. Appl. Phys.* **29**, 508 (1990).
[2.23] K. Harada *et al.*, *Phys. Rev. Lett.* **71**, 3371 (1993).
[2.24] For example, see D. J. Bishop *et al.*, *Science* **255**, 165 (1992).
[2.25] R. N. Kleiman *et al.*, *Phys. Rev. Lett.* **62**, 2331 (1989).

3

Microanalysis by scanning transmission electron microscopy

L. M. BROWN and J. YUAN

3.1 Introduction

The electron–specimen interaction not only provides information about the atomic structure of the materials, but also gives clues about other physical properties of the materials. The science of extracting this valuable information from smaller and smaller volumes of specimens is called microanalysis and is a very fruitful development of modern electron microscopy. Playing a pivotal role in this science is the scanning transmission electron microscope (STEM) which was introduced mainly to provide point-by-point analysis at high spatial resolution. The advantage of scanning is that it enables each pixel to be associated with a data set which might be a diffraction pattern, an X-ray emission spectrum, or an electron energy loss spectrum. The advantage of transmission is that it prevents the beam broadening which degrades the resolution of images formed by scanning a bulk sample. In this chapter, we will review the physical principles of the various imaging and analytical techniques available in the STEM, and assess their usefulness in connection with the research into high-temperature superconductors (HTSC). In many ways, there is a large overlap between the requirement of HTSC research and other branches of the material sciences. Thus we hope that our review can also serve as a brief introduction to STEM for researchers both inside and outside the HTSC community. For specific examples of the application of the STEM in HTSC, we will refer readers to other chapters in this book and the original papers.

3.2 Electron optics of STEM

There are currently two variants of STEM, depending on whether it is specifically designed for scanning microscopy operation or adopted from conventional transmission electron microscopy work. Figure 3.1 shows a

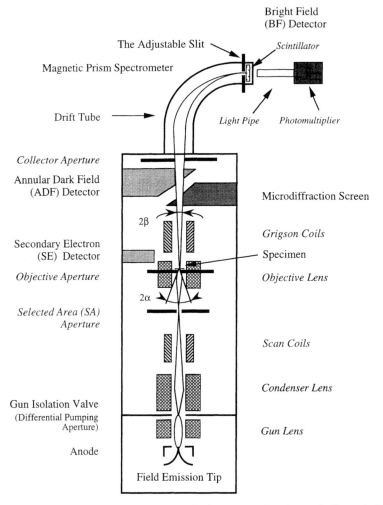

Fig. 3.1. A schematic diagram of electron-optics in a dedicated Scanning Transmission Electron Microscope (VG HB501 STEM).

schematic diagram of the Vacuum Generators (VG) HB501, which belongs to the 'dedicated' variants. It has three lenses, two of them acting essentially as condenser lenses to form a demagnified image of the electron source, which is the tip of a tungsten wire from which electrons are drawn by field emission. The VG HB501 has a UHV chamber with a vacuum of around 10^{-10} Torr for the field-emission source, which is protected from the rest of the column held usually at around 10^{-9} Torr by a differential pumping aperture. The 'gun lens' acts to collect as much current as possible from the tip, and the second condenser lens to transfer the image to the objective lens, which forms a high-quality image of the source on the electron-transparent specimen. There are two important apertures: the objective (probe-forming) aperture, of angle α, and the collector aperture, of angle β. In a VG STEM, the specimen is immersed in the field of the objective lens, causing additional compression of the scattering angle of the transmitted beam, so the effective collection angle is $M\beta$, where M is the compression factor. M has a typical value of 5 but the precise value varies depending on the strength of the objective lens excitation. Various scanning coils are available to control the probe. In particular, so-called 'Grigson coils' can scan the scattered electrons and direct them into a collector aperture and the electron spectrometer.

Usefulness of microanalysis information has also encouraged people to adopt TEM for STEM operation. Figure 3.1 shows a schematic of scanning TEM. The finely focused probe is produced by a combination of strongly excited condenser lens and the prefield of the objective lens. The second condenser aperture acts in the same way as the objective aperture in dedicated STEM to limit the convergence angle so as to control the spherical aberration. The beam deflection coil of the beam illumination system is used to pivot the electron beam about the front focal plane of the objective lens so as to produce a scanning probe at the specimen plane. In principle, like in 'dedicated' STEM, no more lenses are needed for STEM operation. In practice, additional lenses are used to couple the transmitted electron beams to detectors such as in electron energy loss spectroscopy as shown. This complicates the factors controlling the collection angle for the transmitted electrons.

Not shown in Fig. 3.1 are other analytical detectors normally found in STEM, because of lack of space. The principal omission is an X-ray detector that analyses X-rays emitted by the specimen, others include detectors designed to collect cathodoluminescence, specimen current, Auger electrons, etc. The lack of physical space around the specimen is often the deciding factor on which detectors are employed on a particular machine. In general, dedicated STEM has room to employ more detectors. In these respects, STEM is very much about detection systems, and not so much about the probe formation.

Having said that, the electron-optics of STEM is dedicated to the production of a small probe. Before we review the probe formation, it is profitable to ask: what determines the useful probe size for a realistic sample with finite thickness? Part of this question can be answered through an understanding of the physics of electron scattering in solids, and its effect on the beam broadening.

3.2.1 Beam broadening

Monte Carlo calculations [3.1] give the best impression of the beam scattering which controls the interaction volume between the fast electron beam and the sample. They are performed by taking account of the dominant process affecting the beam broadening, namely elastic scattering by atomic nuclei (Rutherford scattering). Figure 3.2 shows an example of such calculations. Only by selecting a thin film sample can one avoid the broadening. For samples thin enough that the beam is not scattered through very large angles, it is easy to show that the diameter of the beam at depth t is proportional to the $3/2$ power of the depth. The original formula due to Goldstein *et al.* [3.2] has been found accurate:

$$d = 0.198 \frac{Z}{E_0} \sqrt{\frac{\rho}{A}} t^{2/3} \text{ (nm).} \tag{3.1}$$

Here, d is the diameter within which 90% of the beam is to be found, Z and A are the atomic number and mass respectively of the element, E_0 is the primary beam energy in kilovolts, and ρ is the sample density in grams per cubic centimeter. This produces beam sizes for 100 keV incident electrons as shown in Table 3.1, with most data taken from [3.2] except for the new estimate for the oxide superconductor $YBa_2Cu_3O_7$ (YBCO). The latter is achieved by substituting Z with the composition averaged $\sqrt{\langle Z^2 \rangle}$ and replacing A with the composition averaged $\langle A \rangle$, from cross-sectional considerations.

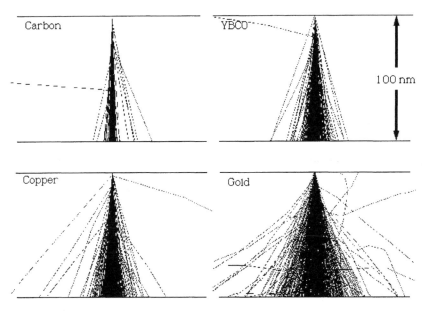

Fig. 3.2. Monte Carlo simulation of beam broadening in a few representative 1000 Å thick films for 100 keV electrons, produced from D. Joy's program SS_mc [3.1] with the help of R. Darji.

Table 3.1. *Estimates of beam broadening,*
b/nm, for 100 keV electrons

	Film Thickness, *t*(nm)				
Element	10	50	100	300	500
Carbon	0.16	1.8	5.1	27	57
Aluminium	0.26	2.9	8.1	42	91
YBCO	0.66	7.4	20.9	109	*
Copper	0.68	7.6	21	112	*
Gold	1.6	17	*	*	*

* Scattering angle $> 20°$.

From Table 3.1 it can be seen that the fine probe, perhaps 0.5 nm at the entrant surface of the foil, is largely wasted even in a film of carbon 50 nm thick, if all the transmitted electrons or the secondary signals produced from them are used as a signal. The beam broadening effect in YBCO is comparable with that of the Cu. The fact that the probe broadens as the 3/2 power of the film thickness is easily understood: the number of collisions is t/λ, where λ is the mean free path for scattering. The inelastic scattering is to very small angles, so it is the elastic scattering which predominates. The electron may be scattered to the left or to the right, but the mean angle regardless of sign will be $(\tau/\lambda)^{1/2}\langle\phi\rangle$, where $\langle\phi\rangle$ is the root mean square angle for single elastic scattering. Then the film thickness multiplied by the mean scattered angle will give the width of the beam at the electron exit surface, and this will be proportional to $t^{3/2}$. Evidently a large accelerating voltage will reduce the broadening, by increasing λ.

Electrons incident along highly symmetric orientations of crystalline materials can channel between atomic columns quite some distance into the materials, before suffering significant large angle scattering (dechannelling). Thus, the beam broadening will be delayed in these special circumstances [3.3].

3.2.2 The role of beam brightness

After the probe size, the current within it is another very important parameter of the STEM. A small probe with a small electron flux may be good on the specification tables of the manufacturers' brochures but is useless in practice. Because electromagnetic lenses suffer from large aberrations, only a small cone of illumination along the central optical axis is useful for information transfer. The important parameter regarding the current in the probe is then determined by the *brightness*, B, namely the current density per unit solid angle. An important property of the brightness is that its value is conserved by perfect optical devices. Suppose that there is a detector of scattered electrons: whatever its nature (elastic scattering, electron spectrometer, etc.) it will have an entrance slit making a solid angle Ω_c (subscript c for collector). Then the fraction of the solid angle covered by the detector will be Ω_c/Ω_i, where Ω_i is the solid angle made by the image of the source, the 'exit pupil' of the lens. If the cross-section for scattering is σ, then the signal in electrons per second – the quantity that determines the shot noise in the signal – is given by $\sigma B\Omega_c/e$. Thus it is the brightness which fundamentally controls the signal-to-noise ratio for a given time of acquisition of the image in any scanning optical system. Of course, one chooses the largest collector aperture consistent with the desired quality of analytical information. (In terms of the semi-angles usually used to define an aperture opening, $\Omega_c = \pi\beta^2$.)

Similarly, the brightness is fundamentally the property of the electron source and can only be altered by changing the accelerating voltage. In the following, we list some typical numbers for the common electron sources used in an electron microscope operating at 100 keV [3.4, 3.5].

> *Thermionic* source brightnesses are about 10^5 A m^{-2} sr^{-1}; maximum current several microamperes.
>
> LaB_6 source brightnesses are about 10^6 A m^{-2} sr^{-1}; maximum current several microamperes.
>
> *Field emission* source brightnesses are about 10^{10} A m^{-2} sr^{-1}; maximum current 0.1 μA.

Thus we see that the field emission source permits many orders of magnitude more counts in any signal than the other sources, if ideal lenses are employed. In all cases, raising the acceleration voltage can increase the intrinsic brightness, by skewing the electron velocity distribution more along the optical axis.

3.2.3 Useful probe sizes

However, electron microscope lenses are far from ideal, and the probe is broadened by spherical aberration (proportional to the cube of the aperture size) and by diffraction (inversely proportional to the aperture size). Balancing the two effects by finding an optimal aperture size (α_0) gives the theoretical minimum probe sizes at the specimen plane. However, the current within the probe size may be too small for useful signal to be collected within an acceptable time. Thus, while many manufacturers will claim smaller probe sizes, in reality the *useful* minimum probe size, one that yields enough analytical information, is generally larger than the theoretical electron-optical limit.

We will illustrate this with an example. Let us imagine that the probe will be used to form a convergent beam diffraction pattern. Then there might be typically 10^6 picture elements, each requiring 10^4 electrons; a total throughput of about 10^{10} electrons. On the other hand, let us imagine an electron energy loss spectrum, with 10^4 channels requiring 10^6 electrons per channel to get an acceptable signal above the background; again a total throughput of 10^{10} electrons. An exposure time of 10 s is about the maximum usually available to avoid too much drift or contamination of the specimen. It follows that one must have total current of about 0.16 nA to achieve acceptable analytical output. Under these conditions, using a typical value for the spherical aberration coefficient, Brown [3.6] calculated the effective probe size achievable using various sources. The results are shown in Fig. 3.3. We see that for small probe-forming apertures the beam diameter is diffraction-limited, but for large apertures it is aberration-limited. The minimum achievable probe size, at a probe-forming aperture semi-angle of about 8 mrad, is about 1 nm for field emission, 10 nm for LaB_6, and 30 nm for thermionic emission from a tungsten filament. These figures are roughly borne out in practice.

3.2.4 Energy spread of electron sources

Being a primary component of an analytical electron microscope, the choice of the electron source in STEM also has relevance to the performance of the

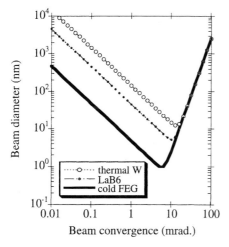

Fig. 3.3. Effective probe size available using various electron sources, as a function of the aperture size of the probe forming lens.

spectrometer attached to it. For example, the ultimate resolution of electron energy loss spectroscopy (EELS) is determined by the energy spread in the fast electron source, for STEM without the use of the monochromator (which is true almost all the time). The thermal spread (ΔE_{ther}) of the electron sources is 2.45 kT_{s} for a Maxwellian electron distribution which applies to thermionic sources. For a tungsten tip operating at 2800 K, ΔE_{ther} is 0.6 eV, and that for LaB$_6$ tip operating at the lower temperature of 1700 K is 0.3 eV. The actual energy spread measured (1–2 eV for tungsten filament) is always larger than these values suggest. The differences have been traced by Boersch [3.7] to electron–electron interactions which occur mainly at various crossovers where the electron density is very high. For a 'cold' field emission gun (FEG) operating at room temperature, the thermal spread is only 0.03 eV, hence negligible. The space–charge (Boersch) effect is also not very important for a tip with such small total emission current [3.8]. The actual measured spread (order of 0.3 eV) is determined by the Fowler–Nordheim distribution which has a sharp Fermi level cut-off [3.9] plus a long tail at high energy. Thus a STEM equipped with a cold field emission gun, such as a VG HB501, can have the additional advantage of resolving many fine features without paying the penalty of reduced electron counts from using a monochromator.

3.2.5 *Coherence of the electron source*

The high brightness of the field emission gun is largely due to the small emission area of the source. According to the Van Cittert–Zernike theorem for any optical system [3.10], the illumination at the probe-forming lens is then highly coherent. This coherent electron probe is formed at the sample surface by adding the wavefunction amplitude from all points in the objective aperture, not their intensity. This has the consequence of producing a probe profile far from Gaussian, containing subsidiary zeros due to interference effects. The analysis given in Section 3.3 is based on geometric optics for incoherent illumination. A proper analysis of the coherent probe formation requires the aid of wave optics. Still the conclusion reached in Section 3.3 regarding the merit of the various electron sources will not be altered, but the detailed pattern of the current flow may be more complicated [3.11]. The small energy spread of the field emission gun also ensures high temporal coherence. Both transverse coherence of the electron probe (linked to source size) and longitudinal coherence (linked to energy spread) allow the possibility of electron inter-ference while propagating through the sample to the detector. The degree of transverse coherence of the illumination at the probe-forming (objective) lens is controlled by the size of the cross-over after the demagnifying (condenser) lens. Since the size of the cross-over is related to the angular size of the source, the transverse coherency like the probe size, is inversely related to the useful current in the probe.

In summary, to produce a useful probe of subnanometer size, a high-brightness gun is essential together with optimised use of the probe-forming aperture to limit the aberration effects. Among the choice of the various electron sources, the field emission gun stands out. The penalty to be paid for a field emission source is the necessity to use ultra-high vacuum techniques. Such instruments were usually restricted to the 'dedicated STEM', but nowadays field emission sources are also popular in analytical TEMs. The improvement in the latter has blurred the distinction between 'dedicated STEM' and 'TEM-STEM'.

3.3 Imaging in STEM

3.3.1 Reciprocity and phase contrast

In the early days of STEM, great attention was paid to the surprising result that even with a point-by-point scanning system, it is possible to observe phase-contrast effects. The reason for this is 'reciprocity', as first pointed out by Cowley [3.12]. Figure 3.4 shows ray diagrams for two idealised instruments, TEM and STEM. It can be seen that if the source and the collector are interchanged, the ray paths for the two instruments are identical. It follows that all phase contrast phenomena observed in the TEM can be observed in the STEM. In particular, Fresnel fringes can be seen. The collector aperture plays the role of the collimating aperture in TEM, and the probe-forming aperture the role of the objective aperture. Reducing the collector aperture in STEM is equivalent to reducing the beam divergence in TEM, thereby improving the coherence of the illumination and producing more Fresnel fringes. Provided drift, beam voltage, and the number of electrons per picture element are the same, the same image is expected from the two techniques, and in particular, the 'resolution' governed by the electron optics alone is the same, approximately $(C_s \lambda^3)^{1/4}$. In STEM, this is simply to be regarded as the optimum probe size achieved by balancing spherical aberration (characterised by the coefficient C_s) against diffraction; this is the minimum in the curves of Fig. 3.3, if an infinite number of electrons are available and there is no brightness limitation. In fact, STEM is a very inefficient way of collecting a phase-contrast image, because the collection is accomplished serially, point-by-point, rather than by parallel exposure of all the picture elements together, as in TEM. The miracle is, that in STEM it can be done at all!

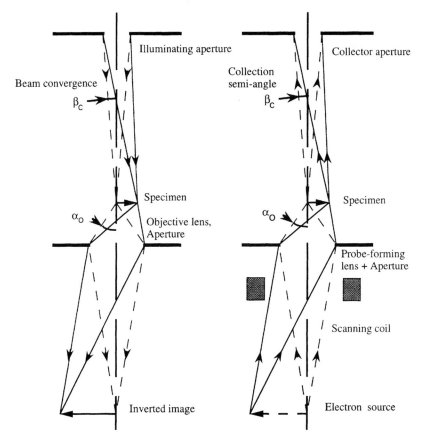

Fig. 3.4. Reciprocity relationship for elastic scattering in TEM (left) and STEM (right).

An important limitation of imaging by STEM is the electronic display utilised: a cathode ray tube has about 10^3 lines, giving 10^6 picture elements, compared with a photographic film which has perhaps fifty times more. Thus information is lost because of the inefficient recording system. Provided one or less resolution element of $(C_s\lambda^3)$ is acquired per picture element in the display, the display does not degrade the resolution. In practice, with a display size of 10 cm on a side and an electron optical resolution of 0.5 nm, a minimum magnification of $10^{-1}/10^3 \times 0.5 \times 10^{-9} = 200\,000\times$ is needed. Below this critical magnification, the STEM image is degraded by the display, and photographic enlargement will not reveal more specimen detail. Above this magnification, all the specimen information capable of being registered by the electron beam is present. Of course, the field of view available in the display is much reduced from that in a photographic image. The STEM is a high-magnification, small field-of-view, imaging system.

3.3.2 ADF/HAADF

The most useful imaging device in STEM is the annular detector, particularly the high-angle annular detector pioneered by Crewe [3.13] and by Howie [3.14]. This is shown schematically in Fig. 3.1. The idea is to use a scintillator in the form of a disc with a hole in it, to allow the forward scattered beam through, but to collect the electrons scattered to high angles. These suffer elastic scattering. If the angle is large enough, even in crystals, the scattering is incoherent. In physical terms, if the spacing of Bragg planes is less than the atomic vibration amplitude, coherent Bragg scattering cannot occur. The incoherent elastic scattering follows the Rutherford scattering law, and is proportional to Z^2, the square of the atomic number. The forward scattering, largely inelastic, is proportional to Z. Thus the ratio of the large-angle signal to the forward signal is proportional to Z: a signal denoted 'Z-contrast' by Crewe. Using this method, Crewe was able to image single heavy atoms [3.15]; and, most recently, using the high-angle version of Crewe's detector, atomic column resolution was achieved in crystals [3.3, 3.16]. Not only that, but the inelastic signal can be passed through a spectrometer, giving column-by-column micro-analysis [3.17]. Most published STEM images are taken with the annular detector, which is very efficient. The images are susceptible to easy interpretation, because they bypass the interference effects which dominate phase-contrast imaging. A further factor in 'column-by-column' imaging is the channelling of the probe upon entering the specimen, due to the attractive potential at the ion cores. This delays the broadening which otherwise results from diffraction, and increases the foil thickness over which column-by-column imaging can be achieved. Many modern examples can be given of Z-contrast imaging. Some authors nowadays do not even reveal that their published images are ADF or HAADF images, so common have the techniques become.

3.3.3 Holography

Still in the development stage are the various forms of electron holography which allow the phase information of the sample to be recovered [3.18]. The STEM has the built-in advantage of a highly coherent electron source because such machines usually have FEG tips. It is mentioned here because of the potential implications for superconductors in terms of imaging the flux distribution in type-II superconductors. We may see development along this line if and when the cryogenics can be sorted out.

3.4 Microanalysis in STEM

Among various signals coming from the volume excited by the focused probe in STEM, we will introduce two forms of spectroscopy that have proved to be most useful in oxide superconductor research, as well as microdiffraction.

3.4.1 Electron energy loss spectroscopy

1. Spectrometers and parallel acquisition systems

The most popular choice of analysing the energy distribution of the transmitted electrons is to pass them through a magnetic sector spectrometer, as shown schematically in Fig. 3.1. To first order, it bends the electrons so that rays from a point on the entrance side are brought together at a point on the exit side if they have the same velocity or energy. Modern spectrometers are designed to be 'double focusing', that is, trajectories out of the plane of the figure are also focused; this is achieved by selecting the angles of the entrance and exit faces of the magnet. They are also designed to eliminate second-order aberrations by curving the faces, so those electrons slightly off axis experience different lengths of trajectory. Two approximate equations are helpful: the first gives the dispersion for a spectrometer which bends the electron beam to a radius R

$$\delta E/E = \delta x/R \tag{3.2}$$

so that if R is about 0.1 m, the dispersion δx is about 1 μm per volt of energy loss for a 100 kV fast electron. The second equation gives the figure of merit for a spectrometer

$$\delta\Omega/(\delta E/E) = 100 \tag{3.3}$$

so that the (solid angle) acceptance aperture $\delta\Omega$ is about 10^{-3} sr for 1 V resolution in 100 kV; this corresponds to a semi-angle for the entrance pupil of about 17 mrad. The formula emphasises the main practical problem of spectroscopy: as the aperture is opened, the signal increases, but the resolution is degraded. The trick is to find an acceptable compromise, bearing in mind radiation damage, drift, and any other exposure-limiting factors.

   Equations (3.2) and (3.3) are approximate formulae for spectrometers of modern design without lenses. The actual value for the VG prism spectrometer used in our HB-501 STEM is about 1.8 μm eV^{-1}. In parallel detection mode, the dispersion achieved is usually further magnified by quadrupole lenses. A single quadrupole produces a line focus. Two such lenses in series can act as a round lens. Modern parallel spectrometer systems use three or four quadrupoles in series to adjust the dispersion and to operate either with a line focus for spectral analysis or a point focus for recording microdiffraction patterns.

The spectrum is converted to light using a transmission phosphor, then converted to electrons using a photomultiplier for serial acquisitions or using a 'position sensitive detector' for parallel acquisition, either a self-scanned photodiode array or a charge-coupled device. The former has been used by Gatan systems [3.19] whereas the latter has been developed by McMullan and co-workers (see [3.20] and papers referenced therein). The latter system has many advantages, not least the ability to produce a full two-dimensional image of the spectrum as well as energy-filtered microdiffraction patterns. For this reason, systems designed for energy-filtered imaging in the conventional microscope also use CCD detection [3.21], and the future will see the predominance of this method. Figure 3.5 shows a schematic diagram of the McMullan system currently in operation.

One of the parameters characterising the EELS detection system is its Detective Quantum Efficiency, DQE, defined as the ratio of the number of counts to the mean square fluctuation in them. A detection system is said to have unit DQE if it is shot noise limited, i.e. the mean square signal variation in a channel is equal to the number of counts within it. However, channel-to-channel gain variations in photodiode arrays, dark current, and detector noise

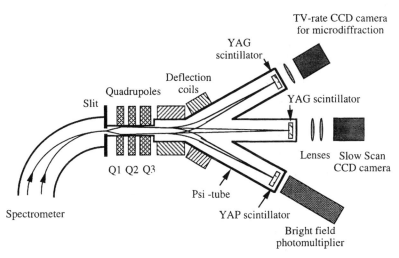

Fig. 3.5. A schematic diagram of the McMullan PEELS acquisition system currently in operation (courtesy of Dr C. Walsh). Simple switching of the post-quadrupole deflection coil can direct the transmitted electrons to separate detectors to form a bright field image as well as a two-dimensional microdiffraction image respectively.

of various sorts, make it difficult to come close to unit DQE [3.19]. The best performance so far achieved with the McMullan system using a cooled slow-scan CCD gives a DQE of about $1/2$; the diode-array systems have values substantially less than this.

The other important parameter characterising the EELS detection system is its dynamical range. Serial acquisition can cope better with spectra containing the zero loss peak (involving elastic and quasi-elastic scattered electrons only), where signal variation is in excess of 10^5. In parallel detection systems, some form of signal attenuation has to be applied to reduce the intensity of the strong signal to the detector designed for single electron collection.

A limitation of the magnetic prism-based EELS detection system is the spectrum drift in the energy dispersion direction. Various sources of instability in the microscope contribute to this effect: High voltage fluctuation, magnetic field creep etc. This places a fundamental limit on the useful exposure time. Experimental efforts in electron spectroscopy consist largely in reducing unwanted electrical noise and specimen drift.

2. Virtual photon field picture of EELS

What is measured in electron energy loss spectroscopy is the optical absorptive properties of the materials. This statement can be made clearer if we adopt the virtual photon field picture pioneered by C. F. Weizsäcker and E. J. Williams [3.22] by considering the time-dependent electric field generated at the atom site by the passage of a fast electron nearby:

$$E(t) = \frac{r(t)}{4\pi\varepsilon_0 r(t)^3} \tag{3.4}$$

where $r(t)$ is the separation of the atom and the fast moving electrons, ε_0 the vacuum permittivity. The energy loss suffered by the fast electrons is exactly equal to the energy absorbed by the atomic electrons when driven by this time-dependent electric field, with a rate given by [3.23]:

$$\text{Im}[\alpha(\omega)]|E(\omega)|^2 \tag{3.5}$$

where $\text{Im}[\alpha(\omega)]$ is the imaginary part of the atomic polarisability and $E(\omega)$ the Fourier components of the field $E(t)$. The form of (3.5) is similar to the optical absorption formula if we assume $E(\omega)$ represents a photon field associated with the fast electron. This has been proved to be very useful for spectral interpretation.

3. Plasmon excitation

In the low-energy (low-frequency ω) region, the resonance between different atomic electrons needs to be considered and can be heuristically included if

Im $a(\omega)$ can be replaced by the imaginary part of the effective polarisability by taking into account the screening of the neighbouring atoms, with a mean-field value of

$$\text{Im}\left\{a(\omega)\left(1 + \frac{2a(\omega)\rho}{3\varepsilon_0}\right)^{-1}\right\}|E(\omega)|^2 \propto \text{Im}[-1/\varepsilon(\omega)]|E(\omega)|^2, \qquad (3.6)$$

where ρ is the atomic density. The expression (3.6) in terms of the dielectric function $\varepsilon(\omega)$ of the material is obtained with the aid of the Clausius–Mossotti relation, and agrees with the result of more detailed derivations. It can be reduced to (3.5) in the limit of low density (atomic case) or a small polarisability (high energy losses).

The factor $1 + (2a\rho/3\varepsilon_0)^{-1}$ is unique to EELS excitation and its pole represents the collective resonance of all the screening atomic electrons, even in the absence of the external field E, i.e. plasmon excitation. In the case of oxide superconductors, of great interest is the carrier density n, in the superconducting plane. This can be related to the intraband plasmon excitation energy (i.e. the pole energy) via the relation:

$$\omega_{\text{p}}^2 = \frac{ne^2}{\varepsilon_0 m}. \qquad (3.7)$$

Here, n is the number of electrons per unit volume contributing to the excitation. Typically, the carrier plasmon excitation in high-temperature oxide superconductors peaks at or below 1 eV [3.24]. This places the plasmon excitation peak too close to the tail of the very intense zero loss peak centred at zero loss, for it to be realistically distinguished in most of the current generation of EELS in STEM. Researchers studying the low-energy loss excitation in high-temperature oxide superconductors by STEM start the spectral analysis from 1 eV upwards. The higher energy peaks in the low loss spectra are then due to the plasmon excitation, involving the whole valence band, which are not that illuminating for superconductivity properties.

4. Kramers–Kronig analysis and interband transition

However, usually EELS can be easily measured over an extended energy range from 1 eV upwards and very accurately. Together with suitable extrapolation down to 0 eV (usually with the aid of optical data), the function $\text{Im}[-\varepsilon^{-1}]$ can be deduced over the whole energy range where it has significant intensity. The causality relation governing linear response functions can be used to relate $\text{Re}[-\varepsilon^{-1}]$ with $\text{Im}[-\varepsilon^{-1}]$:

$$\text{Re}[-\varepsilon^{-1}(\omega)] = \frac{\pi}{2}\int_0^\infty d\omega' \frac{\omega' \text{Im}[-\varepsilon^{-1}(\omega')]}{\omega^2 - \omega'^2}. \qquad (3.8)$$

This is called the Kramers–Kronig (KK) relationship, from which the dielectric function $\varepsilon = \varepsilon_1 + i\varepsilon_2$ can be derived [3.25]. Since ε is also a linear response function, ε_1 and ε_2 are again related by the KK relationship, thus the information contained in the dielectric function can be examined by concentrating on one of the two components of the dielectric function. We choose to work with $\varepsilon_2(\omega)$ because it is what optical (X-ray) absorption spectroscopy measures and can be directly related to the atomic polarisability $\mathrm{Im}[\alpha(\omega)]$ that appeared in (3.5).

Application of the Fermi golden rule allows one to relate ε_2 to atomic transitions in the atoms

$$\varepsilon_2 \propto |\langle \phi_\mathrm{f} | x | \phi_\mathrm{i} \rangle|^2 \delta(E_\mathrm{i} - E_\mathrm{f} - \Delta E) \rho(E_\mathrm{i}) \rho(E_\mathrm{f}) \qquad (3.9)$$

where ϕ_i and ϕ_f are the initial and final states of the wavefunction of the atomic electrons in the solid, $\rho(E)$ defines the density of states of initial and final states. Because of energy conservation during the transition, ε_2 measures the product of the matrix element with the joint density of states:

$$\mathrm{JDOS} = \delta(E_\mathrm{i} - E_\mathrm{f} - \Delta E) \rho(E_\mathrm{i}) \rho(E_\mathrm{f}). \qquad (3.10)$$

The matrix element is usually a weak function of the energy so most of the features in ε_2 can be attributed to the critical points in the joint density of states which can be related to the electronic structure of the material. Such critical points occur where the valence band runs nearly parallel to an unoccupied conduction band.

5. Inner-shell absorption spectroscopy

The most important use of the EELS method in high-temperature oxide superconductors has been to investigate the O 1s and Cu 2p absorption edges. The density of states $\rho(E_\mathrm{i})$ of the core electron level is very narrow so the measured JDOS mainly reflects the density of the final unoccupied states in the single-electron approximation. The matrix element determines the symmetry of the final states reachable from a given core level. In the case of a dipole transition, the final states reached from O 1s and Cu 2p core levels are precisely the O 2p and Cu 3d orbitals whose bonding are critical to the electronic structure near the Fermi level in the CuO_2 sheets in all the high-temperature oxide superconducting materials. In particular, the pre-edge peak at 529 eV in the O 1s absorption spectra can be identified with carriers in the hybridised CuO band in hole-doped superconductors (see Fig. 3.6). This affords a simple measurement of the local carrier density in superconductors with nanoscale resolution when EELS is used in STEM [3.26]. Such high resolution is necessary because of the intrinsic short coherence length in HTSC materials.

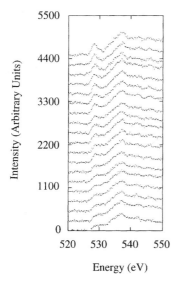

Fig. 3.6. Oxygen 1s absorption (K-)edge observed in $YBa_2Cu_3O_{7-\delta}$ as a function of the oxygen deficiency δ which also controls the doping in the CuO_2 plane [3.26].

6. The nature of the transient field and the validity of dipole selection rules

Expressions (3.5) and (3.6) suggest that a knowledge of the transient field E can allow us to make deductions from the EELS spectra. In our case, we can assume that the fast electron is hardly deflected by inelastic scattering. The separation between the atomic electrons and the swift electron is given by

$$r(t) = b + v(t) \tag{3.11}$$

where the vector b is the closest approach between the swift electron and the atom. The Fourier transform of the electric field (3.4) then gives

$$E(q, \omega) = \frac{2eq}{q^2} \delta(\omega - q \cdot v). \tag{3.12}$$

This expression highlights an important distinction between EELS and optical experiment, namely that the virtual photon field associated with the fast electron is longitudinal ($E \| q$) whereas the real photon field is transverse ($E \perp q$) [3.23]. This difference does not affect the main point of the photon analogy namely that electron energy loss and optical experiments measure the same quantity $\varepsilon(\omega)$ in most (and practically all the) cases. It does mean that the polarisation of the electric field acting on the atomic system is defined differently in both cases, a fact of obvious importance for anisotropic materials.

Examination of (3.12) also suggests that most excitations involve small momentum transfers because of the kinematical factor q^{-2}. The momentum transfer vector is related to the scattering angle through momentum conservation (Fig. 3.7). Small momentum transfer then results in small-angle scattering, justifying the use of a straight-line trajectory to calculate the transient field at the atomic site.

The small-angle scattering also means that atomic transitions are mostly dipole in character, as we have assumed in most of the above discussion implicitly. This may be verified for the O 1s and Cu 2p absorption using typical collection angles used in the experimental set-up. The dipole selection rule holds as long as $qa \leqslant 1$, where a is the mean atomic radius of the core electron wavefunction. Using tabulated Hartree–Fock wavefunctions, the radii are estimated to be about 0.1 nm for both O 1 s and Cu 2p electrons. The wavevector of the 100 keV electron is 1697 nm^{-1}. For a typical collection aperture semi-angle of 10 mrad, the maximum momentum transfer is $\sim 17 \, nm^{-1}$, hence $qa \leqslant 1.7$, i.e. the dipole region is just about covered.

7. Anisotropic electron energy loss spectroscopy

A characteristic of the electronic structure of the oxide superconductor is the two-dimensional nature of the common CuO sheets. This has important implications for EELS interpretation and the dielectric function, particularly for states near the Fermi level. Thus, for example, a transition from the O 1s core level to the hole states in the CuO plane in $Bi_2Sr_2CaCu_2O_8$ material will only occur if the applied electric field lies in the plane of the CuO sheets [3.24]. Because of the nature of longitudinal excitation, the direction of the **E** field is determined by the direction of the momentum transfer.

In EELS, it can be easily demonstrated, through conservation of momentum, that the orientation of the momentum transfer vector **q** is a rapidly varying function of the scattering angle in the important small-angle range where the dipole limit is still obeyed (Fig. 3.7). In normal STEM acquisition mode, the use of a convergent probe inevitably means that spectra acquired contain a range of momentum transfer vectors **q**. At θ_E $(= \Delta E/2E)$, the momentum transfer vector **q** is oriented more than 45° from the beam direction. The value of θ_E is 2.6 (4.6) mrad for O 1s (Cu 2p) absorption at an energy loss of the order of 530 eV (930 eV). This is to be compared with the typical convergence angle of 8 mrad for optimal probe formation. In the case of uniaxial materials such as oxide superconductors, all the spectra contain linear combinations of the two independent components: $\parallel$ and $\perp$ to the crystal c-axis of the material. The exact proportion of the two components in the spectra can be determined by taking into account the kinetics of the inelastic scattering as done in [3.27]

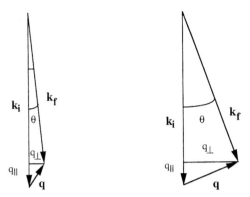

Fig. 3.7. The momentum conservation in inelastic scattering. The momentum transfer is dependent on the scattering angle and its orientation changes rapidly at the important small-angle limit.

and [3.28]. Such analysis is indispensable if quantitative spectral analysis is required.

3.5 X-ray fluorescence spectroscopy

Once atomic electrons are promoted to excited states, they inevitably want to dispose of this excess energy. When the excited atom has electrons ejected from its inner core states, the subsequent filling of the core hole is accompanied by either X-ray fluorescence or Auger electron emission. Fluorescence dominates over Auger processes for deeper core relaxation. Both fluorescence and Auger emissions are characteristic of the excited atoms, hence providing the microscopes with highly sought chemical information. The relatively short mean free path of the Auger electrons in bulk solids (on the order of few nanometers) means that only those produced near the surface layer can escape to the free space to be collected. This restricts the application of Auger spectroscopy to surface study. By contrast, the relatively large mean free paths of X-rays, particularly those of high-energy photons, allows signals to be collected from the whole volume of the thin-film sample probed by the electron beam.

The X-ray signal in STEM is usually collected by compact solid state detectors based on lithium-drifted silicon diodes. The X-ray photon is sorted by energy, hence the alternative names for X-ray fluorescence analysis: Energy Dispersive X-ray Spectroscopy (EDS or EDX). The typical energy resolution for an X-ray photon is of on the order of 150 eV. This is sufficient in most cases for resolving peaks of different elements, but is inadequate for detecting chemical shifts of the atoms which are of the order of a few electronvolts. EDX is thus mainly used for composition analysis. There are a few peculiarities of EDX detectors that are worth mentioning. The Si(Li) diode is operated at liquid nitrogen temperature. To prevent condensation or contamination build up on the cooled silicon crystal surface, the detector is usually isolated from the microscope column by a thick beryllium 'window'. The window is transparent only to X-rays of energy greater than about 1 keV. Thus the conventional EDX is only usable for X-rays generated from elements whose atomic number is from 11 (sodium) upwards. Thinner vapour barriers are used in the so-called 'windowless' detector. These extend the range of the detectable light elements down to and including oxygen, an important element in the high-temperature superconductors (see [3.29] for more detail).

The use of electron beam transparent samples facilitates the interpretation of the EDX result. In such thin foils, the self-absorption of the outgoing X-ray and secondary fluorescence can be ignored except for the very soft X-rays such as those produced by excited oxygen atoms. Hence the intensity of character-istic X-rays produced per incident electron is directly given by the formula

$$I_A(\omega) = f_A \sigma(\omega) n_A \, dt I_0 \tag{3.13}$$

where f is the fluorescence yield, σ the cross-section for the inner shell excitation, n_A the atomic density of the element A, d the effective probe diameter, t the foil thickness and I_0 the total current in the probe. Direct measurement of the individual factors involved can be avoided by using the Cliff–Lorimer ratio method [3.4]:

$$\frac{I_A}{I_B} = k_{AB}^{-1} \frac{n_A}{n_B}. \tag{3.14}$$

The constant k_{AB} only depends on the identity of the two elements A and B as well as the accelerating voltage. It can be calibrated with a standard containing a known ratio of the two elements. By convention, k_{AB} is normally tabulated for a particular instrument with element B being Si. The choice of Si is partially influenced by the large number of different silicates available. It is interesting to note that the microanalysis of thin films is carried out to give atomic number ratios, according to eqs. (3.13) and (3.14) whereas the analysis of the bulk materials is performed in terms of mass ratios. This can sometimes cause confusion.

The simplicity of the ratio method for thin films is to be contrasted with the complexity of microanalysis for bulk samples using EDX. Combined with the high spatial resolution achievable, it makes EDX a very attractive analytical technique. It has been extensively used in oxide superconductor research, for example, in phase identification of powdered samples, and in identifying the homogeneity of the elemental distribution. As more and more heavy elements are introduced into the new types of oxide superconductors, the critical thickness beyond which the absorption effect becomes significant decreases. The validity of eqs. (3.2) and (3.3) then needs to be questioned. The critical thickness is also wavelength dependent, being much smaller for soft X-rays than for hard X-rays. This is one of the contributing factors for the unreliability of oxygen stoichiometry determination using windowless EDX spectroscopy.

3.6 Microdiffraction

In crystalline materials, the elastic scattering will split the incident electron probe along several Bragg angles to form a diffraction pattern in the far field. The detection system on the STEM usually consists of a phosphor screen whose image is relayed to an observer through a TV camera, or by scanning the Grigson coils (see Fig. 3.1). Both are less than satisfactory as they do not have sufficient resolution to reveal detailed structure in the microdiffraction pattern. Various 'home-made' microdiffraction detectors have been built; the McMullan system originally designed for parallel EELS detection is one of the most successful (see Fig. 3.5). To operate as a microdiffraction detector [3.30], the quadrupole excitation is altered to act as an image magnifying device. A post-spectrometer slit is placed at the plane conjugate to the specimen to filter out inelastic scattering. The use of a two-dimensional detector overcomes the beam stability problem encountered in the serial scanning mode of operation. Nowadays, there is also a commercially available system through Gatan Inc., USA [3.21].

3.7 Resolution attainable in analysis

To reiterate: STEM is mainly practised because point-by-point analysis can be carried out. The information available is at least three-dimensional: a two-dimensional image, which is a projection of the specimen, and for each picture element, one or more spectra: either electron energy loss or X-ray emission, or any other spectrum such as optical cathodoluminescence. An important question is, given an incident probe of a certain size, d_0 say, how large is the region from which the spectra originate?

The answer to this question is in three parts. The first part is related to the properties of the electron probe, such as the size of the probe and its broadening, as well as the impact parameter problem, i.e. the characteristic distance beyond which the influence of the electric field from the swift electrons is negligible. The second part is more a specimen problem, namely the spatial extent of the excitation induced in the solid. The final part is related to the selectivity of the detector which will be important for some signals. Thus, the size of the interacting volume is different for different signal and detection strategies.

3.7.1 Impact parameter and X-ray microanalysis

A fundamental limitation on the size of the X-ray generating volume is the impact parameter b, which is the distance from the fast electron trajectory where the electrostatic impulse from the electron is sharp enough to excite a given electronic transition. The existence of this parameter is understandable because the electron beam-induced inner-shell absorption responsible for X-ray generation is mediated by the long-range Coulomb force. This parameter is not infinite because the effect of the electron at far field largely cancels out, leaving the field in the direction transverse to the swift electron motion dominant.

The impact parameter can be estimated simply from the following considera-tion: the transverse electrostatic pulse contains frequencies of the order of v/b, where v is the velocity of the fast electron. Thus a core electron bound with an energy ΔE can only be promoted to empty states above the Fermi level in the conduction band if $r < b = hv/\Delta E$. This fundamental property of the radiation can also be seen by explicitly integrating the resultant electric field at an atomic site a distance b away (as given in eq. (3.12)), over the momentum space d^3q giving:

$$E(\omega) = \frac{1}{(2\pi)^{2/3}} \int d^3 q E(q, \omega) e^{ib \cdot q}$$

$$|E(\omega)|^2 = \frac{\hbar^2}{\pi^2 \varepsilon_0 m v^2 a_0} \frac{\omega^2}{v^2} \left(K_0^2 \left(\frac{\omega b}{v} \right) + K_1^2 \left(\frac{\omega b}{v} \right) \right). \tag{3.15}$$

The functions K_0 and K_1 are modified Bessel functions which decay asymptotically as $K_n(t) \propto t^{-1/2} e^{-t}$. The expression (3.15) shows that the energy density of the evanescent wave decays exponentially with $v/\omega b$. For 100 kV incident electrons, this translates to an approximate relationship be-tween the energy loss and the impact parameter,

$$b/\text{nm} = 40/(\Delta E/\text{V}). \tag{3.16}$$

The parameter b is to be interpreted as the resolution obtainable in a perfect instrument, where the resolution is limited only by the uncertainty principle. For losses in the valence region, where ΔE might be 20 V, b is about 2 nm; for the carbon K-loss at 185 V, b is about 0.2 nm. Only for losses around 200 V and above does b shrink to atomic dimensions.

Beam broadening is the next fundamental effect which degrades resolution.

Measurements of the probe diameter at the exit surface of a silicon thin film, measured by the X-ray output from the sharp boundary of a gold film, give results which agree well with the theory of eq. (3.1). It follows that for X-ray microanalysis, one must think of the probe as having a breadth D compounded of three ingredients: the initial size, d_0, the impact parameter, b, and the broadening due to multiple scattering, d. The simplest estimate is obtained by adding these independent contributions in quadrature, yielding

$$D_x^2 = d_0^2 + d^2 + b^2 = d_0^2 + Ct^3 + b^2 \tag{3.17}$$

where the constant C depends upon accelerating voltage, elemental composition, etc., but not upon foil thickness. Thus for thick foils, D_x is proportional to the $3/2$ power of the foil thickness. For a coherent probe along the channelling direction, the beam broadening term may be more complicated.

Since all inner-shell excitation, irrespective of the way it is excited, leads to the same probability of X-ray generation, when detecting inner-shell excitation by X-ray emission, all the current in the probe will contribute. The spatial extent of the inner-shell excitation is localised on the atoms, hence is negligible by comparison. For a probe with significant intensity in a non-Gaussian tail, the size of the probe d_0 entering eq. (3.17) will not be the size of the coherent part which determines the image resolution, but may be much larger [3.11].

1. EELS and detector aperture function

For analysis by electron energy loss, however, the situation is rather different. Here, the detector is usually limited by a collector aperture in order to reduce energy resolution degradation. This has two effects: firstly, only beams spread by an amount $\beta_c t$ are collected, where β_c is the solid angle subtended by the collector aperture. Thus an estimate of D_e, the resolution obtainable by EELS, is

$$D_e^2 = d_0^2 + (\beta_c t)^2 + b^2 \tag{3.18}$$

We see that for EELS analysis, the resolution is degraded only linearly with foil thickness. Comparison with the values in Table 3.1 shows that if the collector aperture is about 10 mrad, and the energy losses are 200 eV or more, nanometer resolution can be maintained for foils up to about 100 nm thick. This thickness coincides approximately with the mean free path for inelastic scattering, so that spectra from thicker foils suffer degradation of energy resolution anyway.

The above simple analysis is applicable if the detector aperture is large enough to collect almost all the scattered electrons [3.31]. This is relatively easy to satisfy because of the forward nature of the inelastic scattering. In this case the estimate for the impact parameter based on treating the swift electron as a classical particle is valid. However, with a small aperture either centred or displaced from the optical axis, the detector is only collecting inelastic scattering from a small part of the momentum transfer space. This is equivalent to integrating $E(q, \omega)$ in eq. (3.12) only through a small momentum space. The effect is that the detector acts as an effective filter in q-space to select only excitations within the allowed range of momentum transfers. For example, when very small apertures are used, only small momentum transfers are detected, i.e. excitation with a large spatial extent. On the other hand, a displaced aperture only senses high momentum transfer, hence more localised excitations. In this way the choice of the detector arrangement influences the localisation of the signals.

2. Microdiffraction

Another important analytical signal is the microdiffraction pattern. Here, the rays are spread by coherent (Bragg) scattering, so the resolution will be given by

$$D_\mu^2 = d_0^2 + (\theta_B t)^2, \tag{3.19}$$

where, because the scattering is elastic, the impact parameter makes no contribution, but the Bragg angles θ_B spread the beam.

   These equations are generally borne out in practice. EELS and microdiffraction provide high-resolution analytical information, whereas EDX usually gives lower resolution. The most striking demonstration of high resolution attainable in EELS is the observation of changing fine structure at interfaces between thin films and the substrates on which they are grown [3.32, 3.33]. Recent work on grain boundaries in metals confirms that electronic structure of segregants can also be determined [3.34, 3.35]. The interpretation of the EELS is, as usual, tricky. On the other hand, EDX, although conveying less information, is easy to use and the spectra are usually easy to interpret. Accurate analysis of strains in a 10 nm strained layer superlattice has recently demonstrated the spatial resolution attainable in microdiffraction; see [3.30].

3.8 Radiation damage and nanolithography

The main limitation on nanoanalysis using STEM is damage to the specimen. A striking indication of this is to estimate the dose received per picture element per second, which is given by

$$\frac{\mathrm{d}D}{\mathrm{d}t} = \frac{I}{\pi d^2 \rho} \sum_i \frac{E_i}{\lambda_i} \qquad (3.20)$$

where I is the beam current, d the effective probe diameter, ρ the mass density of the sample and E_i and λ_i are energy loss and the mean free path of the inelastic scattering processes. If we only consider the plasmon losses, the dose rate is about 10^{12} Gy per second of exposure, where the Gray is the SI unit of dose: $1\ \mathrm{Gy} = 1\ \mathrm{J\,kg^{-1}}$. Lethal whole-body exposure for humans is about 3 Gy. It is therefore not surprising that analytical information is limited in nearly all practical cases by radiation damage.

In the case of high-temperature oxide superconductors, some research has been made into their radiation resistance. The most sensitive work has been carried out in thin films on substrates where changes in T_c due to irradiation are measured. Because backscattered electrons from the substrate may also contribute significantly to the total damage, the result can only serve as a guide on the irradiation damage sustained in the STEM microanalysis. Nevertheless, the experiments suggest that using a dose of electron irradiation similar to that used for STEM analysis can cause a reduction in T_c of about 30 K, but not changes in normal state resistivity in YBCO [3.36]. This suggests that the superconductivity is more fragile than the electronic structure controlling the normal state transport. The effect is partially reversible upon warming to 300 K, suggesting that some effect is probably due to disordering of the oxygen sublattice, rather than the removal of oxygen atoms. This is consistent with the EELS observation that the oxygen content related pre-edge peak in O 1s absorption does not change in the first few spectral analyses. Thus, it is perhaps not realistic to detect superconductivity itself with the highest resolution available in current generation STEMs [3.26], but it is just about adequate to study the local electronic structure underlining the superconductivity.

In practice, a number of obvious experimental tricks are available to keep irradiation effects under control, such as focusing on one area, and moving to a contiguous area to acquire a spectrum; using very short acquisition times from fresh areas to acquire noisy spectra, and then superimposing them to reconstitute an acceptable spectrum – whose spatial resolution is now governed by how close together in space the individual acquisitions can be made [3.34].

Of course, in some circumstances, particularly in inorganic thin films, the damage appears in the form of holes which can be used to create nanostructures. These are ROMs which are robust, carry a very high information density, and can be organised to have a short readout time. In the long run, such techniques will be used to store information conveniently at much higher density than a compact disc, which is the current state-of-the-art for libraries and computer databases. In the case of oxide superconductors, the damage has been put to use as a way to fabricate weak links with dimensions commensurate with the short coherence of the superconductivity involved (Fig. 3.8). High quality junctions with good Josephson junction characteristics have been successfully manufactured [3.36–3.38].

3.9 Suggestions for further reading

The best comprehensive account of transmission electron microscopy, including STEM, is Ludwig Reimer's book entitled *Transmission electron microscopy* published by Springer-Verlag [3.4]. An indispensable companion is R. F. Egerton's book *Electron energy loss spectroscopy in the electron microscope*, published by Plenum Press [3.39]. An important reference on early developments is the article by the inventor of STEM, A. V. Crewe, entitled 'Scanning transmission electron microscopy' [3.40]. The development of the subject is recorded mainly in conference proceedings, of which the Institute of Physics series entitled 'EMAG' appears every other year.

3.10 Summary

Scanning transmission electron microscopes are capable of producing sub-nanometer probes with useful beam currents of about 1 nA. With a sufficiently

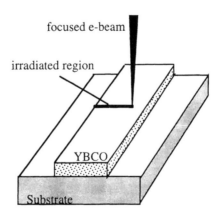

Fig. 3.8. Josephson junction fabricated by a focused electron probe in a STEM.

thin foil (10–100 nm thick), the probe size does not become degraded significantly, so analytical and visual information can be obtained with sub-nanometer resolution. Annular dark field imaging provides a relatively simple image of the microstructure of the superconductor, with chemical sensitivity. Among the analytical techniques available in STEM, EDX and EELS have been reviewed because of their importance in the HTSC research. EELS is the only method that can provide bulk electronic properties of the materials in the samples and is particularly suited for analysing the local electronic structures of defects in the superconductors. The factors affecting the performance of the STEM such as source brightness, beam broadening and radiation damage are examined. We have shown that nanoscale microanalysis and microfabrication as exemplified by the dedicated STEM can play important roles in HTSC research.

3.11 Postscript

A very important recent advance in electron microscope development is in the correction of lens aberration. From the very start of electron microscopy development, the importance of lens aberration as a limiting factor of the spatial resolution of the electron microscope has been recognised [3.41, 3.42]. Ways to correct these aberrations were proposed a long time ago and put into actual microscopes with partial success at least about 30 years ago [3.43]. The main idea is to compensate the positive spherical aberration of the round lens with the negative aberration of multipole lenses.

A recent flurry of activity in aberration correction has been partially prompted by the ease of reliable precision lens control offered by the personal computer and the computer-aided tuning of complex lens systems. Several demonstration models have been developed for TEM [3.44], SEM [3.45] and STEM [3.46] that illustrate the general validity of the concept. The application of spherical aberration correctors is most effective in dedicated STEM because it suffers less from chromatic aberration due to a lack of post-specimen lenses (for example for high-angle annular dark-field imaging).

We foresee that a new generation of dedicated STEM with zero first and second order spherical aberration and minimal chromatic aberration will super-sede the current generation of STEM. The main advantage for microanalysis is higher point-to-point resolution or improved probe current at the same point resolution. As we have seen in the main text, the resolution of the microanalysis is limited more by the signal-to-noise ratio than the theoretical constraint. On the practical side, the elimination of dominant aberration correction will also allow large specimen tilt and/or higher take-off angle for the X-ray detector to be incorporated. In these aberration-corrected electron microscopes, the bright-field image will be very different because of lack of structure in the lens transfer function. Thus a new imaging methodology will have to be developed to extract specimen information from the image contrast. The remaining problem will be the radiation susceptibility of the samples under investigation. Solutions to this problem are also being developed through the use of non-spherical probes. For example, this can be generated as equally spaced fringes either by coherent illumination of two tilted electron sources [3.47] or through dynamical scattering by atomic columns in a crystal [3.48]. Such non-conventional probes can be used to study atomic columns or planar interface structures with much higher spatial resolution without the sample enduring an excessive electron dose.

All these developments are testimony to the vitality of the research in high-resolution microanalysis and will have an important impact in material research such as high-temperature superconductivity.

References

[3.1] D. C. Joy, *Monte Carlo modelling for electron microscopy and microanalysis* (Oxford, Oxford University Press, 1995).

[3.2] J. L. Goldstein *et al.*, *Quantitative X-ray analysis in the electron microscope*, in SEM/1977 (Chicago, Illinois: IIT Research Institute, 1977).

[3.3] S. J. Pennycook & D. E. Jesson, *Phys. Rev. Lett.* **64**, 938 (1990).

[3.4] L. Reimer, *Transmission electron microscopy.* 3rd edn. Springer Series in Optical Sciences, eds. A. L. Schawlow, K. Shimoda, A. E. Siegman & T. Tamir, Vol. 36 (Berlin: Springer-Verlag, 1993).

[3.5] C. Mory, C. Colliex, & J. M. Cowley, *Ultramicroscopy* **21**, 171 (1987).

[3.6] L. M. Brown, *J. Phys. F* **11**, 1 (1981).

[3.7] H. Boersch, *Z. Phys.* **139**, 115 (1954).

[3.8] C. Lea & R. Gomer, *Phys. Rev. Lett.* **25**, 804 (1970).

[3.9] P. E. Batson, *Ultramicroscopy* **18**, 125 (1985).

[3.10] M. Born & E. Wolf, *Principles of Optics*, 6th edn (New York, Pergamon, 1980).

[3.11] C. Colliex & C. Mory, Quantitative aspects of scanning transmission electron microscopy, in *Quantitative electron microscopy*, ed. J. N. Chapman (SUSSP publications, Edinburgh, 1983).

[3.12] J. M. Cowley, *Appl. Phys. Lett* **15**, 58 (1969).

[3.13] A. V. Crewe, High intensity electron sources and scanning electron microscopy, in *Electron microscopy in material science*, ed. U. Valdre (Academic Press, New York, pp. 162–207 1971).

[3.14] M. M. Treacy, A. Howie, & S. J. Pennycook, *Phil. Mag. A* **38**, 569 (1978).

[3.15] A. V. Crewe, J. Wall & J. Langmore, *Science* **168**, 1338 (1970).

[3.16] S. J. Pennycook *et al.*, *Phys. Rev. Lett.* **67**, 765 (1991).

[3.17] N. D. Browning, M. F. Chisholm & S. J. Pennycook, *Nature* **366**, 143 (1993).

[3.18] J. M. Cowley & A. Gribelyuk, *MSA Bulletin* **24**, 438 (1994).

[3.19] O. L. Krivanek, C. C. Ahn & R. B. Keeney, *Ultramicroscopy* **22**, 103 (1987).

[3.20] D. McMullan *et al.*, *Further development of a parallel EELS CCD detector for a VG HB501 STEM* in EUREM 92 Paris (1992).

[3.21] O. L. Krivanek & P. E. Mooney, *Ultramicroscopy* **49**, 95 (1993).

[3.22] E. J. Williams, *Proc. Roy. Soc. A* **139**, 163 (1933).

[3.23] J. D. Jackson, *Classical electrodynamics*, 2nd edn (New York, John Wiley, 1976).

[3.24] N. Nücker *et al.*, *Phys. Rev. B* **39**, 12379 (1989).

[3.25] J. Yuan, L. M. Brown & W. Y. Liang, *J. Phys. C: Solid State Physics* **21**, 517 (1988).

[3.26] N. D. Browning, J. Yuan & L. M. Brown, *Supercond. Sci. Technol.* **4** S346 (1991).

[3.27] N. D. Browning, J. Yuan & L. M. Brown, *Ultramicroscopy* **38**, 291 (1991).

[3.28] N. D. Browning, J. Yuan & L. M. Brown,. *Inst. Phys. Conf. Ser.* **119**, 121 (1991).

[3.29] S. J. B. Reed, *Electron microprobe analysis*. 2nd edn (Cambridge, Cambridge University Press, 1993).

[3.30] W. T. Pike & L. M. Brown, *J. Crystal Growth* **111**, 925 (1991).

[3.31] R. H. Ritchie & A. Howie, *Phil. Mag. A* **58**, 753 (1988).

[3.32] P. E. Batson, *Nature* **366**, 727 (1993).

[3.33] D. A. Muller *et al.*, *Nature* **366**, 725 (1993).

[3.34] L. M. Brown *et al.*, *Microscopy Microanalysis Microstructure* **6**, 121 (1995).

[3.35] J. Yuan *et al.*, *J. Microscopy* **180**, 313 (1995).

[3.36] S. K. Tolpygo *et al.*, *Appl. Phys. Lett.* **63**, 1696 (1993).

[3.37] A. J. Pauza *et al.*, *Physica B* **194**, 119 (1994).

[3.38] J. Yuan, Y. Yan, & A. J. Pauza, *Inst. Phys. Conf. Ser.* **147**, 429 (1995).

[3.39] R. F. Egerton, *Electron energy loss spectroscopy in the electron microscope*, 2nd edn (New York, Plenum Press, 1996).

[3.40] A. V. Crewe, *J. Microscopy* **100**, 247 (1974).

[3.41] O. Scherzer, *Z. Physick* **101**, 593 (1936).

[3.42] O. Scherzer, *Optik* **2**, 114 (1947).

[3.43] J. H. Mm. Deltrap, Ph.D. Thesis, Cambridge University (1964).

[3.44] M. Haider *et al.*, *Optik* **99**, 167 (1995).

[3.45] U. Zach & M. Haider, *Optik* **99**, 112 (1995).

[3.46] O. L. Krivanek *et al.*, *Inst. Phys. Conf. Ser.* **153**, 35 (1997).

[3.47] P. Kruit, *Inst. Phys. Conf. Ser.* **153**, 269 (1997).

[3.48] J. M. Cowley *et al.*, *Ultramicroscopy* **68**, 135 (1997).

4

Specimen preparation for transmission electron microscopy

J. G. WEN

4.1 Introduction

This chapter is intended as a convenience to those readers actively engaged in the investigation of high T_c superconductors by transmission electron microscopy (TEM). A future possible application of the newly discovered high T_c superconductors is their use in electronic devices. The electrical properties of a device strongly depend on their microstructure, since grain boundaries in these materials can behave as weak links as reported by Dimos *et al.* [4.1]. Therefore, TEM is an important tool in the study of the relationship between the microstructure and the electrical properties.

To obtain a TEM sample representative of the as-received sample is not only a technical problem but also a problem of understanding the sample preparation process. Unfortunately, the solution is often strongly dependent on the materials being prepared. For instance, high T_c superconductors easily react with moisture and degrade during sample preparation. Moreover, they easily become amorphous during ion milling. For high T_c thin films, the films are usually softer than the substrate, thus have a much higher ion-milling rate. Taking precautions against these kinds of difficulties makes sample preparation for high T_c superconductors relatively difficult.

There have been a number of review articles on TEM sample preparation techniques [4.2–4.5]. TEM samples of high T_c superconductors are mostly prepared either by crushing, cleaving or ion milling. These methods will be dealt with in Sections 4.2 and 4.3. Minor details frequently determine the success of a technique. To illustrate this, we will describe the normal preparation procedure for both techniques while we will concentrate on some 'tricks' to obtain a good TEM sample reliably and fast. The method described here will mainly be focused on cross-section sample preparation. Finally, two new techniques will be introduced. (i) a simple TEM cross-section ion shadow method which is described in section 4.6; and (ii) how to clean an amorphous overlayer from a TEM sample in section 4.7.

4.1.1 General requirements

As we know, some high T_c superconductors such as YBCO easily degrade due to the moisture in the air. This means that we have to reduce the exposure time of the sample to air and avoid water in the whole preparation process. When the sample is dedicated for cross-section observation, it is recommended to deposit gold or silver directly (*in situ*) on the surface of the film in order to protect it from degradation. Copper grids are often used for support of samples. However, in order to check the chemical composition of high T_c superconductors (containing Cu) Ag, Al, or Ni grids can be used as supports to avoid background contributions from Cu. To preserve the specimen it is best to handle it with vacuum tweezers or a pair of fine tweezers. A well-ground thin sample can safely be moved around by a piece of tissue paper soaked in acetone or ethanol.

4.1.2 What is a good TEM sample?

Besides the basic requirements for TEM observation (e.g. electron transparent, 3 (or 2.3) mm diameter, conducting), several more requirements need to be fulfilled in order to get a good TEM sample. (i) The thin area must be as large as possible to enable us to check the sample on a macroscopic scale. (ii) Cleanliness is necessary to avoid background signals coming from contaminating overlayers or oxide layers. A large background contribution from an amorphous layer can completely disrupt a phase-contrast image [4.6], or result in a completely wrong spectrum when doing analytical TEM. [4.7, 4.8]. (iii) Bending should be avoided, especially in a local area. In a bent region the direction of the zone axis varies, which makes it difficult to image in HREM. (iv) Good electrical and thermal conductivity are required, otherwise the image may move on recording the image due to thermal drift or discharges. (v) For cross-sectional samples, an equal thickness of film and substrate is important in order to interpret the image correctly.

4.2 Crushing and cleaving

Crushing and cleaving are often used in checking bulk samples and cross-sections of thin films. Sometimes these methods avoid unwanted damage caused by ion milling.

4.2.1 Crushing

Crushing is the most simple preparation technique. This technique only needs an agate mortar (and pestle), then the sample can be ground into a fine powder, often immersed in acetone or alcohol. A few droplets of the suspension of ground material is put on a holey carbon film supported by, e.g. a Cu grid. However, for a highly anisotropic sample such as Bi2212, it is extremely difficult to obtain the *c*-axis in-plane by dropping a suspension onto a grid (i.e. with the viewing direction parallel to the *a*- or *b*-axis). In order to overcome this problem, the ion-milling method is often used. But the crushing technique is still applicable if it is slightly modified [4.9]. The procedure can be modified by dry grinding, using a sticky Cu grid (a Cu grid dipped into a sticky solution of cellophane tape in benzol or CH_3Cl_4) and putting this grid into the ground powder. For heating experiments a Si_3N_4 foil can be used because the foil can sustain high temperature up to 500 °C. [4.10, 4.11]

4.2.2 Cleaving Bi2212 with tape

It is rather easy to obtain a c-axis view with a sample such as the Bi2212 single crystal. This crystal can directly be cleaved along the *a−b* plane by a cellophane tape. Put one small piece of Bi2212 single crystal between two pieces of tape and separate the Bi2212; repeat this procedure 10 times. Wash out the small fragments of Bi2212 sticking to the tape using CH_3Cl_4 or benzol. Stir the solution and use a copper grid with 200 or 400 mesh to catch some fragments suspended in the solution. These fragments are usually thin enough to enable observation by TEM. Compared to crushing, the cleaving method offers: (i) large thin areas (several micrometers); (ii) preservation of most defects such as antiphase grain boundaries [4.12] and other defects [4.13] which can be studied by diffraction contrast imaging and (iii) homogenous thickness and no bending. A specimen prepared in this way is especially suitable for symmetry studies using convergent beam electron diffraction (CBED) [4.14, 4.15].

4.2.3 Cleaving a thin-film sample

The cleaving method can be also applied to prepare cross-section and plan-view samples of thin films on substrates [4.16, 4.17]. This cleaving method basically consists of mechanical grinding and successive sample cleaving. Substrates for high T_c superconducting thin films can be cleaved more or less along any directions which implies that a quite sharp wedge can be obtained. Basically a thin film sample is ground from the substrate side to about 50 μm thick and then cleaved by placing the sample on a flexible surface and applying pressure with a sharp instrument. Cleaved pieces with sharp wedges are selected and mounted on a slot grid. The following are key points in order to succeed using this technique. (i) When the thickness of the sample is about 50 μm, finally grind the sample with about 800 grit sandpaper along a direction about 10° to 15° away from a low-index orientation such as [100] or [110], as shown in Fig. 4.1(a). Cleaving the sample along the direction of the low-index plane (for example the [100]) and the sand paper scratches can easily lead to a sharp wedge. (ii) Avoid thermal drift by using conducting silver glue to mount the specimen on a support grid. In addition, gold deposited on both sides of the sample, after grinding and cleaving, can lead to better conductivity.

To make a plan-view sample by cleaving, first cut the sample into a 1 mm wide strip. Then cut the sample at the edge and parallel with this as shown in Fig. 4.1(b). The cleaved sample often contains large thin areas suitable for diffraction contrast imaging and HREM. Since the thin-film surface side is not a cleaved surface, a short time of ion milling of the surface side can reduce sample thickness and remove surface contamination.

As described before, cleaving is sometimes the only way to check the real structure. One example is given here. A (Ca, Sr)CuO$_2$ thin film was grown on a (100) SrTiO$_3$ substrate by RF thermal plasma evaporation. After plasma annealing at 200 °C, the c parameter is found to change from 3.4 Å (infinite layer) to 3.6 Å (a new phase) [4.18]. But if the annealing temperature is above 300 °C, the c parameter changes back to 3.4 Å. After ion milling a film with a c parameter of 3.6 Å, it is also found to change to 3.4 Å. Therefore, the

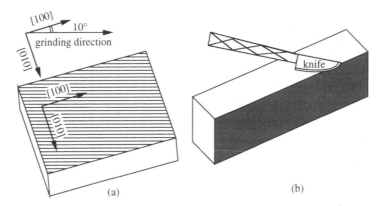

Fig. 4.1. Schematic of cleaving for (a) cross-section and (b) plan-view samples. A final grinding slightly away from the ⟨100⟩ direction with rough sand paper creates many scratches. A sharp wedge is easily formed by cleaving along these scratches.

cleaving method is necessarily introduced for plasma-annealed thin films in order to avoid the c parameter change caused by the ion milling. Figure 4.2 shows a plan-view HREM image along the [001] zone-axis of the substrate and a selected-area diffraction (SAD) pattern after plasma annealing.

Figure 4.3 shows a cross-sectional HREM image along the [010] zone axis of the substrate ([110] zone axis of the film) and the corresponding SAD pattern. From these two images, one can conclude that a $2\sqrt{2}\,a_p \times 2\sqrt{2}\,a_p \times c$ structure with $a_p = 3.90$ Å and $c = 3.6$ Å occurs in the $a-b$ plane of the film [4.18].

4.3 Ion milling

Although crushing and cleaving is often the most favorable preparation technique with respect to the preservation of the structure, many times these

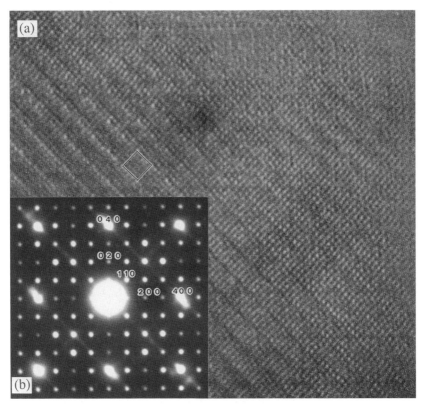

Fig. 4.2. Plan-view HREM image and diffraction pattern for a 3.6 Å $SrCuO_2$ phase prepared by cleaving. The superstructure disappears for an ion-milled sample.

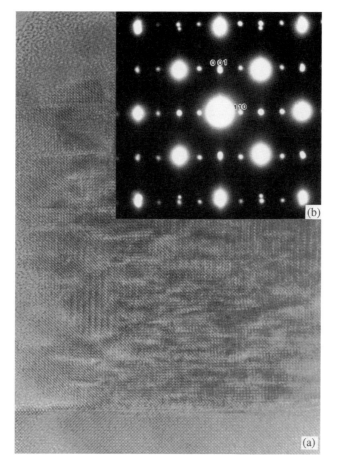

Fig. 4.3. Cross-section HREM image and diffraction pattern along [010] zone axis of a 3.6 Å SrCuO$_2$ phase prepared by clearing. The superstructure disappears for an ion-milled sample.

methods cannot be used, particularly if the overall microstructure or the interface structure is to be checked. An ion-milling procedure is often needed to prepare this type of specimen. Preparing a TEM sample by ion milling generally goes through (i) a mechanical thinning; (ii) an ion thinning down to a thickness less than 1 μm and (iii) an ion polishing. Several groups can even mechanically thin the sample down to electron transparency [4.3, 4.19, 4.20].

4.3.1 Precision thickness control

Thickness measurement in sample preparation is quite important. Here we describe how to control the thickness precisely by observing color fringes.

When the thickness is above 1 μm it is best measured by a micrometer screw or by a micrometer calibrated focus control on an optical microscope. When the thickness is below 1 μm it is convenient to use light interference fringes (colored or monochromatic). Figure 4.4 shows color fringes taken just before perforation occurs. The appearance of colored fringes is due to the interference of incident and reflected light, and occurs when

$$2nt = (2m + 1)\lambda/2 \qquad (m = 0, 1, 2, 3, \ldots). \tag{4.1}$$

Light of wavelength λ becomes extinct which results in the observation of the complementary color. Here n is the refractive index of the substrate, t its thickness and m is an integer. As seen in Table 4.1, the three colors actually observed are yellow, purple, and green (or blue in the 0th order). The thinnest area corresponding to yellow has a thickness below 50 nm, and close to the center is actually thinner than 20 nm (as checked by CBED). This is sufficient for HREM. The thickness difference between adjacent orders of purple is about 100 nm. The appearance of color fringes corresponds roughly to a thickness of 1 μm. It is recommended that one picture is taken of such color fringes as a reference standard. One can immediately see the order of the color fringe, and the corresponding thickness, by comparing with the standard.

4.3.2 Ion-thinning machines

The most commonly used ion thinner is a Gatan 600 DuoMill. It has an option for tilting the incident ion-beam angle (guns) and for rotating the specimen holder. During ion milling of a cross-section sample, the specimen is not

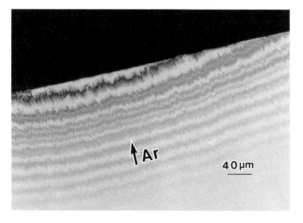

Fig. 4.4. Fringes can be observed before a perforation occurs at the edge of the sample. Based on these fringes, one can exactly know the thickness.

Table 4.1 *Color, wavelength, corresponding complementary color and estimated thickness assuming a refractive index for the substrate of 2*

Color	Wavelength λ (nm)	Complementary color	Estimated thickness (nm)
purple	400	yellow *	50
blue	475	orange	60
green	550	red	70
yellow	600	purple *	75
orange	700	blue	85
red	750	green *	95

* Indicates the color observed.

rotated and is oriented such that the thin-film side is facing away from the ion gun. Only one gun is used. In this way, the substrate is used as an ion-milling block to prevent preferential ion milling of the high T_c superconducting thin film. An oscillating system with different oscillation angles of 30°, 60° and 120° is useful to avoid comb-like structures when one has to ion mill an inhomogeneous sample [4.5]. By inserting a small spring between the ion gun and the high-voltage terminator, it is possible to adjust the ion gun in or out so that the ion beam can be shifted slightly sidewards (about 1 or 0.5 mm).

4.3.3 *End point detection*

If the sample is not transparent to light, the final thickness can directly be controlled by a laser auto-terminator. For bulk and plan-view samples of the high T_c superconductors, we are able to use a laser auto-terminator. But in the case of a cross-section sample, we have to frequently check the thickness (color fringes) by an optical microscope. Before the end point, the ion milling is stopped before producing a hole or when only a very small perforation in the sample has occurred. Excessive ion polishing is to be avoided because, if the perforation gets too large, there is a possibility that sputtered material will be deposited on the opposite side of the specimen.

4.3.4 *Plan-view sample preparation*

It is relatively easy to prepare a plan-view sample. The common way to prepare a plan-view sample is to grind the sample down to less than 10 μm and then ion mill the sample. For a bulk sample, the sample can be ion milled from both sides. In the case of a thin film, a back-thinning process has to be applied where milling is done only from the substrate side. However, material sputtered during ion milling can redeposit on the back surface. One can reduce the resputtering effect as follows: (i) ion milling is stopped as soon as a hole appears; and (ii) a short time of double side ion polishing is applied to remove the resputtered material.

In order to study the microstructure at a specific location such as a bicrystal junction, one can use a dimpler to produce a dimple right on the specified location before ion-milling. When one wants to check the microstructure of a plan-view sample close to the film-substrate interface instead of the surface of the film, one has to control the ion-milling process by checking the color fringes under an optical microscope. First, ion mill from the substrate side until color fringes appear due to the substrate being thin enough. Then control the ion-milling process until a small perforation of only the substrate occurs, and finally apply ion milling from the film side until a real perforation is obtained.

4.4 Cross-section sample preparation

A reliable method to prepare a cross-section was first introduced by J. P. Benedict *et al.* [4.3] where the sample is mechanically ground using a tripod. The tripod can be adjusted by three micrometer feet to control the wedge angle of the sample. The tripod has several advantages: (i) a sample can be ground keeping a homogenous thickness (plane parallel) which enables one to check the sample on a large scale; (ii) because one can obtain a large homogeneous thickness, three or four samples can be ground at the same time. Another important method, like dimpling, can also provide a large thin area as described by H. L. Humiston *et al.* [4.21].

4.4.1 Cutting

The sample is fixed on a clean smooth glass plate with the film side facing the glass. It is mounted by wax with a melting point of around 100 °C. For several samples to be glued together, one should grind the back (substrate) side with 600 grit sandpaper in order to remove any silver glue or similar materials. The sample is cut by a diamond wheel saw or a wire saw. If a certain device is aimed for, one has to consider the loss of sample material since the diamond blade or wire is about 100 μm thick. The sample is cut into strips 0.8–1 mm wide and 5 mm long as shown in Fig. 4.5(a). Before cutting, one should position the sample such that the cutting direction is perpendicular to the

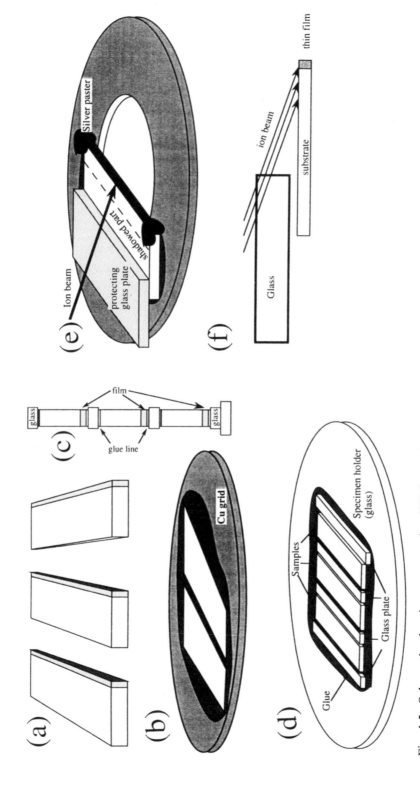

Fig. 4.5. Schematic showing cross-section TEM sample preparation using the tripod. A piece of glass is used to provide a shadow during ion milling, as shown in (e) and (f).

direction along which one wants to view, such as [100] or [110]. If one wants to cut the sample along an exact orientation and/or position, the saw position can be checked by a stereo microscope (magnifying glass).

It is necessary to prepare some clean and smooth pieces of glass 1 mm wide and 5 mm long for further use. These pieces of glass are prepared by gluing several pieces of microscope cover glass (150 μm thick) together with wax and then cutting these into the required size. The glass pieces are cleaned several times in a solution of HNO_3. $HCl = 3:1$ at a temperature of 150 °C and then washed with clean water.

4.4.2 Gluing together

The conventional method of gluing samples together is to glue them face to face as shown in Fig. 4.5(b). We glue the sample with the film facing the same direction, because the edge facing the direction of grinding is found to be better preserved than the trailing edge. A piece of glass (1 mm × 5 mm) prepared in advance is inserted between two samples so that the thin film side is always glued facing a piece of glass. The sample is glued into a stack as shown in Fig. 4.5(c). In order to grind the cross-section sample as thin as possible, the glue gap has to be as narrow as possible. The sample is sandwiched between two pieces of glass in order to protect it from damage and rounding effects as shown in Fig. 4.5(d). The use of a piece of glass allows the area of interest to be seen during the processing so that a polished plane on a prespecified area can be obtained as discussed below. M-Bond 600/610 is chosen as a suitable glue, since it has several advantages compared to conventional epoxy glue, e.g. it is resistant to acetone, and a glue line of less than 1 μm can be obtained. The glue is cured at 120 °C for 2 h, applying a slight pressure to the cross-section stack.

4.4.3 Grinding the first side

The cross-section stack is mounted on a tripod holder using wax. The sample is ground by BN powder (20 μm) on a flat glass plate until a flat surface without large cracks is obtained. After cleaning, the sample is ground using diamond paste with the grain size 6, 3, 1, and 1/4 μm. Different types of cloth are prepared for different diamond grain sizes. These are: (i) a hard woven nylon cloth for 6 μm; (ii) a foam cloth for 3 μm; (iii) a silk cloth for 1 μm; and (iv) a soft long-haired cloth for 1/4 μm. The hard nylon cloth is good to obtain a flat surface while this is not always the case when grinding the sample with the foam cloth (3 μm). On the other hand, the silk cloth and the soft long-haired

Table 4.2 *A list of grinding paste and cloth used. The 'critical thickness' column gives the minimum thickness advisable for the specified grain size (about three times the grain size used)*

Grain size	Cloth type	Critical thickness	Remarks
20 µm BN	flat glass plate	60 µm	grinding and
6 µm diamond	hard woven nylon cloth	20 µm	flattening
3 µm diamond	foam cloth	10 µm	grind/smooth
1 µm diamond	silk cloth	< 10 µm	smooth/polish
0.25 µm diamond	soft long-haired cloth	< 10 µm	polish

cloth produce a round edge. Therefore, two pieces of glass are glued on both ends to protect from this rounding effect. For the first side, it is important to have a flat surface in order to get a homogenous final thickness. The flat surface is also important for reliability during the next stage of gluing, and for accurate thickness estimation. Therefore, it is allowable to change the height of the tripod feet when grinding with 6 µm until a really flat surface is obtained. After finishing grinding with 6 µm, it is better not to change the height of the tripod feet but just to polish the surface smooth by further grinding. The grinding direction is oriented so that the film is the leading side.

4.4.4 Gluing the first side

Experience tells us that good gluing of the polished (first) side on the tripod holder is the key to the whole process. If this gluing is bad, after grinding the sample down to less than 20 µm, bubbles often appear underneath the sample and the final grinding/polishing fails. So the polished surface should first be thoroughly cleaned by acetone, using a lens tissue wrapped on a fine tweezer. By checking the polished surface in reflected light, one can know if there is any dirt left on this side. The surface of the tripod holder has to be cleaned thoroughly also. Make sure that the surface of the holder is clean and smooth each time a new piece of glass (150 µm thick) is fixed onto it with wax. A cyanoacrylate glue is used to glue the polished side on to the holder. This glue is relatively fast curing, soluble in acetone and readily available. Put a small drop of this glue on the holder and put the sample on the holder as soon as possible with the polished side facing the holder. Press the sample in the center slightly with a sharp tool. Make sure that no bubbles are present. Turn the holder upside down, focus and observe in reflecting light the glued surface. If some color fringes from the glued surface can be seen, it is an indication of a good gluing (glue thickness less than 1 μm). Leave the sample in air for the glue to dry. For safety it is recommended to let it dry overnight.

4.4.5 Grinding the second side

After choosing a certain size of diamond paste, one can only obtain a certain minimum critical thickness according to experience. Usually the critical thickness is about three times that of the grain size of the diamond paste. First grind the sample with 25 μm BN paste on a flat glass plate to a thickness of 80 μm. Before continuing, it is necessary to adjust the three micrometers so that the plane determined by the three feet is exactly parallel to the glass plate of the holder. Placing the tripod on an inverted metallographic microscope, one can adjust the three micrometers until all the edge of the glass is at the same focus height. Afterwards, 50 μm and 20 μm thickness can be obtained with 15 μm and 6 μm diamond paste respectively on a nylon cloth. With 3 μm and 1 μm diamond paste on a foam cloth and a woven silk cloth respectively, a thickness of 10 μm and less than 10 μm can be obtained. The sample is finally polished by a soft long-haired cloth with 1/4 μm paste. It is recommended to check the thickness of the sample by an inverted metallographic microscope after finishing each grinding. While grinding with 1 μm diamond paste, one may adjust one of the three micrometers 10 μm out, so that the front side of the film becomes a little thinner than the back side. By doing so, the stack is polished in a wedge shape so that perforation (during ion milling) can be controlled to appear on the film side of the sample. A perforation at the back side of the sample is bad since the substrate cannot then be used as an ion block. In order to obtain a TEM specimen with a large thin area, a homogeneous thickness is the key factor.

4.4.6 Gluing the sample onto a grid

Put the ground sample stack into acetone until it detaches from the specimen holder. The stack is very brittle so it is impossible to pick it up with tweezers. Use a small piece of tissue paper soaked in acetone. The stack will stick to the soaked tissue paper and can thus be removed. The acetone on the tissue paper evaporates quickly and the stack then falls off by its own weight. Clean the sample carefully with a lens tissue wrapped around the tip of a pair of tweezers using acetone. Separate all strips of the cross-section carefully with a sharp knife. The glue is not so strong any more since the cross-section is less than

10 mm. Then, with the knife, cut one strip of cross-section into several segments about 2 mm in length. In order to avoid a piece jumping away during cutting, the filter paper underneath the strip is wetted with acetone or ethanol. In this way all the pieces stick to the filter paper. A small drop of conductive glue (silver glue and epoxy mixed, which becomes hard in 5 min) is painted on one part of a slot grid with a 1 mm hole as shown in Fig. 4.5(e). Hold the grid with a pair of tweezers and position it on top of the strip. During this process, the sample should not be touched by the tweezers. Finally put a small drop of silver glue on the two ends of the strip to ensure good conductivity between the thin film, the grid and the holder. By doing so, normally no liquid nitrogen cooling will be needed during ion-milling.

4.4.7 Ion thinning

The specimen is mounted on the ion-milling holder by silver paste in order to obtain good thermal conductivity. The upper cover of the ion-milling holder is removed so that the ion-milling angle can be lower than 5°. During ion-milling, the specimen is not rotated and is oriented such that the thin film side is facing away from the ion gun. Only one gun is used to ion mill the sample. By doing so, the substrate can be used as an ion-milling block to prevent preferential ion milling of the high T_c superconducting thin film. One can also oscillate the sample through a range of angles. Normally, comb-like features can be created due to surface roughness left from grinding, which is enhanced by ion milling. Angular oscillation during ion milling can reduce this effect. A protective glass plate is placed so that it partially shadows the ion beam on the specimen as shown in Fig. 4.5(e). Therefore, only the area close to the film is ion milled and the rest remains thick. This can prevent bending and fracturing of the specimen when it is very thin.

The protecting glass plate has another advantage which can be exploited when the film is harder than the substrate. For high T_c superconducting thin films, the substrate is more resistant to ion milling than the thin film. Therefore, one can mill from the back side using the substrate as an ion block. If the thin film is harder than the substrate, for instance, a diamond film on a Si substrate, this method has to be modified. By using a protective glass shield as shown in Fig. 4.5(f) the whole substrate can be protected so that the film will start ion milling first. Adjust the ion beam incident angle and the glass plate position until the ion beam is just on the edge of the thin film, using the X-ray fluorescence generated on the substrate. The glass shield being ion milled away becomes thicker and thicker when the ion beam shifts from the thin film to the substrate side so that the hard thin film gets a longer milling time than the softer substrate. By doing so, an interface with a reasonably smooth thickness transition can be obtained.

Typical ion-milling parameters for the ion thinning process are: incident angle 15°, accelerating voltage 4 kV, gun current 0.5 mA. One ion gun is used and the angle is changed from +15° to −15° for each hour. The ion thinning rate using these parameters for $SrTiO_3$ substrate is roughly 1 μm/h with a new gun cathode. When the thickness of the specimen is less than 1 μm, color fringes can be observed in a reflected light optical microscope. It is recommended to check the specimen regularly with an optical microscope. As soon as color fringes are observed, the ion milling conditions can be changed into an ion-polishing procedure.

4.4.8 Ion polishing

As mentioned above, the ion-thinning rate is in the order of 1 μm/h. When an ion-polishing process is applied, a polishing rate of about 150 nm/h can be achieved with the conditions: 3 kV, 0.3 mA, and ±8°. If a special area like a junction is being thinned, the specimen should be checked regularly with TEM when the last color fringe appears. The first yellow band may be as wide as 200 μm where the thickness is less than 50 nm. This enables us to check these areas by high-resolution electron microscopy (HREM). The area where the thickness is below 200 nm is more than 500 μm wide, where we can check the sample by diffraction contrast imaging.

4.5 Examples

Figure 4.6 shows a HREM image of the interface at a YBCO thin film grown on a YSZ substrate. An intermediate layer with thickness of 3–4 nm can be observed. The TEM specimen has the same thickness and orientation in all these three layers. Both the lattice spacing of 4.2 Å, and EDX analysis of the composition (only Ba and Zr using a spot size of 2 nm), indicate that this intermediate layer is $BaZrO_3$. This is a reaction product of YSZ and YBCO. The zone axes of YBCO, $BaZrO_3$, and YSZ are [100], [100] and [110] respectively. At the $BaZrO_3$/YBCO interface, steps occur corresponding to a whole unit cell of YBCO. At the focus used, the brightest line in the figure is the CuO chain plane. This is the starting atomic layer of the YBCO thin film [4.22]. In this figure, a stacking fault corresponding to '125' can be observed.

A ramp-edge Josephson junction is a junction whose performance can be controlled by controlling the thickness of the barrier layer. Figure 4.7 shows a

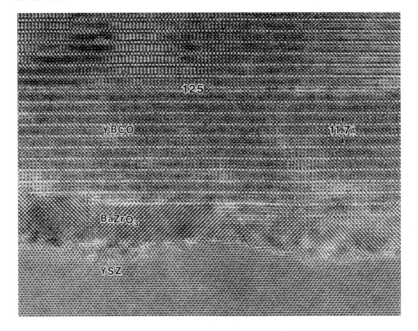

Fig. 4.6. An HREM image of the interface at a YBCO thin film grown on a YSZ substrate. A reaction layer of BaZrO$_3$ can be observed. The zone axes of YBCO, BaZrO$_3$, and YSZ are [100], [100] and [110] respectively.

Fig. 4.7. A cross-section image of a ramp-edge junction. All components of the film are *c*-oriented and are epitaxial with (001) parallel with (001) of the substrate.

cross-section TEM image of such a ramp-type junction. The thickness of the bottom DyBCO and the top DyBCO layers are about 80 nm and 150 nm respectively, while the separating layer is about 60 nm. The barrier layer in this figure has a thickness of about 8 nm. All components of the film are *c*-oriented and are epitaxial with (001) parallel with the (001) of the substrate. The slope of the junction is 20°, in agreement with the ion-milling angle which was used for the ramp production.

Figure 4.8 shows a HREM image of a tip showing the epitaxy of the barrier PrBCO layer in the dimple of the $SrTiO_3$ substrate. The interfaces between the barrier layer PrBCO and the neighboring DyBCO layers are indicated by black arrow-heads. The barrier is about 2 nm thick, and grows epitaxially on the $SrTiO_3$ substrate and the ramp of the bottom DyBCO layer. Therefore, this ramp-edge Josephson junction DyBCO/PrBCO/DyBCO has been fabricated as designed. More experiments show that several grain boundaries in the top YBCO layer are always observed at the tip of the ramp when the substrate is YSZ. Therefore, only substrates with perovskite-like structure can be used as substrates for the fabrication of ramp-edge junctions [4.23].

4.5.1 Precision grinding

Sometimes, one has to check a specific area within a micrometer-sized region, especially when one wants to check the microstructure of a device such as a

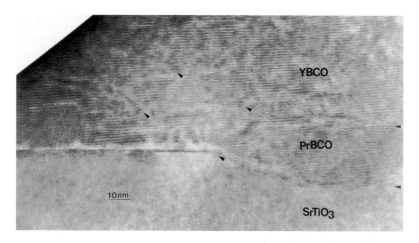

Fig. 4.8. An HREM image of a tip showing the epitaxy of the barrier PrBCO layer (about 2 nm thick) in the dimple of the $SrTiO_3$ substrate. The interfaces between the barrier layer PrBCO and the neighboring DyBCO layers are indicated by black arrow-heads.

Josephson junction after having measured its electrical properties. The best tool to fulfill this purpose is a tripod as described by J. P. Benedict *et al.* [4.3]. We will use a junction fabricated by a focused-ion beam (FIB) as an example to show how this is done. Figure 4.9(a) shows a FIB junction after making electrical measurements, with a width of 5 μm. As shown by the arrow-head, one can see a line corresponding to the FIB junction. While grinding the first side, the grinding plane has to be terminated exactly in this 5 μm region. Figure 4.10 is a schematic diagram showing how the grinding plane is used as a mirror plane, making it easy to check the location of this plane.

The way to obtain this mirror image is to view the sample in plan view from the back side and slightly tilt the sample along the grinding line so that one can see the mirror image. Before approaching the interesting area, it is recommended to adjust the tripod so that the grinding plane is parallel to the FIB junction. Figure 4.9(b) shows the grinding plane just on the junction line. Figure 4.9(c) shows a top-view image after finishing the second side grinding. When the sample is thinned by ion milling, first ion mill 1 or 2 μm off the first grinding side, then ion mill from the other side until the color fringes appear. Figure 4.11 shows a typical microstructure at a FIB junction where three grain boundaries can be observed [4.24]. From this image, we can understand that FIB junction is a grain-boundary type Josephson junction.

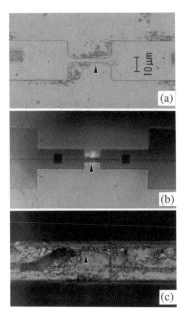

Fig 4.9. We can stop grinding just on the 5 μm wide FIB junction. A top-view image after finishing the second side grinding is shown in (c).

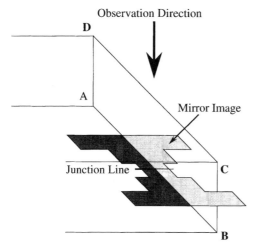

Fig. 4.10. Schematic diagram showing how the grinding plane is used as a mirror plane. By doing so, it is easy to check the location of this plane.

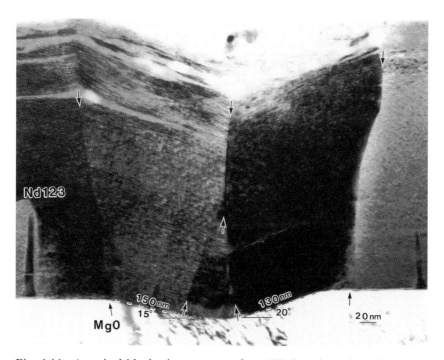

Fig. 4.11. A typical ideal microstructure for a FIB junction where three grain boundaries can be observed. This microstructure corresponds to RSJ-type I-V characteristics.

4.5.2 Cylinder method

When samples being cross-sectioned are very fragile and small in size, such as YBCO thin film on YBCO single crystal [4.25] or radiation damaged Bi2212 single crystals [4.26] which is about 0.1 mm thick, we use a metal cylinder as shown in Fig. 4.12 to strengthen the specimen. Prepare a metal cylinder as shown in Fig. 4.12. When the sample is a YBCO single crystal cut it to the same width as the cylinder's slot. Prepare some dummy silicon samples. When the sample is an irregular Bi2212 single crystal, immediately glue the samples together with M-bond 610. Silicon is chosen in order to use the laser auto-terminator since it is not transparent when thick. The flat surface of the dummy silicon should face the thin film side so that a thin glue line can be achieved. After curing, cut the cylinder into 1 mm thick slices as shown in Fig. 4.12. The remaining steps for thinning the sample are the same as previously described. This method can save samples and increase the rate of success.

It is important for oxide superconductors to homoepitaxially grow *a*- and *c*-axis oriented YBCO thin films on YBCO single crystals in order to really develop device applications. From the point of view of matching between film and substrate, a YBCO single crystal is the best substrate for growing YBCO thin films. The best way to characterize these YBCO thin films is by cross-sectional TEM. Due to the small size of YBCO single crystals and their fragile nature, a cross-section sample has to be prepared by the cylinder method. Figures 4.13 and 4.14 show HREM images of YBCO thin films grown on YBCO (001) and (100) substrates by MOCVD, respectively. In Fig. 4.13, except for one unit-cell at the interface that appears different from normal

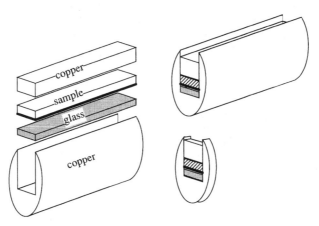

Fig. 4.12. Schematic showing cross-section sample preparation by the metal cylinder method.

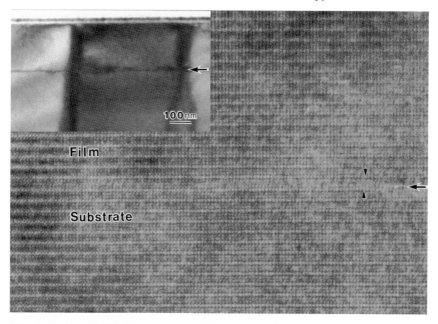

Fig. 4.13. An HREM image of a YBCO thin film grown on YBCO (001) substrate by MOCVD. Only one unit-cell at the interface is slightly different from the others. The inset shows twin boundaries in the substrate extending into the film.

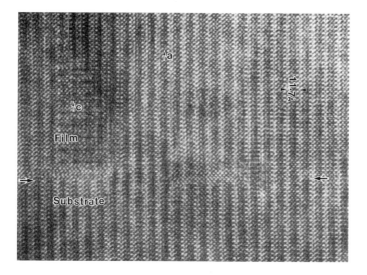

Fig. 4.14. An HREM image of YBCO thin film grown on a YBCO (100) substrate by MOCVD. The YBCO lattice structure in the substrate extends completely into the film. Some *c*-axis YBCO nucleation due to the surface of the substrate helps locate the interface.

YBCO, we cannot distinguish the thin film and substrate. A substrate slightly off (001) can offer better growth than the exact orientation due to the appearance of (100) surface steps. The inset of Fig. 4.13 shows twin boundaries in the substrate that extend into the film, which demonstrates the high quality of the homoepitaxial film. In Fig. 4.14, the interface between a-axis film and substrate sometimes cannot be observed. Some c-axis nucleation at the interface is due to surface degradation, an anisotropic (100) surface is covered by a thin isotropic amorphous layer. A polished (100) or (110) surface can be atomically smooth over a large area (e.g. 10Å roughness over a 5 μm area). One may use this atomically smooth surface to grow a SIS junction by depositing a thin insulating layer and then an a-YBCO layer.

4.6 Ion-shadow method

It is time consuming and precise work to prepare a TEM cross-section sample by the ion milling method as described above. One has to glue the sample, grind the sample down to about 10 μm and then ion mill the sample at a low angle, 15°, to obtain electron transparency. In order to control the final ion milling process, one often has to check whether the sample is ready or not. Here we introduce a new simple and useful preparation method for TEM cross-section sample preparation; the so-called ion shadow method [4.27].

The ion-shadow method was first introduced by Langer & Katzer [4.28] and T. Yoshioka [4.29]. The principle of this method is to place diamond particles or a tungsten wire on the surface of a thin film and then ion mill parallel to the normal of the thin film. Because the diamond particles and the tungsten wire are more resistant to ion milling, the areas covered by diamond particles or the wire will be protected from ion milling, while the other areas are removed by the incident ions. After ion milling for a short time, the areas covered with diamond particles will develop a stalactite morphology. These stalactites are thin enough to enable us to check the microstructure in cross-section. However, the samples prepared by this method can not be checked by HREM, because they are too thick along the beam direction. Here the method has been modified such that the thickness is thin enough for HREM studies.

First, cut a thin-film sample into a strip of thickness 0.8 mm with a diamond cutter. Mechanically grind the sample down to less than 100 μm only from one side. Clean the sample with acetone and put the ground sample on a piece of glass, the ground surface facing the glass as shown in Fig. 4.15(a). Prepare a very thin diamond particle suspension by putting diamond powder (about 0.2 μm to 1 μm in size, commercially available) into acetone or alcohol. Apply a small drop of the suspension on the glass a small distance away from the film as shown in Fig. 4.15(a). The suspension spreads and reaches the thin-film side. Some diamond particles stick to the bottom edge of the thin-film side.

Figure 4.16(a) shows an optical image of the diamond distribution on the thin-film surface. Notice that diamond particles are only found along one edge.

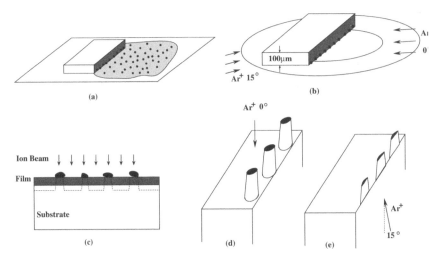

Fig. 4.15. Schematic showing ion shadow cross-section sample preparation method.

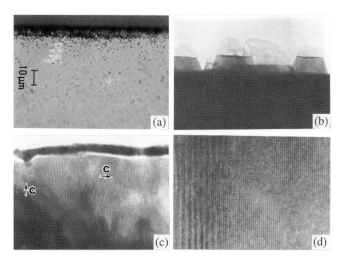

Fig. 4.16. (a) An optical image of the diamond distribution on the thin-film surface; (b) a low-magnification image of a YBCO thin film grown on a SrTiO$_3$ substrate, prepared by this method; (c) an HREM image from one of these thin areas; and (d) a HREM image where one can clearly see the atomic structure.

Glue the sample on a copper slot grid using silver paste, with the diamond covered edge facing the copper grid, as shown in Fig. 4.15(b). During ion milling the sample is not rotated, and is oriented such that the thin-film side faces the ion gun at an angle of 1° to 2°. Typical parameters are accelerating voltage 4 kV and gun current 0.5 mA. After ion milling for 10 min of a lot of stalactites are formed (shown in Fig. 4.15(c) and (d)). The typical diameter of these stalactites is several thousand ångströms. It is too thick to be checked by HREM but is suitable for a SAD study. If one then ion mills from the back side of the film with a 15° angle in the usual way, some of these stalactites can be milled thin enough for HREM, as shown in Fig 4.15(e).

Figure 4.16(b) shows a low-magnification image of a YBCO thin film grown on a $SrTiO_3$ substrate, prepared by this method. Figure 4.16(c) shows an HREM image from one of these thin areas. The dark contrast at the top is a Au layer deposited on the surface before cutting. It indicates that there is no damage at all on the surface of the thin film. A 90° grain boundary can clearly be seen in this image. If one wants to study the microstructure of the interface between the film and the substrate, one can simply ion mill the sample for a longer time, e.g. 15 min at an incident angle of 1° and 2°. Figure 4.16(d) shows an HREM image where one can clearly see the atomic structure.

Compared to the normal sample preparation method, this method takes much less time, and it is not necessary to check whether the sample is ready or not. By using this technique, one can obtain more or less the same quality cross-section samples as with the conventional method even for HREM studies. However, one can not tilt the sample so much. Fortunately, it is still suitable for an epitaxially grown thin film.

4.7 Low-energy plasma cleaning

As discussed above, amorphous layers on the TEM sample surface disrupt a phase-contrast image. Also, carbon contamination due to the focused electron beam is a serious problem for analytical TEM. Recently, theoretical calculations for sputtering and damage production by impinging particles indicate that it is possible to minimize the defect layer to less than 0.5 nm [4.30]. H. W. Zandbergen *et al.* found that low-energy ion plasma can remove surface amorphous layer without damaging the sample crystalline structure. Figure 4.17 shows a schematic diagram of the low-energy ion plasma cleaning setup. An Ar plasma is generated by radio-frequency excitation of neutral Ar gas. A voltage of 50 V is applied to the sample to attract ions towards the sample surface. The low-energy plasma particles clean the sample surface by sputter-

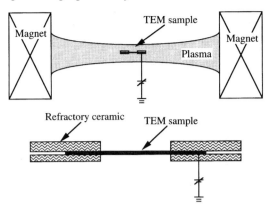

Fig. 4.17. A schematic diagram of the low-energy ion plasma cleaning setup.

ing. During the plasma cleaning the temperature of the specimen is lower than 50 °C and the thinning rate is about 1 Å/min.

Figure 4.18 shows Bi2212 before and after plasma cleaning. Experimental results show that low-energy plasma cleaning can effectively remove amorphous layers. Furthermore, the rate of carbon contamination caused by a focused electron beam can be reduced by a factor of 100 after plasma cleaning.

4.8 Artifacts

Artifacts introduced in the TEM sample preparation of the YBCO superconductor are discussed by H. W. Zandbergen *et al.* and M. S. Louis-Weber *et al.* [4.31, 4.32]. Zandbergen *et al.* found that ion milling can cause grain boundaries or microcracks to be filled with amorphous material. Then it becomes difficult to determine whether an amorphous area (e.g. at a grain boundary) was present before the ion milling or not. Sometimes it can be distinguished by checking whether the amorphous material exists in the thicker areas. In the case of thin films, the amorphization problem is more serious.

Because the high T_c superconductors are more conductive than the substrates, the thin film carries more current and dissipates more heat. When the cross-section area of the film decreases, the temperature in the film due to ion milling increases. Therefore, a good heat conduction between the film, the grid and the ion-milling holder is extremely important to avoid the creation of amorphous areas. Depositing gold or silver on the top of thin film can greatly enhance the conduction between the thin film and the grid. Experiments indicate, that by doing so it is not necessary to use the cooling stage during ion polishing. In the case of an ultra-thin YBCO film, although gold is deposited

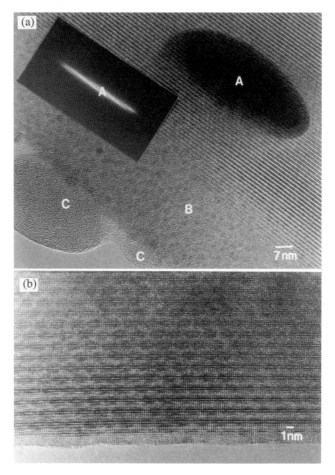

Fig. 4.18. The different contamination rates of Bi2212 (a) before and (b) after plasma cleaning. Notice that the amorphous edge of the Bi2212 at B and C in (a) has nearly disappeared after plasma cleaning in (b).

on top of the thin film and a liquid N_2 cooling stage is used, the ultra-thin YBCO film can sometimes still become amorphous. If the cross-sectional area of the thin film is $S = t_f t_s$, where t_f and t_s are the thickness of the thin film and the average thickness in the viewing direction after ion milling, respectively, then as S becomes smaller the thin film easily becomes amorphous. Therefore, in the case of an ultra-thin film with very small t_f, better conductivity is necessary. Gold coating the unmilled side of the cross-section each time was found to prevent this amorphization. Figure 4.19 shows a low-magnification cross-sectional HREM image of a two-unit-cell YBCO film sandwiched between La_2CuO_4 thin films grown on a $SrTiO_3$ substrate. By coating with gold, the amorphization of the YBCO thin film has been prevented. The

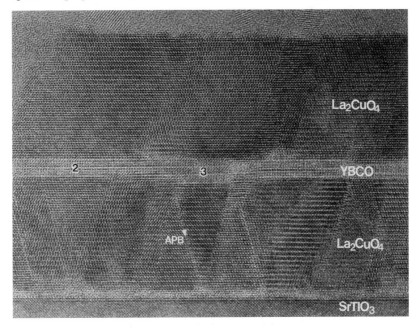

Fig. 4.19. A low-magnification cross-sectional HREM image of a two-unit-cell YBCO film sandwiched between La_2CuO_4 thin films grown on a $SrTiO_3$ substrate. By coating with gold, the amorphization of the ultra-thin YBCO film is prevented.

La_2CuO_4 layer next to the substrate is found to contain many antiphase boundaries, resulting in a rather stepped interface between this La_2CuO_4 and the YBCO thin film. Although it was intended to grow a YBCO layer of two unit-cells, due to the stepped surface of the La_2CuO_4 layer, a variation in the number of unit-cells occurred. However, the CuO chain planes, focused as bright lines in the YBCO thin film, are more or less continuous. This suggests that the YBCO thin film has the ability to smooth out the (001) atomic layer. The top La_2CuO_4 layer on the YBCO thin film contains considerably fewer defects which seems to be caused by the smoothening of the interface by the YBCO film [4.33].

4.9 Case studies

One of the most important research issues related to theory and application of high T_c superconductors is the understanding of the influence of grain boundaries (GBs) on critical current density. Dimos *et al*. [4.1] reported that the critical current density through a GB fabricated on a bicrystal substrate dramatically decreased with increasing misorientation angle. Recently, more and more experiments have shown that the order parameter of high T_c superconductors is a *d*-wave type. This means that the critical current density through a GB strongly depends not only on the misorientation between two grains but also on the orientation of the GB itself down to an atomic scale [4.34]. Therefore, a single-facet GB is necessary to study the influence of GB microstructure on the critical current density.

Many TEM studies show that the GB films grown by physical vapor deposition (PVD) on bicrystal substrates are wavy and the typical facet size is about 50 nm [4.35–4.37]. Therefore, the dependence of critical current density on misorientation angle is difficult to study with these films. Recently, liquid-phase epitaxy (LPE) was successfully used to obtain large single-facet grain boundaries. Fig. 4.20 shows a plan-view image of the GB of a YBCO film grown on a 24° MgO bicrystal. It clearly shows that the GB is a symmetrical

Fig. 4.20. A plan-view HREM image of a symmetrical YBCO grain boundary grown on a 24° MgO bicrystal substrate by LPE. A large single-facet grain boundary over 10 mm was obtained by this method.

one with a single facet and no second phase. The low-magnification image of the GB shows that the single facet of the GB is over 10 mm. Figure 4.21 shows the detailed atomic arrangement of the GB which is very close to the ideal atom positions. The cross-sectional image (Fig. 4.22) also shows that the GB is sharp and straight from the bottom to the surface of the film. These GBs grown by LPE are single faceted over a large area which is quite different from the GB prepared by PVD since both plan-view and cross-section images show the GB grown by PVD is wavy [4.35].

Therefore, these GBs can be used to study the real dependence of the critical current density on GB misorientation angle. Primary measurements on electrical properties for the bicrystal junction grown by LPE indicate that high critical density values and reproducible $J_c \times R_n$ products can be obtained from these GBs grown by LPE. Further studies of J_c dependence on misorientation angle are in progress [4.38].

A large single-facet GB can be obtained when one of the two BB faces is a basal-plane-faced (001) plane [4.39, 4.40]. These kinds of *a/c* GBs were obtained from the needle-like *a*-axis outgrowth grains in the *c*-axis oriented YBCO thin films grown on a MgO substrate. The typical size of a needle-like *a*-axis grain is 30 μm is length and 20–50 nm in width. Two kinds of needle-like *a*-axis grains in *c*-axis YBCO thin films on a (001) MgO substrate were

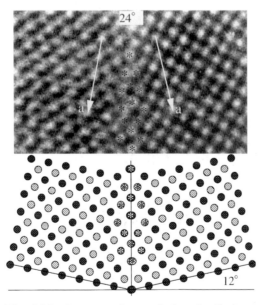

Fig. 4.21. A comparison of the detailed atomic arrangement of the GB between an experimental image and a simulation.

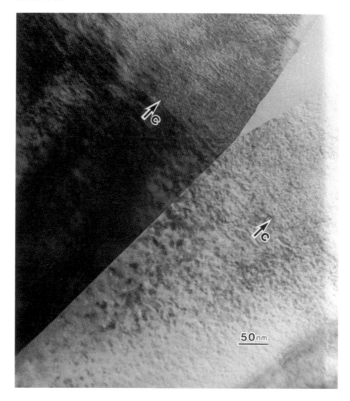

Fig. 4.22. A cross-sectional image of the grain boundary shown in Fig. 4.20 at a medium magnification. Notice the V-shape at the top of the grain boundary. Low-magnification images confirm that the grain boundary is a single facet from the bottom to the top of the film.

found. One type of GB (90° GB) is formed by (001) of the *a*-axis grain and (100) of the *c*-axis film, while another type (45° GB) is formed by (001) of the *a*-axis grain and (110) of the *c*-axis film. Cross-sectional transmission electron microscopy observations for 90° and 45° GBs showed that the *a*-axis grains start right from the MgO substrate. Both kinds of grain boundaries are found to have atomically sharp interfaces over 30 μm, and the distortion due to mismatch is localized only within the range of about one atomic layer. Figure 4.23 shows a plan-view HREM image of a 45° GB. Further HREM observations indicate that the (001) surface atomic layer of the *a*-axis grain is always a BaO layer irrespective of the orientation of the opposite grain [4.40]. These GBs are very suitable for studying the symmetry of superconducting pairing states and *d*-wave pairing characteristics have indeed been observed for the 45° GB [4.41, 4.42].

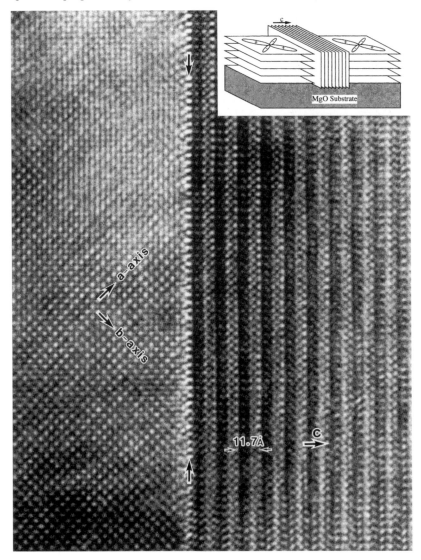

Fig. 4.23. A plan-view HREM image of 45° *a/c* grain boundary. The grain boundary is found to have atomically sharp interfaces over 30 μm and the distortion due to mismatch is localized to within about one atomic layer. The inset shows a schematic configuration of the needle-like *a*-axis grain.

Acknowledgments

It would not have been possible to write this chapter without the effort and co-operation of my research colleague Dr C. Traeholt. I should like, in particular, to thank Dr H. W. Zandbergen and Dr N. Koshizuka for providing financial support and stimulating scientific guidance.

References

[4.1] D. Dimos, P. Chaudhary & F. K. Legoues, *Phys. Rev. Lett.* **61**, 219 (1988).

[4.2] A. Barna *et al.*, *Mat. Res. Soc. Symp. Proc.* **61**, 367 (1986).

[4.3] J. B. Benedict *et al.*, *EMSA Bulletin* **19**, 74 (1989).

[4.4] P. J. Goodhew, Thin Foil Preparation for Electron Microscopy, Chapter 5 in *Practical methods in electron microscopy*, vol. 11, ed. A. M. Glauert (Elsevier, Amsterdam, 1985).

[4.5] C. Traeholt, Ph. D. Thesis, 'Electron microscope investigation of GaAs/$Al_xGa_{1-x}As$ heterostructure and the high temperature superconductor $YBa_2Cu_3O_{7-x}$', Physics Department, Technical University of Denmark (1994).

[4.6] R. Kilaas & R. Gronsky, Ultramicroscopy **16**, 193 (1985).

[4.7] J. W. Steeds, Convergent beam electron diffraction, Chapter 15 in *Introduction to analytical electron microscopy*, ed. J. J. Hren, J. I. Goldstein & D. C. Joy (New York, 1979)

[4.8] S. J. Pennycook & D. E. Jesson, *Phys. Rev. Lett.* **64** , 938 (1990).

[4.9] H. W. Zandbergen *et al.*, *Nature* **332**, 621 (1988).

[4.10] H. W. Zandbergen, A. Pruymboom & G. Van Tendeloo, *Ultramicroscopy* **24**, 45 (1988).

[4.11] H. W. Zandbergen, R. Gronsky & G. Thomas, *Phys. Status Solidi* **105**, 207 (1988).

[4.12] K. K. Fung *et al.*, *J. Phys. Condens. Matter* **1**, 317 (1989).

[4.13] J. G. Wen, Ph.D. Thesis, 'TEM study on the structures and defects of Bi-based high T_c superconductor', Institute of Physics, Chinese Academy of Sciences (1991).

[4.14] K. K. Fung, C. Y. Yang & Y. F. Yan, *Appl. Phys. Lett.* **55**, 280 (1989).

[4.15] J. G. Wen, Y. F. Yan & K. K. Fung, *Phys. Rev. B* **45**, 12561 (1992).

[4.16] H. Kakibayashi & F. Nagata, *Jap. J. Appl. Phys.* **24**, L905 (1989).

[4.17] R. J. Graham, *Ultramicroscopy* **27**, 329 (1989).

[4.18] J. G. Wen *et al.*, *Physica C* **228**, 279 (1994).

[4.19] H. L. Humiston *et al.*, *Mat. Res. Soc. Symp. Proc. Vol.* **254**, 211 (1992).

[4.20] J. Ayache & P. Albarede, *Proc. of ICEM 13-Paris*, **1**, 1023 (1994).

[4.21] H. L. Humiston, B. M. Tracy & M. L. A. Dass, *Mat. Res. Soc. Symp. Proc. Vol.* **254** , 211 (1992).

[4.22] J. G. Wen *et al.*, *Physica C* **218**, 29 (1993).

[4.23] J. G. Wen *et al.*, *Physica C* **255**, 293 (1995).

[4.24] J. G. Wen *et al.*, *Advances in Superconductivity VIII*, (1995) 1097.

[4.25] M. Konnishi *et al.*, *IEEE Transactions on applied superconductivity* **5**, 1229 (1995).

[4.26] W. L. Zhou, Y. Sasaki & Y. Ikuhara, *Physica C* **234**, 323 (1994).

[4.27] J. G. Wen & N. Koshizuka, Presentation at the Spring Meeting of the Japanese Applied Physics Society, Kanagawa, Japan (1995), submitted to *Ultramicroscopy*.

[4.28] E. Langer & D. Katzer, *J. Mat. Sci. Lett.* **13**, 1256 (1994).

[4.29] T. Yoshioka, Presentation at the Metal Society of Japan Seminar, Yokyo, Japan (1994).

[4.30] H. W. Zandbergen *et al.*, *Proc. of ICEM 13-Paris*, **1**, 1003 (1994).

[4.31] H. W. Zandbergen, C. Hetherington & R. Gronsky, *J. Superconductivity* **1**, 21 (1988).

[4.32] M. S. Louis-Weber, V. P. Dravid & U. Balachandran, *Physica C* **243**, 273 (1995).

[4.33] J. G. Wen *et al.*, *Appl. Phys. Lett.* **66**, 1830 (1995).

[4.34] H. Hilgenkamp, J. Mannhart & B. Mayer, *Phys. Rev. B* **53**, 14586 (1996).

[4.35] C. Traeholt *et al.*, *Physica C* **230**, 425 (1994).

[4.36] J. W. Seo *et al.*, *Physica C* **245**, 25 (1995).

[4.37] X. F. Zhamg, D. J. Miller & J. Talvacchio, *J. Mater. Res.* **11**, 2440 (1996).

[4.38] T. Takagi *et al.*, *IEEE Transactions on Applied Superconductivity*, in press (1999).

[4.39] A. F. Marshall & C. B. Eom, *Physica C* **207**, 239 (1993).

[4.40] J. G. Wen *et al.*, *Mat. Sci. Eng.* **B41**, 82 (1996).

[4.41] Y. Ishimaru *et al.*, *Jpn. J. Appl. Phys.* **34**, L1532 (1995).

[4.42] Y. Ishimaru *et al.*, *Phys. Rev. B* **55**, 11851 (1997).

5

Low-temperature scanning electron microscopy

R. P. HUEBENER

5.1 Introduction

Scanning techniques for obtaining topographical information about an object are now widely used in science and technology. Usually, a two-dimensional image is constructed from the signal generated by the scanning process, and this information is restricted to a region close to the sample surface. [5.1]. Here a well focused electron beam is scanned over the surface of the specimen and a response signal such as the emitted secondary electrons or the back-scattered electrons are recorded as a function of the coordinate point (x, y) of the beam focus on the sample surface. We emphasize that this technique essentially yields only information about the composition (atomic, chemical, or metallurgical microstructure) and the geometry of the specimen.

Some time ago, low-temperature scanning electron microscopy (LTSEM) was introduced [5.2] which by now has matured into an important new diagnostic tool. LTSEM extends the temperature of scanning electron microscopy to the regime of liquid helium and liquid nitrogen by providing the necessary sample cooling. However, more importantly, LTSEM yields information on the local *electronic function* and not just the local *structure* with high spatial resolution. In this way it has provided important input for the understanding of the physics of superconducting electronic circuits and devices. Of course, LTSEM is equally important for the evaluation and analysis of low T_c and high T_c superconductors. The principle of LTSEM utilizes the electron beam as a local heat source on the one hand, and the sensitive response of superconductors to small temperature changes on the other hand [5.2]. By scanning the electron beam over the sample surface while the sample is cooled to low temperatures, a local temperature increment is effected. Simultaneously a *functional electronic response signal* is recorded as a function of the coordinate point (x, y) of the electron beam focus on the sample surface. For obtaining the functional response signal, the sample is current or voltage biased during the scanning process, and a corresponding voltage signal $\delta V(x, y)$ or current signal $\delta I(x, y)$ is generated, respectively.

In LTSEM, high spatial resolution is achieved by recording the response of a *global* quantity such as the total current or voltage of the sample to a *local* thermal perturbation. Of course, in each case it is the beam-induced change in the local electric resistance which is detected. The characteristic length scale of the thermal perturbation then determines the spatial resolution limit. The thermal perturbation of a system has been utilized in the past in different contexts. As an example we mention the principle of global thermal perturbation introduced by Eigen [5.3] for studying the physico-chemical kinetics of a system from its temporal response.

In the following we outline the principle of LTSEM and its application as a diagnostic tool for high T_c superconductors. We summarize the results which have been obtained up to now. Recently several reviews of LTSEM have appeared [5.4–5.8], and for details we will refer to these papers.

5.2 Electron beam as a local heat source

Low-temperature scanning electron microscopy is ideally suited for the spatially resolved diagnostics of superconductors because of the extreme sensitivity of the electronic properties of the latter to temperature changes. Usually the resistive transition to the superconducting state is sharp along the temperature axis. Also, the superconducting critical current and the quasi-particle current of superconducting tunnel junctions, and Josephson weak links are sensitively temperature dependent. Furthermore, the superconducting circuits and electronic devices are usually fabricated as a thin-film configuration deposited on a substrate, and, hence, represent ideal objects for local heating by electron-beam irradiation. The local thermal perturbation then results in a local increment $\delta \rho(x, y)$ of the electric resistivity near the coordinate point (x, y) of the electron beam focus on the sample surface. Depending on the bias condition, a voltage signal $\delta V(x, y)$ or a current signal $\delta I(x, y)$ is generated. Here and in the following $\delta V(x, y)$ and $\delta I(x, y)$ always denote the *beam-induced change* of the sample voltage and current, respectively. These signals are also referred to as electron beam induced voltage (EBIV) or current (EBIC), respectively, similar to those widely used in the analysis of semiconductors [5.1]. Because of the thin-film configuration of the superconducting objects under study, a quasi-two-dimensional treatment of the physics underlying the LTSEM signal is usually sufficient, and variations within the superconductor along the z-coordinate perpendicular to the film plane do not need to be considered. However, the coefficient α quantifying the heat transfer between the superconductor and the substrate and the thermal properties of the substrate represent important parameters which must be taken into account.

The study of both low T_c and high T_c superconducting samples below their critical temperature requires sample temperatures in the range from well below the liquid helium boiling temperature of 4.2 K up to more than 120 K. Coverage of this temperature range is achieved by using liquid helium or liquid nitrogen (77 K) as cooling liquids. The most effective cryogenic arrangement consists of a conventional ^{4}He bath cryostat attached to the outside of the sample chamber of the scanning electron microscope. The large tank of liquid helium then serves as a reservoir for a small tank of about 100 cm^3 volume located in the center of the sample chamber. This arrangement allows electron-beam scanning of the thin-film superconductor while the backside of the substrate is in direct contact with liquid helium. If sample temperatures only above 77 K are sufficient, liquid nitrogen can serve as the cryogenic liquid. For measurements at temperatures above 4.2 K a temperature-controlled heating stage is inserted into the small tank. In Fig. 5.1 we show a cross-sectional view of the small helium tank including the mounting of the sample for the temperature range below and above 4.2 K. Further details of the cryogenic apparatus can be found elsewhere [5.6, 5.7, 5.9, 5.10].

The instrumental apparatus for LTSEM is shown schematically in Fig. 5.2. The substrate carrying the sample film is mounted on the temperature-controlled low-temperature stage. While scanning the specimen surface with the electron beam the response signal is displayed synchronously on a video screen. In this way a two-dimensional image of a specific sample property is obtained. Here modern image processing is extremely helpful. In addition to the signals $\delta V(x, y)$ and $\delta I(x, y)$ mentioned above, the standard response signals of scanning electron microscopy such as backscattered (BE) and secondary electrons (SE), the Auger electrons (AE), or the X-rays generated by the electron beam irradiation can be utilized. From the latter signals information on the topology and chemical composition of the specimen is obtained.

The typical beam parameters of commercial scanning electron microscopes are 1–40 keV voltage and 10^{-12}–10^{-6} A current, yielding about 1 nW–40 mW beam power. The scanning speed usually can be varied over a wide range from more than several seconds to less than 100 µs per line. Often signal averaging methods are necessary for improving the signal-to-noise ratio. Here phase-sensitive signal detection together with periodic beam modulation (typical frequency 10–20 kHz) is usually applied. For studying transient phenomena with high time-resolution boxcar averaging in conjunction with periodic short beam pulses can be performed. Commercial beam blanking units

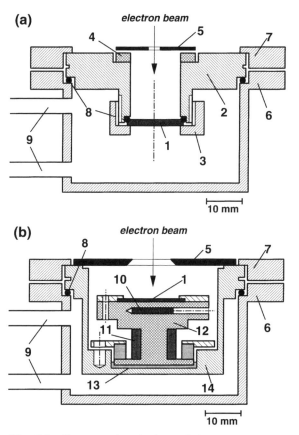

Fig. 5.1. Cross-sectional view of the small liquid helium tank including the sample mounting for the operation below (a) and above 4.2 K (b). 1: sample; 2: sample holder; 3: clamping screw; 4: copper ring for wire heat sink; 5: thermal shield; 6: liquid helium tank; 7: clamping ring; 8: indium seal; 9: liquid helium tubes; 10: temperature sensor; 11: heater; 12: copper block; 13: nylon disk; 14: lid for liquid helium tank. (From Ref. [5.7].)

allow the generation of beam pulses as short as only a few picoseconds. Note that electron-beam pulses of 10 pA current and 10 ns duration on the average only contain about a single electron. Such pulses have been used for studies of the energy resolution of superconducting tunnel junction particle detectors [5.11, 5.12].

The physical principles of LTSEM applied to superconductors were first discussed by Clem & Huebener [5.2]. The minimum beam diameter which can be achieved in commercial scanning electron microscopes is typically about 10 nm. On the other hand, the spatial resolution of LTSEM is limited by the spreading of the beam-induced sample perturbation. In thin-film superconductors it is essentially the *thermal perturbation* (local heating effect) which

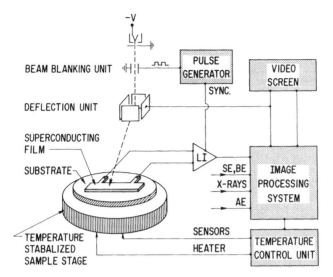

Fig. 5.2. Principle of low temperature scanning electron microscopy of super-conducting thin-film devices and circuits. (From Ref. [5.7].)

determines the spatial resolution obtained with the signals $\delta V(x, y)$ and $\delta I(x, y)$. Thermalization of the beam energy takes place predominantly through Coulomb interaction between the incident and the target electrons, in addition to electron-phonon interactions. The typical time scale of this thermalization process is about 10^{-14} s, and a more detailed discussion can be found elsewhere [5.2, 5.5–5.7]. Here we only concentrate on the local heating effect, since the essential physics of LTSEM is contained in this process [5.2]. For the configuration of a thin-film superconductor of thickness d on a substrate the spatial spreading of the heating effect in the sample film is determined by the thermal healing length

$$\eta = \left(\frac{\kappa \cdot d}{\alpha}\right)^{1/2}. \tag{5.1}$$

The temporal response is given by the thermal relaxation time

$$\tau_\alpha = \frac{C \cdot d}{\alpha}. \tag{5.2}$$

Here κ and C are the heat conductivity and the heat capacity per unit volume of the film, respectively, and α is the heat transfer coefficient describing the heat transfer between the film and the substrate. The range of penetration of the beam electrons into the target material is proportional to about the 1.5 power of the beam energy E_o and inversely proportional to the mass density ρ_m of the absorbing medium. For beam energies up to 10 keV in high T_c and low T_c superconductors the range of penetration is typically 1 μm or less [5.7].

For estimating the thermal healing length for high T_c superconductors from eq. (5.1) we assume an electric resistivity $\rho = 100\ \mu\Omega\,\text{cm}$, $d = 1\ \mu\text{m}$, $\alpha(4.2\ \text{K}) = 1\ \text{Wcm}^{-2}\text{K}^{-1}$ and $\alpha(77\ \text{K}) = 10^3\ \text{Wcm}^{-2}\text{K}^{-1}$ [5.7]. Using the Wiedemann–Franz law for finding κ, we obtain $\eta(4.2\ \text{K}) = 3.2\ \mu\text{m}$ and $\eta(77\ \text{K}) = 0.43\ \mu\text{m}$. Taking the value $C(4.2\ \text{K}) = 3\ \text{mJ cm}^{-3}\text{K}^{-1}$ for YBa$_2$Cu$_3$O$_{7-\delta}$ [5.7], we estimate $\tau_a(4.2\ \text{K}) = 300$ ns from eq. (5.2). Since the ratio C/α is nearly temperature independent, a similar value of τ_a is expected at 77 K. If the beam power P_o is totally dissipated within the specimen film, the temperature rise $T - T_b$ in the film above the bath temperature T_b is given by

$$T - T_b = \frac{P_o}{\alpha\eta^2\pi}. \tag{5.3}$$

Assuming the value $\pi\cdot\eta^2 = 30\ \mu\text{m}^2$ and taking $\alpha(4.2\ \text{K}) = 1\ \text{Wcm}^{-2}\text{K}^{-1}$ we find for $P_o = 1\ \mu\text{W}$ the temperature increment $T - T_b = 3$ K.

The thermal healing length η of eq. (5.1) governs the *static response* to an unmodulated electron beam. If the beam power is modulated at the angular frequency ω, the static healing length η can be replaced by the frequency-dependent thermal healing length [5.2, 5.5]

$$\eta_\omega = \eta\left(\frac{1}{2}\{[1 + (\omega\tau_a)^2]^{1/2} + 1\}\right)^{-1/2} \tag{5.4}$$

representing the characteristic length scale of the modulated sample region. Here η is given by eq. (5.1). In the limit $\omega\tau_a \ll 1$ we have $\eta_\omega \approx \eta$. In the opposite limit $\omega\tau_a \gg 1$ eq. (5.4) approaches the expression

$$\eta_\omega = \eta\cdot\left(\frac{2}{\omega\tau_a}\right)^{1/2} = \left(\frac{2D}{\omega}\right)^{1/2} \tag{5.5}$$

where $D = \kappa/C$ is the thermal diffusivity. The length η_ω determines the *dynamic response* and is referred to as the dynamic thermal healing length. The frequency dependence of η_ω in eq. (5.5) is known as the thermal skin effect, and η_ω is the thermal skin depth. In the high-frequency regime the length η_ω decreases with increasing modulation frequency ω. This results in a corresponding increase of the spatial resolution obtained by LTSEM, if the modulated signal is detected. On the other hand, the signal amplitude decreases with increasing modulation frequency, since the signal-generating volume of the thermally modulated sample region becomes smaller. Therefore, the gain in spatial resolution becomes limited at high modulation frequencies. The validity of the thermal skin effect for LTSEM, as discussed above, has been demonstrated experimentally by Pavlicek *et al.* [5.13]. The range of the modulation frequency has been extended up to 50 MHz.

The ideas we have outlined in this section can be generalized and expressed in terms of a general response theory [5.2, 5.7]. In addition, the principles apply equally well to other scanning techniques such as scanning with a laser beam or a tunnelling microtip. In each case, the size of the scanned area and the achieved spatial resolution depend on the individual method. Scanning with an electron beam utilized in LTSEM appears particularly well suited, in terms of the obtained spatial resolution in the μm range and the size of the scanned area, for the microanalysis of the functional behavior of microelectronic super-conducting circuits and devices.

5.3 Thin films

The analysis of thin films of high temperature superconductors represents an important application of LTSEM. Here we have in mind in particular the detection of deviations of the critical temperature T_c or of the critical current density J_c from spatial homogeneity. The principles for such an analysis have been outlined shortly after the discovery of high-temperature superconductivity [5.14–5.16]. For illustration we consider the case where the thin-film high T_c superconductor consists of three regions each with a different transition temperature T_c. The resistive transition curves of the three regions are shown schematically in Fig. 5.3. (For simplicity we have assumed that the normal-state resistivity in the three regions is the same.) In the following we take the

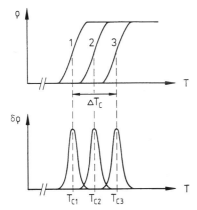

Fig. 5.3. Electric resistivity ρ (top) and beam-induced resistivity change $\delta\rho$ (bottom) versus temperature T for three regions with different transition temperature T_c. (From Ref. [5.14].)

midpoint of the transition curves as the T_c value of each region. We denote the total spread of the T_c values by ΔT_c and the beam-induced local temperature increment by δT. With the further assumption $\delta T \ll \Delta T_c$ the beam-induced resistivity change $\delta\rho = (\partial\rho/\partial T)\cdot\delta T$ is shown in the lower part of Fig. 5.3 for the three regions as a function of the operating temperature for LTSEM. We see that the response signal $\delta\rho$ is strongly different for the three regions reaching a sharp maximum at different temperatures. The maximum of the response signal $\delta\rho$ is close to the temperature where the transition curve of the corresponding region reaches its steepest slope. According to these arguments the beam irradiation effects a local resistivity increment $\delta\rho$ which sensitively depends upon the T_c value of the irradiated region and the operating temperature. In this way LTSEM can provide information on the spatial variation of the local critical temperature in thin-film high T_c superconductors. In the scheme outlined above we have assumed that the absorbed beam power and the thermal coupling of the film to the substrate is uniform over the total film area investigated. If spatial inhomogeneities exist due to one or both reasons, they fold into the electric response signal $\delta\rho$.

Next we turn in more detail to the detection of the beam-induced local resistivity change, $\delta\rho$. For simplicity we start with the *one-dimensional case* shown schematically in Fig. 5.4 with a series connection (part a) and a parallel connection (part b) of regions with different values of T_c. For practical reasons we only consider current-biased operation. The total current and voltage is denoted by I and V, respectively. The direction parallel and perpendicular to the current flow is denoted as the x- and y-direction, respectively.

For the series connection (Fig. 5.4(a)) the local beam-induced resistivity increment $\delta\rho(x)$ results in a voltage signal

$$\delta V(x) \sim \delta\rho(x). \tag{5.6}$$

In the parallel connection (Fig. 5.4(b)) the beam irradiation causes the current reduction $-\delta I(y) \sim -\delta\rho(y)$ in the irradiated part. As a consequence the

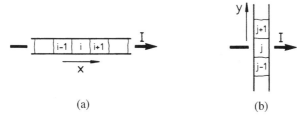

(a)   (b)

Fig. 5.4. Series connection (a) and parallel connection (b) of regions with different values of the transition temperature T_c (one-dimensional case). Along the x- and y-direction the regions are marked with the index i and j, respectively. (From Ref. [5.14].)

current in the unirradiated part is increased by $\delta I(y) \sim \delta\rho(y)$. Assuming that the irradiated part is much shorter than the unirradiated part, the beam induced voltage signal is in good approximation

$$\delta V(y) = \frac{\partial V}{\partial I} \cdot \delta I(y), \tag{5.7}$$

where the derivative $\partial V/\partial I$ is taken as that of the whole film without irradiation. From eqs. (5.6) and (5.7) we see that in both cases the voltage signal yields information on the local resistivity change $\delta\rho$.

As a typical example we show in Fig. 5.5 the LTSEM voltage image

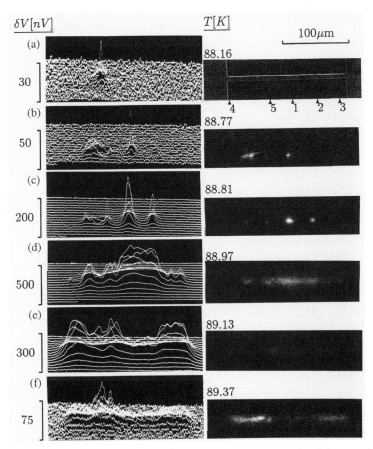

Fig. 5.5. Voltage signal $\delta V(x, y)$ of an epitaxial *c*-axis oriented YBa$_2$Cu$_3$O$_{7-\delta}$ film of 200 nm thickness for different temperatures at $I_B = 11.2\ \mu A$ bias current, on the left in *y*-modulation and on the right in brightness modulation. The micrograph on the upper right identifies the sample geometry in the images. The arrows refer to the results presented in Fig. 5.6. The temperatures are the following: (a) 88.16 K, (b) 88.77 K, (c) 88.81 K, (d) 88.97 K, (e) 89.13 K and (f) 89.37 K. (From Ref. [5.17].)

$\delta V(x, y)$ of an epitaxial c-axis oriented $YBa_2Cu_3O_7$ film of 200 nm thickness and 10 μm width on a (100)-oriented $SrTiO_3$ substrate [5.17]. The image $\delta V(x, y)$ is presented for the bias current $I_B = 11.2$ μA (corresponding to the current density $J_B = 5.6\cdot10^2$ A cm^{-2}) and for different temperatures between 86.16 and 89.37 K. The images on the left show $\delta V(x, y)$ plotted vertically for a series of horizontal line scans (y-modulation). On the right $\delta V(x, y)$ is displayed using brightness modulation, where the bright regions indicate the locations with a large value of the signal $\delta V(x, y)$. On the right at the top of Fig. 5.5 we show a micrograph of the sample configuration for identifying the geometry of the images. In Fig. 5.5 the first signal appears at 88.16 K near the left end of the sample. As the temperature is increased additional signal-generating regions appear. At $T = 88.97$ K the voltage signal reaches a maximum, and at higher temperatures it decreases again with increasing temperature, as one would expect from the decrease of the slope of the $\rho(T)$-curve. From the results of Fig. 5.5 we see that the resistive transition of the whole film is spread only over a temperature range less than 1 K. However, distinct spatial variations of T_c can clearly be resolved within this narrow temperature interval. The temperature dependence of the voltage signal for the five locations marked by the arrows in the micrograph of Fig. 5.5 is shown in Fig. 5.6. From the results of Figs. 5.5 and 5.6 we conclude that variations of T_c

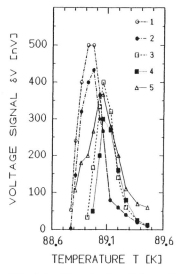

Fig. 5.6. Magnitude of the voltage signal $\delta V(T)$ vs. temperature for the five locations on the sample indicated in Fig. 5.5 by the arrows on the micrograph. Bias current $I_B = 11.2$ μA. (From Ref. [5.17].)

of 200 mK in epitaxial $YBa_2Cu_3O_{7-\delta}$ have been detected by means of LTSEM with high spatial resolution.

The *two-dimensional* case is clearly more difficult. Even for a current-biased sample the current density in the perturbed region does not have to remain constant, and it is only the integral of the current density over a complete cross-sectional area which stays constant. A quantitative LTSEM analysis of T_c variations is possible only under some restrictive conditions. An example would be the case where the current flow is essentially one-dimensional (along the x-direction) and the equipotential lines are predominantly parallel to the y-direction. However, qualitative results can always be obtained in the more complicated general case, and in many cases spatial inhomogeneities of T_c can be detected.

The concepts for detecting spatial variations of T_c can be extended for analyzing inhomogeneities in the local critical current density, $J_c(x, y)$. Above the critical current, I_c, the electric resistivity of a superconductor strongly increases with increasing current and eventually becomes nearly current-independent when the normal state is approached. During this transition the temperature is assumed to be constant. A detailed discussion of this non-ohmic behavior can be found elsewhere [5.18]. In Fig. 5.7 we show such transitions schematically for three sample regions with different values of the critical current. Due to the beam-induced temperature increment δT each curve is shifted slightly to the left, and the corresponding resistivity change $\delta\rho$ is sketched in the bottom part of Fig. 5.7. Again, if the temperature change δT is sufficiently small, the curves $\delta\rho(I)$ display sharp maxima near the current

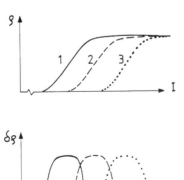

Fig. 5.7. Electric resistivity ρ (top) and beam-induced resistivity change $\delta\rho$ (bottom) versus current I for three regions with different critical currents. (From Ref. [5.16].)

where the steepest slope is reached in the curves $\rho(I)$. For current-biased operation a voltage signal $\delta V(I)$ is generated which depends on the current and has a maximum slightly above I_c. The onset of the voltage signal $\delta V(I)$ images the critical current I_c at which resistance starts to appear.

The procedure for imaging the spatial distribution $J_c(x, y)$ is analogous to that discussed above for imaging the distribution $T_c(x, y)$. At a fixed temperature $T < T_c$, starting from a small value the bias current is increased in small steps. After each step the sample is scanned with the electron beam. A nonzero voltage signal $\delta V(x, y)$ first appears for the coordinate point (x) representing the cross-sectional area with the smallest critical current $I_c(x)$. As the bias current is increased stepwise, more and more regions with higher values of $I_c(x)$ generate a voltage signal. In this way the distribution $I_c(x)$ can be imaged. Of course, adequate spatial resolution is obtained only if the beam-induced reduction $|\delta I_c|$ is small compared to the total spread ΔI_c along the sample: $|\delta I_c| \ll \Delta I_c$. Again, extending this analysis to the case of variations of J_c along the y-direction is difficult, due to possible redistributions of the current density along this direction. However, in special cases meaningful results on the two-dimensional distribution $J_c(x, y)$ can be obtained. A more detailed discussion can be found elsewhere [5.7, 5.14, 5.16]. Finally, we emphasize again that the principles for imaging $J_c(x, y)$ and $T_c(x, y)$ by means of LTSEM are highly similar. Both principles are based on the strong nonlinearity of the curves $\rho(I)$ and $\rho(T)$ in the critical region near $I = I_c$ and $T = T_c$, respectively.

A typical image of $J_c(x, y)$ for a polycrystalline film of $YBa_2Cu_3O_{7-\delta}$ (thickness $= 1.1$ µm; width $= 30$ µm; length $= 520$ µm) obtained at 53.4 K is shown in Fig. 5.8. The images (a)–(e) were observed for increasing bias current ranging between 722 µA and 8.66 mA. Bright regions indicate the locations where a beam-induced signal $\delta V(x, y)$ (typically 20 µV) has been generated during the scanning process. It is in these bright regions where the bias current exceeds the local critical current I_c in the irradiated sample. From Fig. 5.8 we see that with increasing bias current more and more resistive regions appear and that the local I_c values vary by more than a factor of ten.

Figure 5.9 shows a series of voltage images $\delta V(x, y)$ obtained for an epitaxial c-axis oriented $YBa_2Cu_3O_{7-\delta}$ film of 60 nm thickness, 70 µm width, and 400 µm length. The images (a)–(d) were observed for increasing bias current ranging between 12.5 and 25 mA. The bright lines stretching across the film indicate local regions with strongly reduced critical current density. In this case the local weakening of the sample had been caused apparently by small scratches in the substrate.

The images presented in Figs. 5.8 and 5.9 were generated using a constant

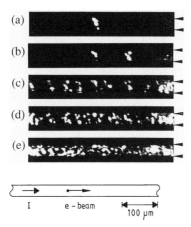

Fig. 5.8. Images obtained at 53.4 K with the following values of the bias current: (a), 722 µA; (b) 912 µA, (c) 1.24 mA, (d) 2.17 mA, (e) 8.66 mA. The Y-Ba-Cu-O film extends horizontally beyond the field of view, as indicated schematically at the bottom. The arrows on the right of each image mark the location of the film boundaries. (From Ref. [5.15].)

bias current and recording the beam-induced voltage signal $\delta V(x, y)$ during the scanning process. (Usually the beam is modulated at about 20 kHz frequency and $\delta V(x, y)$ is detected with a lockin-amplifier.) In addition to this method, a different measuring technique can be applied quite successfully, which we refer to as *maximum critical current detection* [5.20]. In this case the sample current is increased at a constant rate with the electron beam irradiating the sample at the location (x, y). Then, the current value at which the sample voltage exceeds a specified threshold value, V_{ref}, is detected and stored using a voltage comparator and a sample-and-hold unit. The voltage comparator triggers the sample-and-hold unit and a switch between the periodic sweeper and the current source, whenever the sample voltage exceeds V_{ref}. The periodic sweeper and the sample current are then reset to zero. This cycle is repeated periodically up to a rate of 10^4 measurements per second. Subtracting the critical current without electron beam irradiation from the value obtained with beam irradiation, the electron-beam induced change $\delta I_c(x, y)$ is found. A two-dimensional critical current image $\delta I_c(x, y)$ is generated by scanning the electron beam across the sample surface and measuring I_c continuously. This maximum critical current detector has been particularly useful for studying the spatial distribution of the pair tunneling current density in superconducting tunnel junctions [5.20] and Josephson weak links. LTSEM performed with high T_c Josephson junctions will be discussed in the following section.

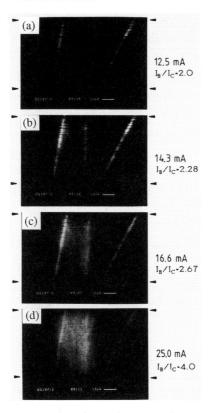

Fig. 5.9. Voltage images of a 70 μm wide superconducting line showing the spatial distribution of the critical current density at 83 K. The line extends horizontally beyond the field of view. The arrows mark the film boundaries. (From Ref. [5.19].)

5.4 Superconducting devices

5.4.1 Microbridges

Microbridges fabricated from high-temperature superconductors are presently being studied for electronic applications such as Josephson weak links, bolometers, etc. Here the detection of spatial inhomogeneities of T_c and J_c becomes crucial, and LTSEM finds important applications as has been discussed in the last section. A dramatic consequence of spatial inhomogeneities in narrow superconducting lines carrying an electric current is the possible development of a hotspot. A self-heating hotspot is a dissipative structure which can be generated in an electrical conductor with an S-shaped temperature dependence of its resistance [5.21, 5.22]. A detailed analysis of the heat balance equation for describing a hotspot in a thin-film superconductor has been reported by Skocpol *et al.* [5.23]. A hotspot shows a distinct temperature profile passing through T_c at both ends and reaching a maximum temperature larger than T_c in the center. Imaging of a hotspot by means of LTSEM can be understood qualitatively as follows. Noting that the hotspot is maintained by the current applied to the sample, the electron beam irradiation only acts as a small perturbation of the system. The beam induced local temperature increment causes a small increase in the electric resistance and a corresponding voltage signal $\delta V(x, y)$ only near the locations where the temperature profile passes through T_c. Further outside the hotspot the sample temperature is too low for any additional electric resistance to appear due to the beam irradiation. On the other hand, in the region within the hotspot with $T > T_c$ the resistance remains nearly unchanged for the small temperature excursions due to the beam irradiation. The sensitive region at both hotspot boundaries extends over a distance given by the thermal heating length η in both directions. Therefore, we expect a peaked voltage signal $\delta V(x, y)$ of width 2η at both hotspot boundaries. Experiments have exactly confirmed these expectations [5.24, 5.25]. Further details and references can be found elsewhere [5.5, 5.6, 5.25].

5.4.2 *Josephson junctions*

In the applications of high temperature superconductors in cryoelectronics Josephson junctions play a central role. Therefore, in recent years a strong effort has been concentrated on the development of high T_c Josephson junctions, and presently a few major types of junctions appear promising [5.26]. Soon after the discovery of high-temperature superconductivity it became apparent that the classical type of tunnel junction with a dielectric layer of only a few ångströms thickness placed between two planar superconducting electrodes is unfeasible for the cuprate superconductors. The reason for this difficulty is the extremely short superconducting coherence length and the large anisotropy in the cuprates, which would require a control of the material composition of the junction on an atomic scale. For applications as Josephson weak links various types of grain boundaries artificially fabricated within the otherwise epitaxial superconducting film have emerged as the most promising junctions. Here LTSEM has already provided important input regarding the physics of these junctions and in particular the identification of the various causes for malfunction. In the following we illustrate the role of LTSEM in the diagnostics of high T_c Josephson junctions mainly for one type of junction, namely the bicrystal single grain boundary junction.

The bicrystal single grain boundary Josephson junction is fabricated by depositing an epitaxial *c*-axis oriented cuprate film on a substrate bicrystal where the crystallographic orientation of both halves is rotated relative to each other by a specific misorientation angle. In this way during the epitaxial growth of the cuprate film a single grain boundary is generated in the film showing Josephson behavior [5.26, 5.27]. This grain boundary junction is usually treated as quasi-one-dimensional, since the thickness of the superconducting film is less than the Josephson penetration depth λ_J. Within the film plane, x and y denote the direction perpendicular and parallel to the grain boundary, respectively. If a magnetic field B is applied in z-direction to a junction of width less than the Josephson penetration depth, the phase difference $\varphi(y)$ of the superconducting wave function between both sides of the junction increases linearly with the y-coordinate

$$\varphi(y) = \frac{2\pi d_m B}{\phi_0} \cdot y \tag{5.8}$$

and the pair current density $J_s(y)$ is sinusoidally modulated along the junction

$$J_s(y) = J_c(y)\sin\left(\frac{2\pi d_m \cdot B}{\phi_0} \cdot y\right). \tag{5.9}$$

Here d_m is the effective magnetic thickness of the junction, ϕ_0 the magnetic flux quantum, and $J_c(y)$ the maximum Josephson current density. A complete oscillation of the pair current density is called a Josephson vortex and contains one magnetic flux quantum. This behavior is shown schematically in Fig. 5.10 [5.28].

During electron beam scanning the maximum Josephson current density

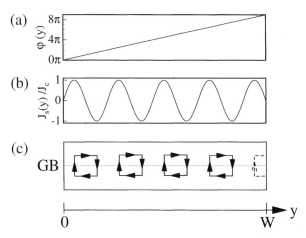

Fig. 5.10. Variation of (a) the phase difference $\varphi(y_0)$ and (b) the normalized supercurrent density $J_s(y_0)/J_c$ along an ideal narrow Josephson junction. The location of the Josephson vortices is schematically sketched in (c) (GB denotes grain boundary). (From Ref. [5.28].)

$J_c(y)$ is changed by the amount $\delta J_c(y)$ in the perturbed region. If the phase-difference function $\varphi(y)$ were to remain unaffected, the beam-induced change of the junction critical current would be sinusoidally modulated as a function of the y-coordinate according to eq. (5.9). Such modulation has been observed for quasi-one-dimensional tunnel junctions, and the different vortex states have been imaged using the maximum critical current detector described at the end of section 5.3 [5.20, 5.29]. In principle, the same technique can also be used for imaging the vortex states in high T_c Josephson weak links. However, for weak links only a small threshold voltage (typically less than 10 μV) can be used because of the non-hysteretic behavior of the current–voltage character-istic. This is in contrast to superconducting tunnel junctions, which are hysteretic and where the threshold voltage can be close to the sumgap voltage (typically a few millivolts). Therefore, for investigating the distribution of the pair current density in high T_c weak-link Josephson junctions it is advanta-geous instead to bias the junction at $I = I_B$ slightly above the critical current and to measure the beam-induced voltage signal $\delta V(y)$, which is given by

$$\delta V(y) = \left(\frac{\partial V}{\partial I}\right)_{I=I_B} \cdot \delta j_s(y). \tag{5.10}$$

Here $\delta J_s(y)$ is the beam-induced change of the supercurrent density at the coordinate point y. A typical LTSEM voltage image obtained in this way is shown in Fig. 5.11 for a bicrystal $YBa_2Cu_3O_{7-\delta}$ grain boundary junction of 23 μm width. The image is recorded at 83 K and displays the pair current density distribution of the 4−5 vortex state. (It is this state which is also shown

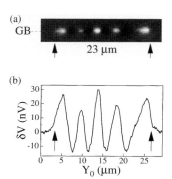

Fig. 5.11. (a) LTSEM voltage image of the 4−5 vortex state in a 23 μm wide $YBa_2Cu_3O_{7-\delta}$ grain boundary junction (GBJ) at $T = 83$ K. The edges of the GBJ are indicated by the arrows. The position of the grain boundary is marked by the broken line. Bright and dark regions correspond to sample regions yielding a positive and negative voltage signal, respectively. (b) Single line scan along the grain boundary with the voltage signal plotted vertically during the horizontal scan. (From Ref. [5.28].)

schematically in Fig. 5.10). In addition, such images prove that the LTSEM signals are restricted to a narrow region of about 1 μm width along the grain boundary, and that the Josephson weak link is, indeed, located at the grain boundary.

For simplicity, we have assumed in our discussion that the extension of the Josephson weak link along the y-direction does not exceed the Josephson penetration depth λ_J. Further, we have ignored the non-local effects arising from the beam-induced change in the phase difference function $\varphi(y)$. Discussions of the deviations from these simplifying assumptions are given elsewhere [5.6, 5.7, 5.20, 5.28].

In our discussion of the vortex states in an applied magnetic field and of the Josephson physics in Figs. 5.10 and 5.11 we have assumed that the maximum pair current density $J_c(y)$ is spatially homogeneous. Of course, LTSEM is perfectly suited to detect deviations from the homogeneity of $J_c(y)$ (due to microshorts, etc.). As an example we show in Fig. 5.12 the voltage image $\delta V(x, y)$ of a $YBa_2Cu_3O_{7-\delta}$ bicrystal grain boundary junction (film thickness $= 0.5$ μm, width $= 20$ μm) measured at 14 K and zero magnetic field [5.30]. The junction was current biased slightly above the critical current I_c. The recorded voltage signal, which is directly proportional to the critical current density of the grain boundary, is plotted vertically during the horizontal line scans. Distinct variations of $J_c(y)$ can be seen along the grain boundary. Self-excited resonances can be generated in the grain boundary junction acting

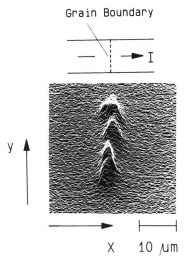

Fig. 5.12. Voltage image $\delta V(x, y)$ of a $YBa_2Cu_3O_{7-\delta}$ grain boundary junction taken with a bias current of 30 mA at 14 K. (From Ref. [5.30].)

as a cavity for electromagnetic waves. Already LTSEM has contributed significantly to our understanding of the physics of these resonances [5.30, 5.31].

In addition to the analysis of single Josephson junctions, LTSEM finds
important applications in the analysis of junction arrays. The development of
high T_c Josephson arrays is presently in its infancy. We illustrate the analytic
power of LTSEM with a recent example, taken again from the bicrystal single
grain boundary technology [5.32]. It was the goal of the experiment described
in the following to determine the critical current value of each individual
junction in a one-dimensional series array of 30 junctions. The array consisted
of an epitaxial $YBa_2Cu_3O_{7-\delta}$ film meandering as a narrow line across the grain
boundary of the substrate bicrystal (see Fig. 5.14(b)). The imaging principle
utilized the sharply peaked beam-induced voltage signal δV appearing at the
critical current I_c when the beam is focused on an individual junction and the
bias current I_B is increased monotonically. A typical example is shown in Fig.
5.13. By scanning over all 30 junctions of the array and stepwise increasing the
bias current, the image presented in Fig. 5.14(a) was obtained. Here the voltage
signal δV is shown as a function of the bias current (plotted downwards on the
vertical axis) for the different junctions appearing along the horizontal axis.
Dark and bright regions correspond to regions with large and small voltage
signal, respectively. An SE image of the array is seen in Fig. 5.14(b) for
geometric identification. The electron beam was always scanned along the
grain boundary (black line in (b)). From this image the critical current for each
individual junction could be obtained. The data of Fig. 5.14 were measured in
zero applied magnetic field. It turned out that the relatively large spread of I_c

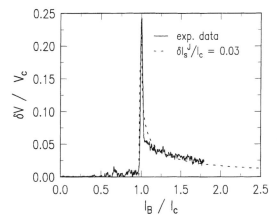

Fig. 5.13. Calculated and experimentally measured electron-beam induced
voltage signal versus the normalized bias current. The experimental data were
obtained with $YBa_2Cu_3O_{7-\delta}$ bicrystal grain boundary junctions at $T = 30$ K.
The theoretical curve was calculated using $\delta I_s^J/I_c = 0.03$. (From Ref. [5.32].)

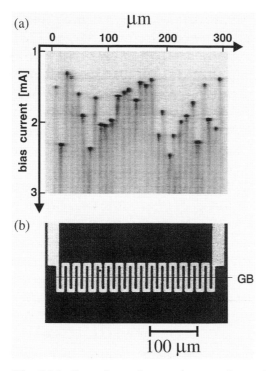

Fig. 5.14. Secondary electron image of a series array of 30 $YBa_2Cu_3O_{7-\delta}$ bicrystal grain boundary junctions (b) and LTSEM voltage image (a) showing the bias current dependence of the electron-beam induced voltage signal at $T = 35$ K. The electron beam is always scanned along the grain boundary [black line in (b)]. The scans recorded at different biased current values are displaced vertically. Dark and bright regions correspond to regions with large and small voltage signal. (From Ref. [5.32].)

for the different junctions seen in Fig. 5.14(a) was caused by trapped magnetic flux within the array structure, affecting the individual junctions differently. Finally, the correct zero-magnetic field value of I_c for each junction was determined by performing the experiment for different applied magnetic fields and finding the maximum I_c for each junction expected when the trapped flux is compensated by the applied field. With this technique one was able to show that the spread of I_c for all junctions of the array was less than 20 % [5.32]. This example clearly shows the importance of LTSEM also for development of the technology of high T_c Josephson arrays.

5.5 Cryoelectronic circuits

Superconducting circuits for microelectronic applications usually represent a more or less complex structure consisting of connecting lines, active or passive devices, etc. Clearly, LTSEM is a powerful tool for checking the functional behavior of this structure and, in particular, for identifying locations of malfunction. For the latter task the spatially resolving capability of LTSEM becomes a necessity. In the following we discuss a few selected examples.

The multilayer structure of the circuits often requires the *crossing of lines* with good electric insulation between them. Such line crossings potentially represent locations with weakened superconductivity. The identification of such weak spots is relatively easy [5.33]. *Sharp bends* in narrow superconducting lines also show a local reduction of the critical current because of the shifting of the supercurrent density towards the inner side of the corner. This nonsymmetric distribution of the current density has been observed recently by means of LTSEM for 90° corners in lines of 150 μm width and 150 nm thickness fabricated from epitaxial c-axis oriented $YBa_2Cu_3O_{7-\delta}$ [5.34].

In many high frequency applications microwaves must be transmitted through superconducting circuits. LTSEM is also highly promising for spatially resolved studies of the *microwave properties* of superconducting devices and circuits [5.35, 5.36]. An apparatus for LTSEM under microwave irradiation has been described by Doderer *et al.* [5.37]. Up to now most of these experiments have been performed with low T_c superconductors. Voltage images of standing microwave patterns due to unintended reflection and of locations where the microwaves are leaking out have been reported [5.35, 5.36]. It is expected that similar experiments with microwave circuits fabricated from high-temperature superconductors will gain importance in the future.

Acknowledgements

A critical reading of the manuscript by R. Gross and R. Gerdemann is gratefully acknowledged.

References

[5.1] L. Reimer, *Scanning Electron Microscopy*, Springer, Berlin, (1985).
[5.2] J. R. Clem & R. P. Huebener, *J. Appl. Phys.* **51**, 2764 (1980).
[5.3] M. Eigen, *Disc. Faraday Soc.* **17**, 194 (1954).
[5.4] R. P. Huebener & H. Seifert, *Scanning Electron Microscopy III*, 1053 (1984).
[5.5] R. P. Huebener, *Rep. Prog. Phys.* **47**, 175 (1984).
[5.6] R. P. Huebener, *Advances in electronics and electron physics*, P. W. Hawkes, Academic Press, New York, Vol. **70**, 1 (1988).
[5.7] R. Gross & D. Koelle, *Rep. Prog. Phys.* **57**, 651 (1994).
[5.8] R. P. Huebener, *Superconducting quantum electronics*, V. Kose (Springer, Berlin, p. 205, 1989).
[5.9] H. Seifert, *Cryogenics* **22**, 675 (1982).

[5.10] R. Gross *et al.*, *Cryogenics* **29,** 716 (1989).

[5.11] F. Hebrank *et al.*, *IEEE Trans. Appl. Supercond.* **3**, 2084 (1989).

[5.12] F. Hebrank *et al.*, *J. Low Temp. Phys.* **93**, 647 (1993).

[5.13] H. Pavlicek *et al.*, *J. Low Temp. Phys.* **56,** 237 (1984).

[5.14] R. P. Huebener, R. Gross & J. Bosch, *Z. Phys. B - Cond. Matter* **70**, 425 (1988).

[5.15] R. Gross *et al.*, *Nature* **332**, 818 (1988).

[5.16] R. P. Huebener & R. Gross, *Scanning Microscopy* **3** , 703 (1989).

[5.17] D. Koelle *et al.*, *Physica C* **167**, 79 (1990).

[5.18] R. P. Huebener, *Magnetic flux structures in superconductors* (Springer, Berlin, 1979).

[5.19] R. Gross *et al.*, *Physica C* **162-164**, 1603 (1989).

[5.20] J. Bosch *et al.*, *J. Low Temp. Phys.* **68,** 245 (1987).

[5.21] R. Landauer, *Phys. Today* **31**, 23 (1978).

[5.22] M. Büttiker & R. Landauer, in *Nonlinear phenomena at phase transitions and instabilities*, T. Riste (Plenum Press, New York, p. 111, 1982).

[5.23] W. J. Skocpol, M. R. Beasley & M. Tinkham, *J. Appl. Phys.* **45**, 4054 (1974); *J. Low Temp. Phys.* **16**, 145 (1974).

[5.24] R. Eichele, H. Seifert & R. P. Huebener, *Appl. Phys. Lett.* **38**, 383 (1981); *Z. Phys. B* **48**, 89 (1982).

[5.25] R. Eichele, L. Freytag, H. Seifert, R. P. Huebener & J. R. Clem, *J. Low Temp. Phys.* **52**, 449 (1983).

[5.26] R. Gross, in *Interfaces in superconducting systems*, S. L. Shinde & D. Rudman, Springer, New York, (1994).

[5.27] P. Chaudhari *et al.*, *Phys. Rev. Lett.* **60**, 1653 (1988).

[5.28] G. M. Fischer *et al.*, *Science* **263**, 1112 (1994).

[5.29] J. Bosch *et al.*, *Phys. Rev. Lett.* **54**, 1448 (1985).

[5.30] J. Mannhart *et al.*, *Science* **245**, 839 (1989).

[5.31] A. Beck *et al.*, *Appl. Supercond. Conf.* (1994).

[5.32] R. Gerdemann *et al.*, *J. Appl. Phys.* **76**, 8005 (1994).

[5.33] K.-D. Husemann *et al.*, *Appl. Phys. Lett.* **62,** 2871 (1993).

[5.34] H. Töpfer *et al.*, *IEEE Trans. Magnetics* **31**, 810 (1995).

[5.35] D. Quenter *et al.*, *Appl. Phys. Lett.* **63**, 2135 (1993).

[5.36] T. Doderer *et al.*, *IEEE Trans. Appl. Supercond.* **3**, 2724 (1993).

[5.37] T. Doderer *et al.*, *Cryogenics* **30**, 65 (1990).

6

Scanning tunneling microscopy

M. E. HAWLEY

6.1 Introduction

The scanning tunneling microscope (STM) is the youngest member of the electron microscopy family, developed only a little over ten years ago. The STM has its own unique list of assets and capabilities to apply to the study of high T_c superconducting materials that distinguishes it from the other family members. The data obtainable by STM can duplicate, surpass, or complement those extracted by the other electron microscopes. The STM has the advantage of having a higher vertical resolution than the scanning electron microscope and can achieve atomic resolution without the extensive and potentially damaging sample preparation techniques required for transmission electron microscopy. A disadvantage is that STM measurements are limited to the near surface region. Its realm is truly the atomic-to-nanometer world of the surface.

In addition to the extremely high vertical resolution (less than a 1 Å) routinely attainable by scanning tunneling microscopy and the often limited sample preparation required as noted above, the STM's additional advantages lie in (1) its sensitivity to both local electronic and structural properties, (2) the variety of measurements possible, (3) the low, generally nondestructive, energy range in which it operates, and (4) its environmental flexibility, i.e. its ability to operate under a wide range of temperatures and atmospheric conditions.

The STM's roots lie in electron vacuum tunneling spectroscopy. In the context of measuring electronic properties, it is more correctly described as a spectrometer, for it is the electronic properties of the surface that are being probed in the STM experiment. The correspondence between the electronic and topographic properties is responsible for the microscope label.

This chapter will start with a very brief background on electron vacuum tunneling and a description of the STM's conceptual basis and historical evolution, particularly as it relates to the study of high T_c superconductors specifically. It will only introduce the concepts behind tunneling spectroscopy. Following, other sections will include a description of the instrumental components, design, and operation with a list of references for more in-depth discussion. It will then address the application of the STM technique to the study of the structural and electronic properties of high T_c single-crystal and thin-film forms. Because most books on STM devote the bulk of their discussion to ultra-high-vacuum (UHV) STM, descriptions of other STM measurement methods and related proximity probe techniques, of which there is an ever growing number, they make little mention of its operation under ambient conditions. The latter has, however, played a more day-to-day role in the study of high T_c thin films driven by the need to optimize and control the structural quality of the films intended for the development of multilayer device structures. Therefore, the bulk of the chapter will be devoted to this aspect of the application of STM to the study of high T_c materials with only a short section on single-crystal and low-temperature spectroscopy of high-temperature superconductors, important for understanding the intrinsic electrical properties of these perovskite materials. A brief description of 'other scanning probe microscope (SPM) techniques' will follow. The chapter will conclude with a short mention of typical artifacts of which the novice in SPM should be aware.

For more in-depth discussion of various STM topics a number of good books [6.1–6.5] are available. Two of them [6.6, 6.7] have chapters devoted to superconductors. A couple of general reviews of STM are found in refs. [6.8, 6.9] These source materials should be referred to whenever the present chapter's discussion gives only a broad brush approach to a topic. Each of these sources gives a rich list of references in addition.

6.2 Tunneling theory and historical perspective on tunneling spectroscopy

Quantum mechanical tunneling is a result of the finite potential barrier at the metal–vacuum interface. The electronic wave function ψ and its first derivative $\partial\psi/\partial z$ are continuous across this interface (or finite potential discontinuity), the electron wave function decaying exponentially, $e^{-\kappa z}$, in the forbidden region where the barrier exceeds the total electron energy. In this context, κ is approximately related to the apparent work function or mean local tunneling barrier, $\kappa^2 = 2m(\phi - E)/(h/2\pi)^2$. Thus, the tunneling current, I_t, or transmission probability also decays with barrier width, z [6.10–6.13]

Two metals, with electron wave functions ψ_1 and ψ_2 respectively, separated by a vacuum distance greater than say 20 Å, are non-interacting with different work functions ϕ_1 and ϕ_2. This situation is portrayed in Fig. 6.1(a) as idealized

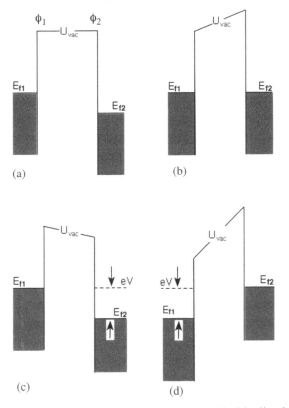

Fig. 6.1. Representation of four possible idealized potential barriers between two metal electrodes for electron vacuum tunneling using a trapezoidal barrier. (a) Two non-interacting electrodes separated by a vacuum. The difference in Fermi levels, E_f, is equal to the difference in work functions, ϕ, of the two materials. (b) The potential barrier after the electrodes are brought within tunneling distance and allowed to come to equilibrium at a common Fermi level. The variation in field within the vacuum space is due to the difference in work function. (c) and (d) Barrier after a voltage, eV, is applied between the electrodes raising the Fermi level of metal 1 relative to metal 2 and vice versa, respectively.

square potentials with the Fermi level at $T = 0$. When the two electrodes are brought within tunneling distance and, therefore, thermal equilibrium, Fig. 6.1(b), the result is an averaging of the Fermi level and a varying barrier within the barrier region. If a field, eV, is then imposed between the two metals, the Fermi level of one electrode is raised with respect to the other, and the barrier shape will depend on which direction the field is applied (Figs. 6.1(c) and (d)). Filled electronic states in one electrode have access to empty states in the other electrode, and a net tunneling current results, the current increasing linearly with voltage. Fig. 6.2(a) depicts the electron wave function as it extends

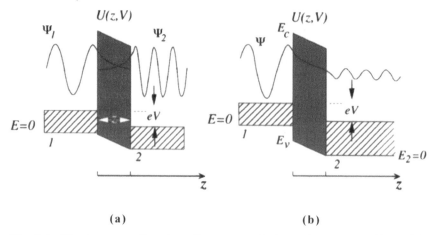

(a)   (b)

Fig. 6.2. Electron wave functions for two metal electrodes separated by a thin insulating layer showing quantum mechanical tunneling with exponential decay of the wave function in the barrier; (a) the wave function of electrode 1 extending smoothly across the entire tunnel junction structure; (b) separate wave functions for electrodes 1 and 2, ψ_1 and ψ_2 respectively [modified from reference 6.10] The wave functions have been offset from the Fermi level for clarity.

smoothly across the metal–vacuum interface, decays exponentially in the gap, extends smoothly across the second vacuum–metal interface, and propagates into electrode 2. Figure 6.2(b) includes the electron wave function of the second electrode. In the absence of a field at finite temperatures (with thermal smearing) electrons can tunnel in either direction with no net current flow. As discussed above, the presence of an applied field leads to current flow.

Around 1960, Giaever [6.11] conceived the idea of taking advantage of the tunneling mechanism to probe the electronic density of states (DOS) of the electrodes; he created planar junctions consisting of either a normal metal–insulator–superconductor (NIS) or superconductor–insulator–superconductor (SIS). He discovered that if he cooled the junctions below the superconducting transition temperature, T_c, in the presence of a pair-breaking magnetic field, $V(I)$ was linear, but in the absence of the magnetic field the I–V characteristics were nonlinear with a gap in the electron excitation spectrum at low bias. He equated this gap with the superconducting energy gap Δ. These experiments were the beginning of tunneling spectroscopy. Figure 6.3 shows three tunnel junction configurations (normal metal–insulator–normal metal (NIN), NIS, and SIS) along with their corresponding current–voltage characteristics. From the BCS density of states relationship, $I_{ns} = R_n^{-1}[(eV)^2 - \Delta 2]^{1/2}$, the gap could be extracted from the I–V characteristics. Giaever speculated that the ideal barrier was not the self-limiting oxides grown to separate the planar

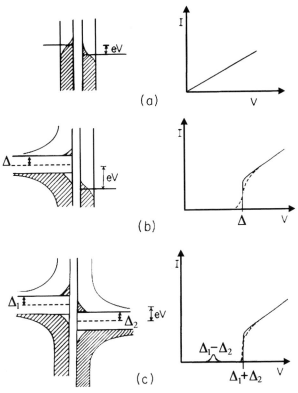

Fig. 6.3. Semiconductor representation for (a) NIN, (b) NIS, and (c) SIS tunnel junctions showing the DOS vs. energy. The expected current/voltage characteristic for each type of junction is included on the right hand side. In each case the Fermi level of metal 1 is raised by eV with respect to metal 2. The dashed lines indicate the characteristics at $T > 0$, and the solid lines indicate the current for $T = 0$.

electrodes, but rather a vacuum. This conceptual idea gave birth to the parallel development of point-contact tunneling and the first variant of the STM.

If one makes one of the electrodes a metal wire with a sharp tip at the end and the other electrode a planar sample separated by vacuum space, one has the basis for both point-contact electron vacuum tunneling spectroscopy and scanning tunneling microscopy. The latter will be discussed in more detail in the next section on instrumentation.

Low temperature (LT) point-contact spectroscopy preceded LT-STM as a means of measuring the energy gap on materials where an ideal oxide barrier could not be fabricated. Typically, noble metal tips were used. Figure 6.4 shows an experimental set-up for one of the earlier designs [6.12] The $I(V)$ and dI/dV measurements taken with this apparatus for one of the first polycrystal-

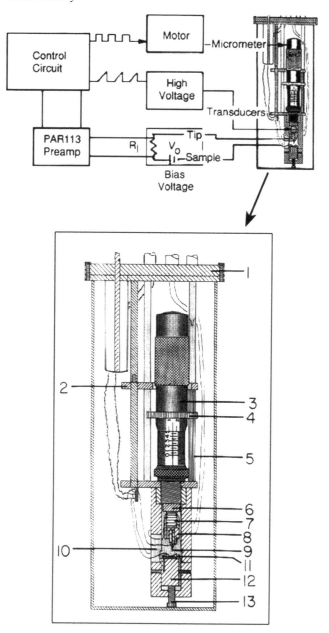

Fig. 6.4. One of the first typical point-contact configurations. The apparatus is mounted in a vacuum can suspended in a superinsulated Dewar filled with liquid He. Shown (insert) are the (1) vacuum can flange, (2) stainless steel (ss) support structure, (3) differential micrometer, (4) ss gear, (5) pinion rod, (6) spindle, (7) piezoelectric transducers, (8) electrode holder, (9) tip, (10) sample cell, (11) sample platform, (12) coarse sample positioning screw, and (13) coarse sample positioning screw to set sample–tip gap [modified from ref. 6.12]

line sample of the La-based superconductor are shown in Fig. 6.5 [6.13] Some
of the other earlier point-contact tunneling spectroscopy references are listed in
[6.14–6.19] Point-contact tunneling continues to be used as a means of measuring the superconducting properties of high T_c single crystals.

The forerunners of point-contact tunneling and STM were the ultramicrometer (Young) and the topografiner, the latter so named by its originators,
Young, Ward & Scire [6.20] The ultramicrometer was a non-scanning device in
which the tip–sample spacing was controlled by differential thermal expansion.
It was quickly followed by the topografiner which also used a tungsten field
emission tip mounted on a piezoelectric element. Two additional orthogonally
placed piezo drives were used to scan the surface. A feedback mechanism
controlled and maintained (constant current mode) the spacing between tip and
sample, typically tens of nanometers but as near as 12 Å off the surface in
static mode. The voltage to the tip piezo element became the resulting image
topography. The instrument, in non-scanning mode, was used to make the first
observations of metal–vacuum tunneling and, in scanning mode, to image the
first surface topography. Unfortunately, it was plagued by vibration problems.
The conceptual scheme for the instrument, however, possessed most of the key
ingredients of the STM; the resolution was low due to the vibrational instabilities that prevented operation at high-resolution tip–sample spacings.

The essential elements necessary for the successful design of an STM were
the selection and fabrication of suitable probes, the use of sub-ångström piezo-
based positioning and feedback capability, and vibration isolation (electromagnetic, mechanical, thermal, and acoustic). Once these were achieved, the field
of STM grew dramatically. In addition, STMs were developed after the
computer revolution and, as a consequence, are computer-driven instruments
with all the assets of computer-driven operation and digital data storage and
manipulation available, see Fig. 6.6. The first successful atomic resolution
images were obtained on Si in UHV with an STM designed by Binnig *et al.*
[6.21] One of the first reports of a charge density wave measured at low
temperature by STM was published by Giambattista *et al.* [6.22].

6.3 General instrumental description

Today the tunneling microscope is a member of a whole new class of com-
puter-based instruments (SPMs) employing scanning proximity probes combined with feedback systems to control the probe's vertical position relative to
its lateral position on the sample under investigation. Conceptually and
operationally they are elegantly simple and versatile. Features common to all
proximity microscopes are the use of (1) very fine sharp tips, (2) small,

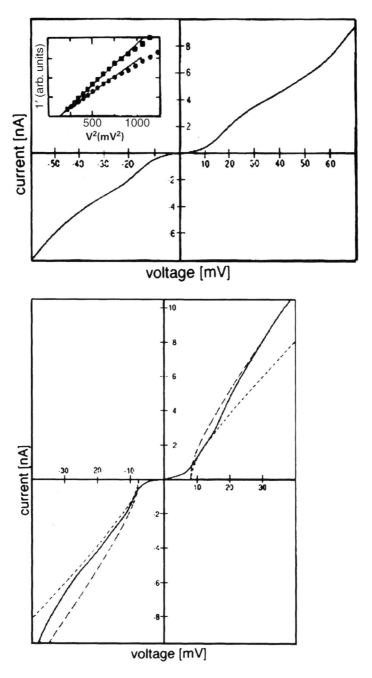

Fig. 6.5. Typical of the first $V(I)$ characteristics for a polycrystalline $La_{1.85}Sr_{0.15}CuO_{4-y}$ sample taken with a Au tip [modified from reference 6.13].

SCANNING TUNNELING MICROSCOPE

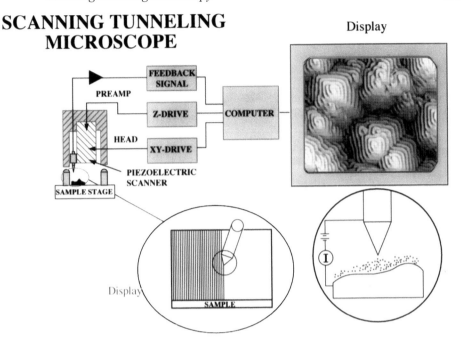

Fig. 6.6. Diagrammatic representation of the scanning tunneling microscope showing the important components.

nondestructive forces, (3) piezoelectric scanners to raster either the sample or tip back and forth with respect to each other, (4) an automated computer-controlled positioning of the probe in contact with or within few ångströms of the sample surface, (5) a feedback mechanism to maintain a fixed sample-tip separation, and (6) digital data collection and storage. The overall design of the SPMs is modular in nature, consisting of a computer workstation with on- and offline software, a power supply/controller, and plug-in uni- or multifunctional scanning probe microscopes. Each microscope is distinguished by the type of measurement(s) of interest, i.e. topography, magnetic domains, etc. Much of the software and hardware (scanners and control electronics) can be shared by the various microscopes. Switching from one type of microscope to another is accomplished by changing a software microscope selection switch and plugging the appropriate microscope into the controller. The user then selects the operating conditions, e.g. scan rate, gains, size of data array, and voltage bias (for the STM), and feedback criteria, i.e. tunneling current in the case of STM. The resolution and maximum scan range, nanometers to micrometers, is determined by the choice of piezoelectric scanner.

Tube-shaped piezoelectric scanners, now commonly in use, are typically partitioned into $+/- x$, y, and z electrodes. As a voltage is applied to one of the electrodes, the piezoelectric material expands or contracts depending on the polarity of the voltage. The piezoelectric material, $PbZrO_3$–$PbTiO_3$, is calibrated in units of ångströms per volt. The amplitude or peak voltage of the triangular waveform signal to the x, y electrodes determines the lateral dimensions of the scanned area. The frequency of the signal determines the scan rate. The in-plane x and y loci of points is controlled by voltage applied to the x and y piezo electrodes, calibrated against standards, e.g. the atomic spacing of graphite for high resolution or the width and height of features on commercially produced gratings for scanners used on large scan areas and rough topography. Other common scanners used are made up of a tripod arrangement of piezoelectric elements and more complicated voltage signals to accomplish the same process. In this case, a separate z piezoelectric element is used to control the tip–sample separation.

The coarse separation between the probe and the sample is adjusted visually under an optical microscope. The online software controls the physical operation of the instrument including the automated tip approach. The final separation between the probe and the sample is typically set utilizing a stepping motor, ramp voltages to the z piezo electrode, and a stopping test. The usual scheme involves monitoring and comparing a tunneling current or cantilever deflection, for example, while the probe is moved toward the sample by a voltage ramp (typically equivalent to a 25 to 50 nm approach) sent to the z electrode of the piezoelectric scanner. If the user-selected set-point value is not detected, the z voltage is reset to zero retracting the tip to its original position, and the stepping motor is used to move the probe closer to the sample by a distance less than that obtained with the voltage ramp. This scheme is repeated until the set-point value is reached on a piezo-controlled approach. Hysteresis in the motor-driven movement makes the stepping motor alone unsuitable for the final placement of the tip vertical position above the sample.

After the tip is in measurement position, the SPM software directs the controller to send high and low frequency, triangular-wave voltage signals to the x and y electrodes of a tube piezoelectric scanner, respectively, to control the scanning process. Some are multimode microscopes that allow simultaneous measurement of two properties, e.g. topography and current, over the same scan area allowing direct comparison of structure and property.

The most common STM mode of operation is the constant current mode, where the bias voltage ($\pm$ tens of millivolts to several volts) is held constant while the feedback circuit connected to the z piezo electrode moves the tip up and down to maintain a fixed pre-selected tunneling current (distance). At each point along each scan line in the x, y raster pattern, the current is measured and compared to the operator-chosen set-point current (typically tens of picoamperes (pA) to several nanoamperes (nA)). A difference between the measured tunneling current and the reference current setting generates an error signal, ∂I. The resulting error signal, ∂I, is converted by the online software to a correction voltage, ∂V, for the z piezo electrode. This ∂V is superimposed on the existing z piezo electrode voltage causing the tip to move up or down as it scans across the sample. Either variations in local sample height (structure or roughness) or conductivity/DOS will cause the current to vary as the tip is scanned over the surface. The STM can be used, therefore, either as a microscope to study atomic, nano-, and microstructures or as a spectrometer to probe the electronic properties of the sample under investigation. The height of vertical detail in the offline, computer-generated microtopograph is related to the ∂I.

Constant height mode of operation is reserved for small scan areas and atomically smooth surfaces. The use of little or no feedback during scanning in constant height mode makes it possible to scan at much higher rates than in constant current mode, which is limited by the feedback circuitry.

Online software controls not just the hardware operation, but the measurement process with data, e.g. z current or piezo-voltage value with its corresponding x, y coordinates, stored digitally as a three-dimensional array for later offline analysis and graphical presentation. Different scanners are available for scan areas of a few square nanometers to > 100 μm^2; vertical extensions are generally limited to under 3 to 4 μm. The collection and storage of data in digital form allows easy offline access later. The offline data manipulation software typically includes filtering and analytical capabilities as well as false-color two- or three-dimensional graphical representation of the data. The vertical detail displayed is determined by the nature of the feedback criteria that were used to collect the data. The analytical capabilities include, for example, determination of RMS roughness, one- and two-dimensional Fourier analysis, and lateral and height information.

6.4 Specific details for STM design

The active components of the STM are the two electrodes, a sharp metallic probe, and a conductive (or semiconductive) sample, one of which is moved relative to the other in a x–y grid utilizing piezoelectric tubes, disks, or rods for precise positioning with ångström accuracy. The STM tip (electrode 1) is typically tungsten (UHV conditions) or Pt/Ir (ambient conditions), although other noble metal tips have been used. The second electrode is the conductive or semiconductive sample to be analyzed or manipulated.

This section will address specific design considerations from the subset of components and references that make up the instruments. The list is by no means complete; the reader should only use this section as a starting point.

6.4.1 Tips and tip preparations

The quality of the data obtained from these techniques is intimately controlled by the nature, shape, and quality of the probing tip. Depending on the resolution desired, the choice of tip will determine the limit of information one can extract from the data. Either chemically etched or cold-worked Pt/Ir STM tips are used to image in air. (Tungsten tips have a native oxide coating that makes them unsuitable for use except in UHV after cleaning.) Either type of tip can achieve atomic resolution; however, cut tips are more susceptible to imaging artifacts due to the presence of multiple tips. On the other hand, etched tips are often plagued by the presence of residual salts left over from the etching process and not completely removed by the post-etch cleaning procedure. Tips are commercially available or one can refer to a number of sources for information about fabricating etched tips [6.23–6.29]. The mechanically formed Pt/Ir tips can be made by cutting wire with a pair of stainless steel scissors.

6.4.2 Low temperature

Although a commercial variable-temperature UHV STM (Omicron) with cooling capability is available, dedicated commercial low-temperature microscopes are virtually nonexistent. For low-temperature electron tunneling spectroscopy, a devoted STM probe has to be designed and built by the researcher, using cryogenic-compatible components. No matter what the specific details of the design that are chosen, a couple of design considerations are common to all: the choice of the correct materials and thermal and mechanical stability. The rest of this section will briefly address these issues.

The choice of materials can be broken down into those that are used to make mechanical and electrical connections to the STM from ambient temperature and those used at low temperature for the STM itself and connections to it. Since the sample and tip need to be cooled, then stably held at that fixed depressed temperature, the STM has to be placed in a controlled cryogenic environment. A superinsulated Dewar is used to house the STM to eliminate vibrations due to boiling liquid nitrogen used in the heat shield of glass Dewars. The STM can be immersed directly in the cryo-fluid, enclosed in a separate vacuum can suspended in the fluid, or transferred from a UHV environment where the sample can be prepared without exposure to ambient conditions into a vacuum enclosure in the cryo-fluid.

To prevent thermal drift or fluctuations, the STM must be thermally isolated from ambient conditions by the careful design and selection of low-thermal-conductivity materials. These materials are used for the ambient-to-cryogenic connections for the support structure, coarse approach mechanism, and the electrical connections to the piezoelectric elements and the STM electrodes. Thin-walled stainless steel tubing with cooling fins is used to support the STM vacuum can or for the vacuum enclosure and as a conduit for electrical and mechanical connections to the tunneling instrument. Electrical leads can go either through the cryogenic liquid into the vacuum space via a low temperature vacuum seal or through the vacuum space in the stainless steel tubing supporting the STM housing. Obviously, if the wires go through the cryogenic liquid heat conducted down the Cu leads is removed by the coolant, but if the wires are fed through vacuum they must be thermally anchored at the cryo-temperature before connection to the STM. Thin stainless steel, chromel or other lower-thermal-conductivity wires can be used between ambient temperature and the thermal anchor point (a cold finger, for example) to further reduce heat conduction to the STM.

The final electrical connections to the STM can be done with copper wires. A small amount of helium is used as an exchange gas to anchor the temperature of the whole assembly to the cryogenic fluid. The body of the STM can be made out of copper, which will respond quickly to temperature changes for variable temperature measurements and provide a uniform temperature environment for the tunnel junction. One has to estimate the differential thermal contraction of the component parts to make sure that a tunnel junction separation set at room temperature is sufficiently large to prevent tip crash on cooling. Other materials like Macor^{TM} or Invar^{TM}, which closely match the thermal expansion properties of the piezoelectric transducers, are used as well but take more time to thermally stabilize. Some references are given in [6.30–6.43]

6.4.3 UHV design

Only a limited amount of work on high T_c materials has been performed in the UHV environment. However, a few references to UHV design are included [6.44–6.51]. A few commercial instruments are available either as a complete system (JEOL, for example) or as an add-on to a custom UHV chamber containing other characterization techniques and/or *in situ* sample preparation (cleaning, annealing, cleaving, pealing) or film deposition capabilities.

6.4.4 Piezoelectric scanners/feedback systems

All SPMs depend on piezoelectric scanners, usually fabricated from lead titanate/lead zirconate ($PbZrO_3$/$PbTiO_3$, PZT)-based materials. Selection of the particular PZT depends on the use, i.e. low temperature [6.52] atomic resolution, or need for long-range scan linearity [6.53–6.55]. They can be designed in many configurations including stacks, tubes, inchworms, or inertial drives. Binnig & Smith [6.56] originated the tube scanner that is in common use today, especially in air-based commercial instruments. Typically one can expect a reduction in expansion/contraction to as little as about one-tenth the room temperature value on cooling to 4 K. The scanners' expansion ranges from tens to hundreds of ångströms per volt, and they have to be calibrated at room temperature [6.57] or low temperature [6.58–6.60] depending on the operating temperature.

Constant-current mode of operation requires the use of a feedback circuit to maintain a tip–sample separation. One drawback is the long data accumulation time limited by the feedback circuit characteristics and resonant frequencies of the scanners. Attempts to design feedback to increase scan rates is an ongoing effort [6.61–6.62] and piezoelectric scanners are designed to maximize the resonance frequencies.

6.4.5 Vibration and thermal isolation

Vibration and thermal isolation are critical factors in the design and placement of the scanning proximity probe instruments. These factors, along with sample quality, will determine the ultimate signal-to-noise ratio and resolution. The approach to vibration isolation is multidimensional since SPMs are sensitive to all vibrational disturbances: mechanical, acoustic, and electromagnetic. The vibration isolation approach is multifold and determined by the source and frequency range with which one is dealing [6.63–6.65]. Anything that will vary the tip–sample spacing or degrade the quality of the tunneling signal on the time scale of the measurement will be a source of trouble. Whether 'home-built' or commercial, all SPM instruments require some vibration isolation, which will depend on the level of built-in isolation (such as suspension system, quiet voltage sources, and electrical shielding) and where they are placed. The sources of trouble range from building vibrations to vacuum pumps to electronic line noise and should be dealt with to guarantee optimum operation of the microscopes.

Simple mechanical isolation techniques include placing the SPM on an air-isolation table or a platform with low-frequency characteristics suspended by bungee cords. UHV-SPMs generally have a built-in eddy current damping/

spring suspension system or take advantage of materials with dissimilar vibration characteristics, such as Viton^{TM} and steel, to built platforms consisting of a stack of alternating decoupled layers of the two materials. The 'soft' suspension system is combined with a rigid SPM design that minimizes vibrations that can effect the tip–sample separation. If mechanical pumps are used, they should be turned off or decoupled during measurements. For low-temperature measurements, vibrations, e.g. pendulum motion, due to boiling cryogenic fluids can be eliminated by the use of a simple spring clamp consisting of hose clamps in a cloverleaf arrangement that will hold the SPM vacuum can rigidly within the Dewar.

Acoustic noise that can excite natural resonance frequencies in the SPM and electrical noise on the piezoelectric transducer voltages or other electronic circuitry will result in an unstable tunnel junction. Particular attention has to be paid to 60-cycle line noise, preventing ground loops, and to the use of electromagnetic shielding. A frequency analyzer is helpful for analyzing the source of noise on the tunnel junction, for example from an ion gauge. A combined electromagnetic/acoustic screen room can be used to house the SPM to help isolate it from the environment. Operationally, reduced feedback gains will prevent overshoots and ringing to improve data quality.

Thermal drift, not a problem in UHV, is particularly troublesome for instruments operated in air or at low-temperatures without proper thermal isolation. Because the piezoelectric materials have a low thermal conductivity and are housed in materials that have been carefully thermally matched, like Macor^{TM} or Invar^{TM}, the whole scanner assembly will drift over a long period of time if heated or cooled, intentionally or unintentionally. A screen room or simply a cover will protect the SPM from temperature fluctuations due to air-conditioning fans cycling on and off. The low-temperature thermal isolation was discussed in the section on low-temperature designs above.

6.5 Related techniques

Over the last ten years, the STM has given birth to a whole family of scanning proximity probes [6.66]. The nature of the feedback criteria, i.e. the property being measured and used to generate the vertical detail, along with the type of sample being examined, distinguishes which SPM is used.

6.5.1 Atomic force microscope (AFM)

Unlike the STM, the AFM does not require the sample to be conductive. The operation of the AFM, Fig. 6.7, is similar in principle to that of a profilometer in

ATOMIC FORCE
MICROSCOPE

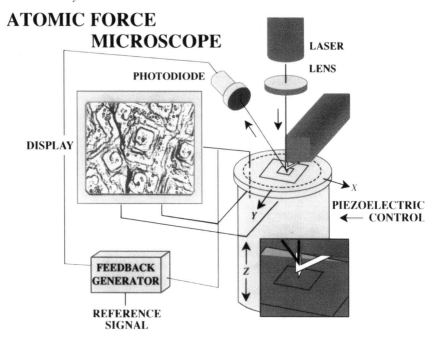

Fig. 6.7. Diagrammatic representation of the atomic force microscope show-ing the important components. A position-sensitive diode is used to monitor changes in the cantilever deflection as the sample moves under the tip.

the sense that a sharp stylus is dragged across the surface of a sample while its vertical deflections are monitored and recorded. See Fig. 6.7 for a schematic of the AFM. In the case of the AFM, however, a three-dimensional surface topograph is generated. Additionally, the stylus, or in this case the tip, is attached to a floppy spring (force constant 0.1–1 N/m) to reduce the forces exerted on the sample to 10^{-9} to 10^{-11} newtons (N). The soft spring makes the probe more responsive to atomic surface corrugation increasing detection sensitivity while minimizing sample damage/modification. The weak spring usually consists of a long, thin spring board or hollow triangular cantilever of a Si-based material to take advantage of existing microelectronic fabrication techniques. The tip, attached to the bottom of one end of the cantilever, is fabricated from either SiO_2 or Si_3N_4. Since the tip is made from a relatively hard material, it is not usually subjected to forces large enough to cause probe deformation.

The cantilever position is held fixed in the AFM laser head while the sample is scanned under the tip. See Fig. 6.7. The feedback system relies on an optical detection system that monitors the cantilever vertical deflection, ∂z, as the tip follows the contours of the sample surface. In the constant force mode of operation, as in the STM constant current case, ∂z is translated into a correction voltage to the piezoelectric translator's z electrode, which results in a vertical displacement of the sample to maintain a constant cantilever deflection. Most commercial AFMs use a low-power red laser and a position-sensitive photodiode detector. The laser beam is focused onto the back of the cantilever apex. A mirror redirects the deflected laser beam from back of the cantilever surface to the detector. Both split and four-position photodiodes are used, with the latter sensitive to torsional twisting of the cantilever in addition to vertical deflections, ∂z. They can be used either in a contact (on the surface) or in a non-contact (tens of nanometers above the sample) mode. Interactions between the AFM tip and the sample include a wide variety of forces: Van der Waals interaction, ionic repulsion, magnetic and electrostatic forces, adhesion and frictional forces, capillary water adhesion, and plastic and elastic responses of the surface. Contact-mode imaging is dominated by topography while non-contact, lift, tapping, or other mixed-mode imaging techniques can selectively reduce lateral force, surface damage, or probe longer-range forces with minimal reduction in resolution.

The first commercially available AFM tips were composed of Si_3N_4 (111) faceted pyramids. Newer 'sharpened' Si_3N_4 tips with an additional shaping step are capable of higher resolution than those fabricated by the original techniques. High-aspect-ratio SiO_2 tips and specialized tips have been developed to allow more accurate measurement of sharp step rise angles, particularly important for high T_c devices. A choice of cantilevers is available with different force constants. Tip development continues to produce tips for special applications. Among those now available are magnetically coated tips for imaging magnetic properties/domains with different coercivity and moments. Though the AFM resolution is still not up to that of the STM, the use of improved tips and greater sample flexibility are making the AFM more useful for many applications. AFM and all its closely related techniques will be the instruments of choice for evaluating high T_c based devices. A description of two of these techniques follows.

6.5.2 *Electric force microscope (EFM)*

The EFM and the magnetic force microscope, MFM, are variants of the AFM. Briefly, EFM and scanning potentiometry utilize a conductive AFM tip to measure static charge distributions or surface potentials, respectively. Both are non-contact methods. In the former case, an oscillating tip is scanned alternately on the surface then at a user-chosen height above the surface. Two images are obtained, the topograph and a electric field gradient map. The force exerted by charges changes the cantilever oscillating frequency, which can be detected as a change in amplitude, phase, or frequency. In the case of scanning potentiometry, the tip is not oscillated when it is off the surface, rather an a.c. voltage is applied to the tip while a d.c. field is put on the sample. This is a null method where an additional d.c. voltage is applied to the tip to balance any variation in potential on the sample. The changing voltage applied to the tip becomes the height contrast in the image. This technique can be a useful method of characterizing high T_c devices.

6.5.3 *Magnetic force microscope (MFM)*

A MFM can be operated in a two-pass way in much the same manner as the EFM. Like the EFM, the MFM can also be operated in two methods, with or without oscillation of the tip in the lift pass across the sample. In the former case, one can get variations in the magnetic stray fields on the surface directly. However, this method is not very sensitive to small stray fields. The second method is analogous to EFM and results in a magnetic field gradient image.

One example is the application of MFM to image magnetic domains or features while simultaneously collecting the corresponding topographical information. This method has been successfully used at low temperatures to image single vortices in YBCO films [6.67]. Because the penetration depth and therefore the vortices' magnetic diameter is hundreds of nanometers, the MFM method may prove to be more successful in identifying the pinning centers in the high T_c films than the STM.

6.6 The study of high T_c materials by STM

6.6.1 *Spectroscopy and atomic imaging*

Measurement of atomic structure and tunneling spectra require a clean surface. The best results are obtained by peeling/cleaving the sample in vacuum, an inert-gas environment, or, in the case of low-temperature STM, in the STM cryostat vacuum can. The more two-dimensional perovskite crystals peel much like graphite. If the sample can't be peeled or cleaved *in situ*, it must be mounted in the STM immediately. Atomic resolution in air, even on the Bismuth samples is not possible. The surface deteriorates very quickly. However, low-temperature STM atomic imaging and spectroscopy [6.10] have been carried out on $Bi_2Sr_2Ca_1Cu_2O$ (BSSCO), $La_{1.85}Sr_{0.15}CuO_4$, and $YBa_2Cu_3O_7$ single crystals by a number of workers [6.69–6.78]. Depending on the quality of the crystals, these studies, besides determining the cleavage plane under investigation, should answer fundamental question about the nature of superconductivity in these type II perovskite superconductors through localized current–voltage and conductance measurements (see tunneling discussion above). Such studies will answer questions about possible gap anisotropy through tunneling measurements on different crystallographic faces. The measurements so far have still left the question of the nature of the gap open for controversy. A review of the data is presented in ref. [6.79]. Materials and experimental problems, coupled with the mobility of the vortices and the small coherence length (in ångströms) in these materials, will make it difficult to take advantage of the STM's high spatial resolution to resolve single vortices. Measurements of the magnetic stray fields with MFM, mentioned briefly above, have been more successful.

The outstanding question in regard to films is the identification of the pinning mechanism for the Abrikosov vortices that is responsible for the superior current-carrying capacity of the films over the single-crystal form. The films possess a large number of structural defects that are good candidates for pinning of the vortices. The rich variety of defects can be seen in the microstructural data collected in air, presented in the section on growth studies. Some successful spectroscopic studies of energy-gap variations have achieved good spatially resolved tunneling spectra in film studies [6.80–6.82] but, because of the problem mentioned above having to do with quality of film, surface contamination, dirty tips, and the fundamental problem of the short coherence length, data from films will be even more controversial than those taken on single crystals.

6.6.2 High T_c thin film growth studies

A large body of work has accumulated on the use of air STM (and AFM) to study the microstructure and growth mechanisms of high T_c films. The early work on films was plagued by poorly conducting surfaces. Boulders on the film surface, prevalent in pulsed laser deposition, usually guaranteed immediate tip destruction. Some success could be achieved by limiting the scan size to areas free of debris providing one of the large particles wasn't encountered within the scan region. The first really high-quality images were obtained relatively recently [6.83–6.90] Improvement in film quality has largely been responsible for the wide-based success reflected in the large number of papers on thin-film studies. A limited amount of work has been reported on the growth of $SmBa_2Cu_3O_7$ [6.91–6.93], $DyBa_2Cu_3O_7$ [6.94], $Bi_2Sr_2Ca_1Cu_2O$ [6.95], and $PrBa_2Cu_3O_7$ [6.96]. The largest contribution to this collection of work has been on YBCO. This includes studies of films grown by chemical-vapor deposition [6.97, 6.98] co-evaporation [6.99, 6.100] pulsed laser ablation deposition

[6.101–6.113] rf-sputter deposition [6.114–6.122] and growth on novel substrates like Mg_2TiO_4 [6.123] or metals [6.124, 6.125]. The interest in growth mechanisms has generated work on ultrathin films [6.126–6.129] and with a gradient growth [6.130].

Certain conditions are common to all high T_c materials: (1) they have anisotropic structures, (2) and, therefore, anisotropic growth behavior, (3) they are grown on lattice mismatched substrates, (4) they are usually fabricated by vapor-phase epitaxy, (5) their microstructure and crystallographic growth direction with respect to the substrate normal is highly substrate- and deposition-parameter sensitive, (6) the growth mechanism, e.g. dislocation mediated, is dependent on growth parameters and crystallographic growth direction (Figs. 6.8 and 6.9) their in-plane growth orientation likewise is growth dependent.

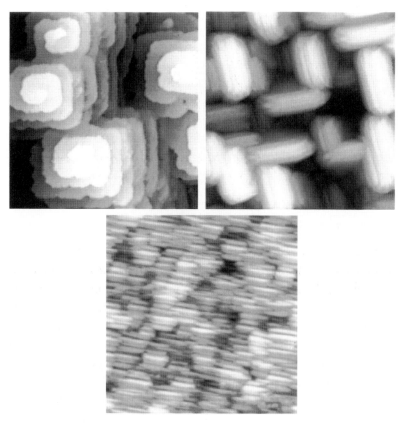

Fig. 6.8. Images of three different sputter-deposited YBCO films showing the striking difference in morphology. The upper left-hand image shows a c-axis normal film grown on CeO_2 buffered sapphire, the upper right-hand image is an a-axis film grown on $LaAlO_3$, and the bottom film is a film grown on $SrTiO_3$ [6.110].

Fig. 6.9. Image of a 2 μm × 2 μm area of a 450 nm thick, sputter-deposited YBCO film grown on MgO. Note the growth spirals including doubles with opposite spin directions (arrows), the one-unit-cell-high terraces, and other defect structures.

The substrate lattice mismatch leads to what is assumed to be a Volmer–Weber growth mechanism resulting in the granular structure observed in the 4500 Å thick c-axis normal YBCO film shown in Fig. 6.9. The vertical scale in the image is greatly exaggerated, as is usual in STM or AFM images, to emphasize the rich detail in the data. The size of the islands that make up the film are dependent on the growth rate and film thickness as is the width of the terraces (Fig. 6.10). The high resolution of the STM not only easily resolves the island structure, but the finer layered structure of each island and, in this case, the fact the islands grew by a screw-dislocation-mediated mechanism. The latter is a result of the relatively low superstaturation controlled by the growth parameters. The one-unit-cell-high terraces are clearly visible along with the terrace width in this single, well-faceted island (Fig. 6.11). The offline software for calculating roughness yielded a 48 Å root-mean-square roughness (RMS) for this film. Another image taken from this film, Fig. 6.12, shows the

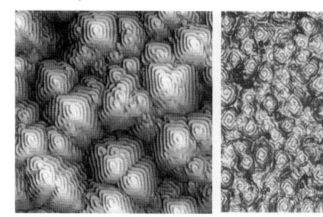

Fig. 6.10. Two different sputter-deposited YBCO films grown on MgO. The left-hand STM image (2 μm × 2 μm) is taken from the same film (450 nm, RMS = 4.9 nm) as Fig. 6.9 while the right-hand AFM image is taken from a thinner film (100 nm, RMS = 2.9 nm), hence the smaller grains. Both images are presented in the edge-enhancement mode to bring out the details, particularly in the data for the second image, i.e. the 1.2 nm high steps and narrow (≤ 20 nm wide) terrace widths. The 45° rotation of the grains in the right-hand image is due to a rotation of the substrate with respect to the STM scan direction.

richly detailed defect structures that the STM reveals. One can also see the interconnections between the *c*-axis grains in Fig. 6.13 as has been observed by others [6.131].

The YBCO out-of-plane crystallographic growth direction is strongly influenced by the growth parameters. At the extremes of temperature one can obtain nearly pure *a*- or *c*-axis perpendicular films of 123 within the temperature range from about 640 °C to 780 °C. The lower temperature favors the *a*-axis growth. At intermediate substrate deposition temperatures the resulting films are mixed *a*- and *c*-axis material with the *a*-axis material concentrated usually at the surface. Depending on the thickness of the film, one can capture with the STM the beginning of *a*-axis growth; see Fig. 6.14. In fact, the STM and AFM images will show the presence of the *a*-axis material at amounts below the X-ray diffraction sensitivity. In Fig. 6.14 you can clearly see the difference structure, i.e. growth mechanism for *a*-axis versus *c*-axis growth. The combined high resolution of the STM to reveal the nanostructural detail of the film morphology and the extensive offline analytical software make the scanning tunneling microscope an invaluable tool to compliment other traditional characterization tools.

Fig. 6.11. The growth spiral, well-faceted terrace edges, and unit-cell-high terrace steps are clearly visible in this STM image of a single 400 nm square c-axis normal (110) in-plane rotated grain from the film in Fig. 6.9. The 45° rotation from the growth direction of the majority of the grains results in large-angle grain boundaries around its perimeter (bottom left side of image).

6.6.3 Surface manipulation/lithography

A choice of the wrong operating conditions can result in engineering on the sample surface. Requiring too much tunneling current at too low a bias will force the tip into the sample surface, where it will dig its way over the scan area. The tunneling tip will faithfully follow the operator's instructions, seeking the required set-point current until it finds it. It is always a good idea to check to see if the tunneling conditions are appropriate. That can be done by reducing the tunneling requirements, i.e. raising the bias and lowering the current, and scanning a larger area will reveal a sunken region in the original scan area. STM images of high-temperature superconducting films are often free of the small particulates, probably insulating phases, seen in AFM images

Fig. 6.12. Another image of the previous film showing a large number of extended defect structures oriented along the [110] direction.

[6.132]. Figure 6.15 includes both STM and AFM images taken on the same film. The resolution in the AFM image is clearly lower than that of the STM, but it does reveal surface particulates that are held and probably non-conducting.

However, the ability to do engineering can also be and has been exploited to create features, i.e. lines or squares, on the surface of samples [6.133–6.140]. So far this has had little applied value for these films but could be used in conjunction with other related techniques such as EFM to study the affect of damage on local electronic effects.

6.6.4 Multilayer devices

At early stages of device fabrication and on surfaces that are conducting, the STM can be used to examine the grain structure, rotation of grains, measure device step heights and angles, and, as mentioned above, check surface

Fig. 6.13. Close-up STM image of the intergrain boundary showing the interlacing of the grains.

roughness. One such partially finished device structure is shown in Fig. 6.16. The 45° rotation between the YBCO film grains on the top step grown on CeO_2 and below the step on the bare MgO substrate is clearly seen in this STM image.

For most device applications, however, the conductivity requirement limits the STM's usefulness. The AFM has taken over the bulk of the characterization of devices and is used routinely to evaluate each of the steps in the growth and processing of multilayer devices. The AFM is used to measure a ramp or step junction angle, height, and smoothness, and to evaluate milling and etching damage to substrates or films, as seen in Fig. 6.17. In the lower half of the figure is an example of the type of offline characterization one can do on a multilayer device. The particular step shown in the figure is about 290 nm high. Although the step looks sharp in the line scan across the step, the angle is only about 8.3°. In the low-resolution image in the upper left the various layers that make up the device can be clearly seen along with the roughness of the ion-milled surface at the base of the step. In the upper right side of the figure is a

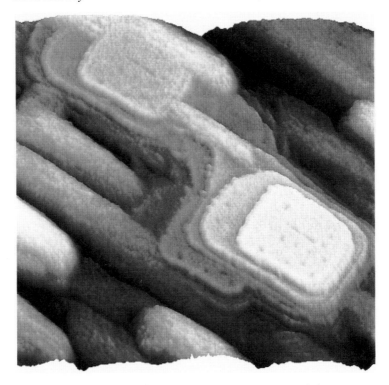

Fig. 6.14. The STM captured the beginning of *a*-axis growth on the under-lying *c*-axis material in this image of YBCO grown on NdGaO₃.

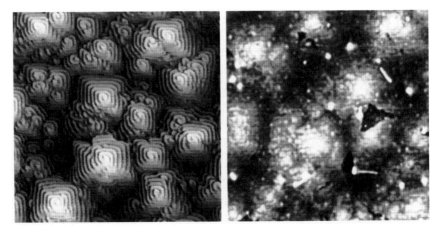

Fig. 6.15. STM (left) and AFM (right) images of the film in Fig. 6.9. Note the poorer resolution of the AFM image and the particulates not present in the STM topograph. The STM tip has 'cleaned' the surface.

Fig. 6.16. An STM image of a partially finished multilayer step edge junction showing the 45° rotation between the YBCO grains on the CeO_2 upper terrace and the grains grown directly on the MgO substrate.

higher-resolution image of the step edge area revealing more of the microstructural detail. The EFM can and is being used to examine potentiometric variations across finished devices, such as the one in the figure, with and without applied external fields. Both AFM and EFM are powerful tools to use in this application in conjunction with other conventional methods.

One other example of the use of STM and AFM to evaluate high T_c devices is shown in Fig. 6.18, which is a composite of three $SrTiO_3$ bicrystal YBCO junctions. The arrows in the images show the progression in angle of rotation from (a) to (c) of 24° to 36.8° to 45°. The presence of *a*-axis grains in different amounts on the two halves of the junctions is probably indicative of inequality of the surface of the bicrystal halves. In the 45° junction, the *a*-axis grains were absent from the immediate junction area and might be symptomatic of messed up epitaxy in the near bicrystal boundary. A higher-resolution image of the 36.8° junction is shown in Fig. 6.19. The meandering boundary between the two YBCO film halves and the spiral grains which overgrow the substrate junction to create this meandering boundary are shown.

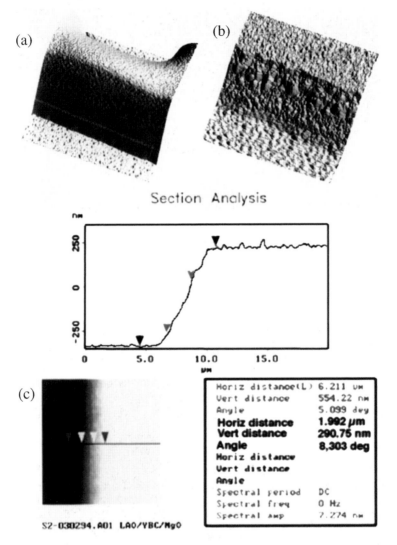

Fig. 6.17. An evaluation of a multilayer step junction. Image (a) is a low-resolution topograph of the entire step area. Image (b) is a higher-resolution close-up of the step detail. Image (c) shows a line scan across the step and the step height and angle measurements.

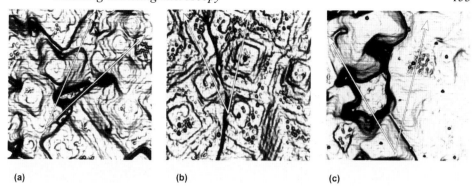

(a)   (b)   (c)

Fig. 6.18. AFM images of three different YBCO/SrTiO$_3$ bicrystal junctions; (a) 24°, (b) 36.8°, and (c) 45°. Note the differences in distributions of *a*-axis grains on the surfaces of these three films.

Fig. 6.19. A higher-resolution AFM image of one of the SrTiO$_3$ bicrystal YBCO junctions in Fig. 6.18 showing the 36.8° rotation of spiral grains across the junction.

6.7 Artifacts

No discussion of the STM and AFM experiments would be complete without a description of artifact effects and how they are dealt with in collecting data and interpreting results. The best known artifacts are due to vibrations, multiple tips effects, tip self-imaging, adsorbates on the tip surface, low conductivity (STM), distortions due to loose STM tips, and image distortions due to thermal drift of the piezoelectric scanner or bending and torsional twisting of the AFM tip cantilever as a result of adhesion between the tip and sample surface. Atomic-resolution imaging is plagued by the manifestations of vibrations that can be confused with atomic corrugations. Thermal drift is also minimized by the isolation box that contains the combined electrical and acoustic isolation. The microscope probes and controller used in this experiment were operated within such a complete mechanical, electromagnetic, acoustic, and thermal drift isolation environment. Tip self-imaging, particularly prevalent in AFM imaging, is of particular concern when imaging surface crystallites whose three-dimensional profile is not unlike that of the tip's own pyramidal shape. Imaging with the STM if the sample is conductive or using a higher-aspect-ratio AFM tip can help distinguish real features from those due to tip self-imaging. In the present experiment two samples were imaged with both the STM and AFM, and the resulting data compared. TappingMode^{TM} imaging, in which the cantilever is oscillated while scanning, significantly reduces the interaction between the tip and sample and, therefore, image distortions due to flexing and twisting of the cantilever.

The AFM cantilever, despite its designed high resonant frequencies, is still a mechanical device and as such is susceptible to vibrations. Probes are capable of 'cleaning the surface' of low- or non-conducting material (STM) or loosely bound material (AFM). The STM cannot differentiate holes from small non-conductive surface particles while the AFM tip will self image when it encounters tall sharp surface protrusions. Neither is yet capable of giving detailed chemical identity information. Conclusions about composition have to be drawn from tunneling direction, bonding distances, and atomic positions of materials whose composition is known. Most often this information is obtained from another technique.

6.8 Summary and future directions

These exciting new technologies present challenging new directions for future development. Collectively they represent more of a concept, that of carrying out an experiment using a variable proximity probe to do a user-defined measurement, than any one specific instrument. This concept is the driving force behind the development of these new capabilities. The variety of possible experiments is a measurement of the elegance of the concept. The configuration that is chosen, i.e. the nature of the experiment and the environment in which it is carried out, is determined by the measurement and variables of interest to the scientist. That, in turn, is determined by the system under study and the important blanks that need to be filled in.

These instruments can address both fundamental and applied research problems. Whether one is trying to solve a device development riddle, understand the complicated vapor-phase thin-film growth process, or delve into the fundamental issues of the nature of superconductivity there is a configuration that can suit that problem. Materials science or physics, whatever the issue, structure or property or their relationship to each other, the SPM capabilities, high resolution, and ability to simultaneously measure structure and property make these instruments valuable additions to the electron microscope family.

Each of the SPMs can make unique or complimentary contributions to the study of high T_c materials. For high-resolution measurements of the LDOS whether at ambient or low temperature, the STM is unsurpassed by its cousin SPMs. The LT-MFM may be better suited to image the vortex lattice and correlate it with microstructure. The STM can directly map the surface variations of the superconducting gap.

Although the resolution of the STM is higher, the AFM, taking advantage of the new high aspect-ratio tips, gives perfectly respectable thin-film microstructure images. It can do so without the need to worry about the immediate-surface conductivity that limits STM use for the same purpose. The AFM will also reveal the presence of non-conductive particulates which are swept away by the STM tip as it cleans the surface. The AFM is particularly useful for routine evaluation of high T_c films and multilayer device fabrication steps. The EFM could be particularly useful in evaluating and trouble-shooting device problems.

Development of new capabilities is moving more and more in this direction, taking advantage of the power of these techniques to probe local non-averaged structure related properties (spectroscopy) simultaneously with beautiful topographic data. The drive is to show the one-to-one correspondence between structure and property. One such example is a new technique that will map surface capacitive variations that, like the EFM, could be used to evaluate high T_c devices.

High on that wish list for the future is the design and construction of dual or interchangeable low-temperature AFM/STM probes to study the temperature- and structure-dependent electrical and magnetic properties of high T_c materials. The low-temperature STM tunneling experiments can ultimately provide some of the answers to the fundamental nature of superconductivity in the high T_c materials.

References

[6.1] H.-J. Guntherodt & R. Wiesendanger, eds., *Scanning tunneling microscopy I: general principles and applications to clean and absorbate-covered surfaces*, Springer Ser. in *Surf. Sci.* **20** (Springer-Verlag, Berlin 1992).

[6.2] J. A. Stroscio & W. J. Kaiser, eds., *Scanning tunneling microscopy*, Methods of Experimental Physics **27** (Academic Press, Inc., San Diego, 1993).

[6.3] D. A. Bonnell, ed., *Scanning tunneling microscopy and spectroscopy: theory, techniques, and applications* (VCH Publishers, Inc., New York 1993).

[6.4] R. J. Behm, N. Garcia, & H. Rohrer, eds., *Scanning tunneling microscopy and related methods*, NATO ASI Ser. E: Appl. Sci. (Kluwer Academic Publishers, Dordrecht, 1990).

[6.5] J. Chen, *Scanning tunneling microscopy* (Oxford University Press, Oxford, 1995).

[6.6] P. J. M. van Bentum & J. van Kempen, ref. [6.1] chapter 8, p. 207.

[6.7] Y. Kuk *et al.*, ref. [6.1] chapter 8, p. 268; H. F. Hess, chapter 9, p. 427.

[6.8] G. Binnig & H. Rohrer, *Rev. Mod. Phys.* **56**, 615 (1987).

[6.9] P. K. Hansma & J. Tersoff, *J. Appl. Phys.* **61**, R1 (1987).

[6.10] See for example E. L. Wolf, *Principles of electron tunneling spectroscopy*, International series of Monographs on Physics **71** (Oxford University Press, New York, 1985), chapters 1 & 2; ref. [6.2] chapter 1; ref. [6.1] chapter 1; L. Solymar, *Superconductive tunneling and applications*, Chapman & Hall, London (1972); C. B. Duke, Tunneling in solids: Suppl. 10 of *Solid state physics*, eds. F. Seitz & D. Turnbull, Academic Press, New York (1969); E. Burstein & S. Lundqvist, eds., *Tunneling phenomena in solids*, Plenum, New York (1969); J. A. Stroscio & R. M. Feenstra, ref. [6.2] chapter 4, p. 95; J. Tersoff & N. D. Lang, ref. [6.2] chapter 1, p. 1.

[6.11] I. Giaever, *Phys. Rev. Lett.* **5**, 147 (1960); I. Giaever, *Phys. Rev. Lett.* **5**, 464 (1960).

[6.12] M. Hawley, *Design and assembly of a low temperature vacuum tunneling apparatus and preliminary measurements of the energy gap of an organic superconductor*, Ph.D. thesis, The Johns Hopkins University (1986).

[6.13] M. E. Hawley *et al.*, *Phys. Rev. B.* **35**, 7224 (1987).

[6.14] U. Poppe, *Physica* **108B**, 805 (1981).

[6.15] H. Srikanth *et al.*, *J. Phys.* **36**, 621 (1991).

[6.16] R. Wilkins *et al.*, *Phys. Rev. B* **41**, 8904 (1990).

[6.17] M. E. Hawley *et al.*, *Phys. Rev. Lett.* **57**, 629 (1986).

[6.18] J. R. Kirtley *et al.*, *Phys. Rev.* **35**, 7216 (1987).

[6.19] H. van Kempen *et al.*, in Prog. in High Temperature Superconductivity **24**, *Proc. of the European conf. on high T_c thin films and single crystals*, Sept. 30–Oct. 4, p. 147 (1989).

[6.20] R. D. Young, *Rev. Sci. Instrum.* **37**, 275 (1966); R. Young, J. Ward & F. Scire, *Phy. Rev. Lett.* **27**, 922 (1971); R. Young, J. Ward & F. Scire, *Rev. Sci. Instrum.* **43**, 999 (1972); R. D. Young, *Phys. Today*, Nov. 1971. p. 42.

[6.21] G. Binnig *et al.*, *Appl. Phys. Lett.* **40**, 178 (1982); G. Binnig *et al.*, *Physica* **109 & 110B**, 2075 (1982); G. Binnig *et al.*, *Phys. Rev. Lett.* **49**, 57 (1982).

[6.22] B. Giambattista *et al.*, *Phys. Rev. B* **37**, 2741 (1988).

[6.23] G. Rohrer, ref. 3, chapter 6, p. 156.

[6.24] V. Weinstein *et al.*, *Rev. Sci. Instrum.* **66**, 3075 (1995).

[6.25] H. Bourque, & R. M. Leblanc, *Rev. Sci. Instrum.* **66**, 2695 (1995); S. J. Oshea, R. M. Atta, & M. E. Welland, *Rev. Sci. Instrum.* **66**, 2508 (1995).

[6.26] L. Libioulle, Y. Houbion, & J. M. Gilles, *Rev. Sci. Instrum.* **66**, 97 (1995).

[6.27] A. Cricenti *et al.*, *Rev. Sci. Instrum.* **65**, 1558 (1994).

[6.28] T. Ohmori *et al.*, *Rev. Sci. Instrum.* **65**, 404 (1994).

[6.29] V. Weinstein *et al.*, *Rev. Sci. Instrum.* **66**, 3075 (1995).

[6.30] F. Gao *et al.*, *J. Vac. Sci. Technol. B* **12**, 1708 (1994).

[6.31] V. Weinstein *et al.*, *Rev. Sci. Instrum.* **66**, 3580 (1995); C. S. Chuang & T. T. Chen, *Chin. J. Phys.* **32**, 289 (1994).

[6.32] A. R. Smith & C. K. Shih, *Rev. Sci. Instrum.* **66**, 2499 (1995).

[6.33] X. Chen, E. R. Frank & R. J. Hamers, *Rev. Sci. Instrum.* **65**, 3373 (1994).

[6.34] S. J. Stranick, M. M. Kamna & P. S. Weiss, *Rev. Sci. Instrum.* **65**, 3211 (1994).

[6.35] S. H. Tessmer, D. J. Vanharlingen & J. W. Lyding, *Rev. Sci. Instrum.* **65**, 2855 (1994).

[6.36] J. W. G. Wildoer *et al.*, *Rev. Sci. Instrum.* **65**, 2849 (1994).

[6.37] R. Gaisch *et al.*, *Ultramicroscopy* **42–44**, 1621 (1992).

[6.38] H. Bando *et al.*, *Ultramicroscopy* **42–44**, 1627 (1992).

[6.39] A. D. Kent *et al.*, *Ultramicroscopy* **42–44**, 1632 (1992).

[6.40] R. R. Albrecht *et al.*, *Ultramicroscopy* **42–44**, 1638 (1992).

[6.41] J. G. Adler *et al.*, *J. Vac. Sci. & Technol.* **9**, 992 (1991).

[6.42] C. A. Lang, M. M. Dovek & C. F Quate, *Rev. Sci. Instrum.* **60**, 3109 (1989).

[6.43] D. P. E. Smith & G. Binnig, *Rev. Sci. Instrum.* **57**, 2630 (1986).

[6.44] S. Horch *et al.*, *Rev. Sci. Instrum.* **65**, 3204 (1994).

[6.45] S. Horch *et al.*, *Rev. Sci. Instrum.* **66**, 3717 (1995).

[6.46] M. Wortge *et al.*, *Rev. Sci. Instrum.* **65**, 2523 (1994).

[6.47] R. Schulz & C. Rossel, *Rev. Sci. Instrum.* **65**, 1918 (1994).

[6.48] A. J. Leavitt *et al.*, *Rev. Sci. Instrum.* **65**, 75 (1994).

[6.49] P. M. Thibado, Y. Liang & D. A. Bonnell, *Rev. Sci. Instrum.* **65**, 3199 (1994).

[6.50] L. I. McCann, R. M. Smalley & M. A. Dubson, *Rev. Sci. Instrum.* **65**, 2519 (1994).

[6.51] J. B. Xu *et al.*, *Rev. Sci. Instrum.* **65**, 2262 (1994).

[6.52] J. Siegel *et al.*, *Rev. Sci. Instrum.* **66**, 2520 (1995).

[6.53] J. Akila & S. S. Wadhwa, *Rev. Sci. Instrum.* **66**, 2517 (1995).

[6.54] M. Hues *et al.*, *Rev. Sci. Instrum.* **65**, 1561 (1994).

[6.55] M. E. Taylor, *Rev. Sci. Instrum.* **64**, 154 (1993).

[6.56] G. Binnig & D. P. E. Smith, *Rev. Sci. Instrum.* **57**, 1688 (1986).

[6.57] L. E. C. Vandeleemput *et al.*, *Rev. Sci. Instrum.* **62**, 989 (1991).

[6.58] K. G. Vandervoort *et al.*, *Rev. Sci. Instrum.* **64**, 896 (1993).

[6.59] K. G. Vandervoort *et al.*, *Rev. Sci. Instrum.* **65**, 3862 (1994).

[6.60] A. M. Simpson & W. Wolfs, *Rev. Sci. Instrum.* **58**, 2193 (1987).

[6.61] A. I. Oliva *et al.*, *Rev. Sci. Instrum.* **66**, 3196 (1995).

[6.62] R. Curtis, M. Krueger & E. Ganz, *Rev. Sci. Instrum.* **65**, 3220 (1994).

[6.63] S. Park & R. C. Barrett, ref. 2, chapter 2, p. 31.

[6.64] M. Schmid & P. Varga, *Ultramicroscopy* **42–44**, 1610 (1992).

[6.65] D. A. Bonnell, ref. 3, chapter 2, p. 17.

[6.66] H. K. Wickramasinghe, ref. 2, chapter 3, p. 77.

[6.67] A. Moser *et al.*, *Phys. Rev. Lett.* **74**, 1847 (1995); H. J. Hug *et al.*, *Physica C* **235**, 2695 (1994); S. Behler *et al.*, *J. Vac. Sci. & Technol.* B **12**, 2209 (1994).

[6.68] C. W Yuan *et al.*, *J. Vac. Sci. Technol.* B **14**, 1210 (1996).

[6.69] S. Tanaka *et al.*, *J. Phys. Soc. of Japan* **64**, 1476 (1995).

[6.70] H. Hancotte *et al.*, *Physica B* **204**, 206 (1995).

[6.71] C. Renner *et al.*, *Physica B* **194**, 1689 (1994).

[6.72] T. Hasegawa *et al.*, *Supercond. Sci. & Technol.* **4**, S73 (1991).

[6.73] E. L. Wolf *et al.*, *J. Supercond.* **7**, 355 (1994); H. J. Tao *et al.*, *Phys. Rev. B* **45**, 10622 (1992).

[6.74] H. L. Edwards *et al.*, *Phys. Rev. Lett.* **73**, 1154 (1994); H. L. Edwards *et al.*, *Phys. Rev. Lett.* **69**, 2967 (1992); H. L. Edwards *et al.*, *J. Vac. Sci. Technol. B* **12**, 1886 (1994).

[6.75] T. Hasegawa *et al.*, *Physica Scripta* **T49A**, 215 (1993).

[6.76] K. Ichimura & K. Nomura, *J. Phys. Soc. Jap.* **62**, 3661 (1993).

[6.77] A. Balzarotti, *Nuovo Cimento D* **15**, 459 (1993); A. Balzarotti *et al.*, *Physica C* **200**, 251 (1992).

[6.78] M. Tanaka *et al.*, *Jap. J. Appl. Phys.* **32**, 35 (1993); M. Tanaka *et al.*, *Physica C* **18**, 1909 (1991).

[6.79] C. Calandra & F. Manghi, *J. Electron Spectroscopy & Related Phenom.* **66**, 453 (1994).

[6.80] M. Nantoh *et al.*, *J. Appl. Phys.* **75**, 5227 (1994); M. Nantoh *et al.*, *J. Supercond.* **7**, 349 (1994); M. Kawasaki *et al.*, *Jap. J. Appl. Phys.* **32**, 1612 (1993).

[6.81] M. Koyanagi *et al.*, *Jap. J. Appl. Phys.* **31**, 3525 (1992); M. Koyanagi *et al.*, *Jap. J. Appl. Phys.* **34**, 89 (1995); S. Kashiwaya *et al.*, *Physica B* **194**, 2119 (1994).

[6.82] P. Rice & J. Moreland, *IEEE Trans. Magnetics* **27**, 5181 (1991).

[6.83] H. P. Lang, T. Frey & H.-J. Guntherodt, *Europhysics Lett.* **15**, 667 (1991); F. Baudenbacher *et al.*, *Physica C* **185**, 2177 (1991).

[6.84] I. D. Raistrick *et al.*, *Appl. Phys. Lett.* **59**, 3177 (1991); M. Hawley *et al.*, *Science* **251**, 1587 (1991).

[6.85] J. Morel *et al.*, *Appl. Phys. Lett.* **59**, 3039 (1991).

[6.86] D. P. Norton *et al.*, *Phys. Rev. B* **44**, 9760 (1991).

[6.87] C. Gerber *et al.*, *Nature* **350**, 279 (1991).

[6.88] R. M. Silver *et al.*, *IEEE Trans. Magnet* **27**, 1215 (1991).

[6.89] L. E. C. Vandeleemput *et al.*, *J. Cryst. Growth* **98**, 551 (1989).

[6.90] J. K. Grepstad *et al.*, *Physica C* **153**, 1453 (1988).

[6.91] J. Summhammer *et al.*, *Physica C* **242**, 127 (1995).

[6.92] W. Schindler *et al.*, *Thin Solid Films* **250**, 232 (1994).

[6.93] B. Staublepumpin *et al.*, *Physica C* **235**, 679 (1994).

[6.94] Chandrasekhar *et al.*, *Appl. Phys. Lett.* **60**, 2424 (1992).

[6.95] X. Zhu *et al.*, *J. Vac. Sci. & Technol. B* **12**, 2247 (1994); X. Zhu *et al.*, *Physica C* **216**, 153 (1993).

[6.96] Y. Bando *et al.*, *Chin. J. Phys.* **31**, 903 (1993).

[6.97] L. Luo *et al.*, *Appl. Phys. Lett.* **62**, 485 (1993).

[6.98] H. Sakai, Y. Shiohara & S. Tanaka, *Physica C* **228**, 259 (1994).

[6.99] Y. Tazoh & S. Miyazawa, *Appl. Phys. Lett.* **62**, 408 (1993).

[6.100] D. G. Schlom *et al.*, *J. Cryst. Growth* **137**, 259 (1994).

[6.101] K. H. Wu *et al.*, *Chin. J. Phys.* **31**, 1091 (1993); J. Y. Juang *et al.*, *Physica B* **194**, 385 (1994); J. Y. Juang *et al.*, *Physica B* **194**, 385 (1994).
[6.102] D. H. Lowndes *et al.*, *Appl. Phys. Lett.* **61**, 852 (1992).
[6.103] H. P. Lang *et al.*, *Helvetica Phys. Acta* **65**, 864 (1992; H. P. Lang *et al.*, *Physica C* **194**, 81 (1992); H. Haefke *et al.*, *Thin Solid Films* **228**, 173 (1993); H. P. Lang *et al.*, *J. Alloys & Cmpds* **195**, 97 (1993); R. Sum, H. Lang & H.-J. Guntherodt, *Appl. Surf. Sci.* **86**, 140 (1995).
[6.104] M. Aindow & M. Yeadon, *Phil. Mag. Lett.* **70**, 47 (1994).
[6.105] H. Koinuma *et al.*, *J. Phys. & Chem. of Solids* **54**, 1215 (1993).
[6.106] M. Schilling, F. Goerke & U. Merkt, *Thin Solid Films* **235**, 202 (1993); M. A. Harmer, C. R. Fincher & B. Parkinson, *J. Mater. Sci.* **27**, 4871 (1992).
[6.107] H. Olin *et al.*, *Ultramicroscopy* **42**, 734 (1992).
[6.108] S. J. Pennycook *et al.*, *Physica C* **202**, 1 (1992).
[6.109] A. Catana *et al.*, *J. Alloys & Cmpds* **195**, 93 (1993).
[6.110] M. McElfresh *et al.*, *J. Appl. Phys.* **71**, 5099 (1992); R. E. Muenchausen *et al.*, *Physica C* **199**, 445 (1992).
[6.111] M. Guillouxviry *et al.*, *J. Cryst. Growth* **132**, 396 (1993); C. Rossel *et al.*, *Physica C* **223**, 370 (1994).
[6.112] J. Z. Liu *et al.*, *J. Supercond.* **8**, 103 (1995); L. Li *et al.*, *Physica C* **235**, 635 (1994).
[6.113] C. S. Chuang, H. T. Chen & T. T. Chen, *Physica C* **220**, 203 (1994).
[6.114] I. Maggioaprile *et al.*, *Ultramicroscopy* **42**, 728 (1992).
[6.115] J. R. Sheats & P. Merchant, *Appl. Phys. Lett.* **62**, 99 (1993).
[6.116] J. H. Xu *et al.*, *Appl. Phys. Lett.* **64**, 1874 (1994); J. H. Xu *et al.*, *Mat. Lett.* **21**, 357 (1994).
[6.117] M. E. Hawley *et al.*, *Ultramicroscopy* **42**, 705 (1992); I. Raistrick & M. Hawley, *Physica D* **66**, 172 (1993).
[6.118] S. K. Mishra *et al.*, *Indian J. Pure & Appl. Phys.* **30**, 685 (1992).
[6.119] D. J. Lichtenwalner *et al.*, *J. Vac. Sci. & Technol. A* **10**, 1537 (1992).
[6.120] T. I. Selinder *et al.*, *Thin Solid Films* **229**, 237 (1993); T. Suzuki *et al.*, *Physica C* **235**, 623 (1994).
[6.121] C. Chen & G. Oya, *IEICE Trans. Electronics* **E77**, 1209 (1994).
[6.122] G. S. Shekhawat *et al.*, *Supercond. Sci. & Technol.* **8**, 291 (1995).
[6.123] R. Sum *et al.*, *J. Alloys & Cmpds* **195**, 113 (1993); D. Hesse *et al.*, *J. Alloys & Cmpds* **195**, 109 (1993); H. P. Lang *et al.*, *Physica C* **202**, 289 (1992); H. Haefke *et al.*, *Appl. Phys. Lett.* **61**, 2359 (1992).
[6.124] S. E. Russek *et al.*, *Appl. Phys. Lett.* **64**, 3649 (1994).
[6.125] R. Chatterjee *et al.*, *Appl. Phys. Lett.* **65**, 109 (1994).
[6.126] M. Bauer, F. Baudenbacher & H. Kinder, *Physica C* **246**, 113 (1995).
[6.127] J. Burger *et al.*, *Appl. Phys. A* **59**, 49 (1994).
[6.128] N. Savvides & A. Katsaros, *Physica C* **226**, 23 (1994).
[6.129] S. Zhu *et al.*, *Appl. Phys. Lett.* **62**, 3363 (1993).
[6.130] H. Haefke *et al.*, *Appl. Phys. Lett.* **15**, 3054 (1992).
[6.131] M. V. H. Rao *et al.*, *Mater. Res. Bull.* **28**, 271 (1993).
[6.132] R. E. Thomson *et al.*, *Appl. Phys. Lett.* **63**, 614 (1993).
[6.133] X. Yang *et al.*, *Annal. Physik* **1**, 575 (1992); U. Geyer, G. Vonminnigerode & H. U. Krebs, *J. Appl. Phys.* **76**, 7774 (1994).
[6.134] R. E. Thomson, J. Morel & A. Roshko, *Nanotechnology* **5**, 57 (1994).
[6.135] S. Chen *et al.*, *Physica B* **194**, 391 (1994).
[6.136] I. Heyvaert *et al.*, *Appl. Phys. Lett.* **61**, 111 (1992).

[6.137] J. A. Virtanen *et al.*, *J. Appl. Phys.* **70**, 3376 (1991).
[6.138] M. A. Harmer, C. R. Fincher & B. A. Parkinson, *J. Appl. Phys.* **70**, 2760 (1991).
[6.139] K. Terashima *et al.*, *Appl. Phys. Lett.* **59**, 644 (1991).
[6.140] H. Heinzelmann *et al.*, *Appl. Phys. Lett.* **53**, 2447 (1988).

7

Identification of new superconducting compounds⋅ by electron microscopy

G. VAN TENDELOO and T. KREKELS

7.1 Introduction

It is clear that electron microscopy is not the most favourable technique for structure determination of new (superconducting) phases; X-ray diffraction and particularly neutron diffraction do a far better job in the *ab initio* structure determination. Electron microscopy and electron diffraction are extremely powerful however to determine the local structure; i.e. to detect deviations from the average structure, as determined by X-rays or neutrons. In this way several new phases have been first identified by electron microscopy; some of them have been later made into bulk superconductors. In other cases the identification of isolated defects in an existing material have inspired chemists to produce new superconducting materials; this was, for example, the case for the occurrence of double HgO_δ layers in a one-layer Hg-1223 superconductor.

In the first part of this contribution we will focus on the well known $YBa_2Cu_3O_{7-\delta}$ superconductor; this material allows a large number of substitutions without drastically altering its structural aspects, but with sometimes completely different physical properties. In the second part, we will concentrate on the more recent Hg-based superconductors and illustrate the extreme importance of the different electron microscopy techniques in the development of new superconducting compounds.

7.2 Oxygen vacancy order in the CuO plane of $YBa_2Cu_3O_{7-\delta}$

From a microstructural point of view, $YBa_2Cu_3O_{7-\delta}$ is an interesting compound. It can assume variable oxygen contents ($0 \leqslant \delta \leqslant 1$), ordered into various ordering schemes as observed abundantly by electron microscopy [7.1–7.16], and more recently by X-ray [7.17–7.19] and neutron diffraction [7.3]. Another important feature of the compound is its susceptibility to elemental substitutions, resulting again in a variety of oxygen ordered phases.

It is well-established that, at room temperature and atmospheric pressure, within the range $0 \leqslant \delta \leqslant \delta_t$ YBa$_2$Cu$_3$O$_{7-\delta}$ is orthorhombic and within the range $\delta_t \leqslant \delta \leqslant 1$, it is tetragonal. Values for δ_t vary around 0.65 [7.2, 7.20–7.24]. In experiments where δ is fixed the transition temperature T_t increases linearly with δ [7.25] from 110 °C at $\delta = 0.67$ to 680 °C at $\delta = 0.34$.

7.2.1 Tweed and twinning

The tetragonal–orthorhombic phase transition that occurs upon cooling from high-temperature oxygenation stages, induces a spontaneous strain, proportional to the orthorhombicity of the material. The strain is released by the introduction of a lamellar texture called twinning that consists of slabs of two orientation variants of the orthorhombic phase. Experimental as well as theoretical work seems to point out that twinning is a reaction to a *macroscopic* orthorhombic strain. This is in contrast to the case of tweed that seems to be triggered by *local* orthorhombic deformations.

In electron diffraction, twinning can easily be recognized by a spot splitting parallel to the unsplit central row, perpendicular to the mirror plane (see below). It is important to note that twinning as well as tweed are effects that are occurring in the orthorhombic range of the phase diagram. By techniques not offering a very local probe (X-rays or neutrons), and thus providing only average structural information, tweed-textured material has often been described as tetragonal.

7.2.2 Tetragonal phase

In the tetragonal phase, the oxygen atoms in the CuO layer are distributed quasi-randomly, with an equal occupation of the O1 and O5 sublattices. The sublattices consist of the oxygen sites between every two copper atoms in the plane, the O1 sites lying along one direction (defined as a), the O5 sites along the perpendicular direction. (See Fig. 7.2). The formation of short CuO segments along both basic directions of the CuO chain plane [7.26] is likely to cause the local orthorhombic strains that lead to the pre-transition tweed texture. The tetragonal phase is easily recognized in the [001] zone electron diffraction patterns, where a perfectly square lattice of basic reflections is observed. The true tetragonality of a pattern is sensitively demonstrated by the absence of spot splitting, associated with the twinning.

7.2.3 Short-range ordering

The presence of short-range order is revealed in electron diffraction patterns by the occurrence of diffuse intensities. The diffuse streaks observed in the square [001] zone diffraction pattern of Fig. 7.1 are due to the presence of irregularly spaced, short segments of CuO chains. This is derived on the basis of the cluster theory of Ridder *et al.* [7.50, 7.28] where knowledge of the geometrical locus of diffuse scattering allows determination of the shape of the atom clusters giving rise to the scattering. CuO chains form in the $CuO_{1-\delta}$ plane by the occupation of the O1 sites and a depletion of the O5 sites or vice versa. Short-range order is commonly observed in highly oxygen-deficient samples and can be induced in oxygen-rich materials by a thermal disordering by an in situ (inside the electron microscope vacuum) heating treatment (see below).

From tilting experiments it can be deduced that the observed streaks are due to reciprocal $(100)^*$ and $(010)^*$ planes of diffuse intensity, intersecting the Ewald sphere, which is supporting the assumption that the scattering is due to linear defects. (See also [7.4, 7.27, 7.28].) This state of order is referred to in [7.50] as a *transition state,* a short-range ordered precursor to a final long-range ordered state.

7.2.4 Ortho-I

The orthorhombic phase that corresponds to the ordered $YBa_2Cu_3O_7$ phase ($\delta = 0$), is termed Ortho-I. The CuO plane in the Ortho-I phase has all O1 sites

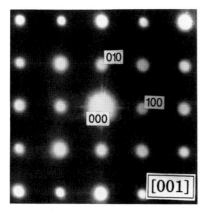

Fig. 7.1. Electron diffraction pattern along the [001] zone, in a tetragonal sample (from the observation of unsplit spots). *In situ* heating and cooling causes disorder reflected in the presence of diffuse scattering along the basic directions.

filled and all O5 sites vacant. CuO chains run along b (Fig. 7.2(a)). As a result, for the Ortho-I phase $a_p \leqslant b_p$ ($a_p = 3.83$ Å; $b_p = 3.87$ Å [7.1]). Figure 7.3 shows a [001] zone diffraction pattern of the Ortho-I phase. The image was obtained on a twinned piece of material, which is revealed by the split spots in the pattern. Since material of both variants was selected by the selected area aperture, the [001] pattern is an overlap pattern of the [001] zone patterns of both orthorhombic variants. Both patterns are identical rectangular patterns (since a_p^* and b_p^* are slightly different, subscript 'p' indicating indexing with reference to the basic tri-perovskite), rotated about 90° with respect to each other. Since the magnitude of the spot splitting is determined by the difference between the lattice parameters a_p and b_p, the spot splitting is a measure for the

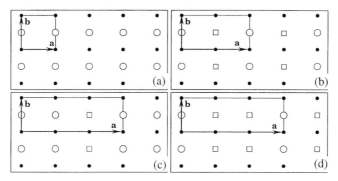

Fig. 7.2. (a) Schematic representation of the CuO plane Ortho-I ordering; (b) model of the Ortho-II ordering. Along a CuO and Cu vacancy chains alternate; (c) and (d) are representations of ordering leading to tripled a-parameters, where (c) represents the Ortho-III phase with an oxygen content of 6.667, and (d) represents a symmetric phase where roles of vacant and filled chains are reversed, that occurs at an oxygen content of 6.333.

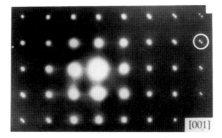

Fig. 7.3. The [001] diffraction pattern of the Ortho-I phase. Splitting of superstructure spots is caused by orthorhombic twinning (see text). Streaks appear when vacancies order in an irregular stacking of chains, as a precursor to Ortho-II or Ortho-III phases.

orthorhombicity of the phase. In the tetragonal limit, the spot splitting is absent; with increasing oxygen content the orthorhombicity will increase and so will the spot splitting. The spot splitting is therefore a direct measure for the local oxygen content in the material.

7.2.5 Ortho-II

The Ortho-II phase, with an ideal composition $YBa_2Cu_3O_{6.5}$, has every other CuO chain evacuated of oxygen (Fig. 7.2(b)). The range over which the Ortho-II phase is observed stretches over $0.3 \leqslant \delta \leqslant 0.5$. The new unit mesh that can be defined on the ordered $CuO_{1-\delta}$ plane has dimensions $a_{II} \approx 2a_p$, $b_{II} \approx b_p$, $c_{II} \approx c_p$ (based on [7.1]).

The Ortho-II phase is easily recognized in [001] ED patterns (Fig. 7.4) as well as in [001] or [010] zone HREM images (Fig. 7.5). In ED patterns superstructure spots appear at positions $h + \frac{1}{2}k\,l$, characteristic for a doubling of the a-parameter. Often however, as shown in Fig. 7.4, in such patterns also the b-parameter seems doubled by the appearance of spots at positions $h\,k + \frac{1}{2}l$ as well. These spots are due to twinning of the orthorhombic material, the resulting diffraction pattern being the overlap of the two orientation variants [7.29–7.31].

In [7.32] it is shown that a disordered array of line defects gives rise to Lorentzian-shaped diffraction spots, with a position associated with the mean spacing of the scattering elements. When chains in the CuO plane are irregularly spaced with a mean spacing of $2a_p$, Lorentzian shaped diffraction spots at positions $h + \frac{1}{2}k\,l$ are to be expected. When the spreading on the average periodicity increases, the superstructure reflections will become more elongated (Fig. 7.4(b)). The superstructure reflection shape will further be influenced by the Ortho-II domain shape; small domain dimensions leading to elongated reflections. Intensity and sharpness of the superstructure reflections can thus be considered as indicative of the domain size and the ordering quality of the phase. When the average period becomes $3a_p$, Lorentzian peaks will be observed at positions $h + \frac{1}{3}k\,l$ and $h + \frac{2}{3}k\,l$. (Fig. 7.4(c)). In [7.33] similar results are obtained for a Magneli homologous series of CuO chain ordered phases. Whereas a perfect Ortho-II domain has an oxygen deficiency $\delta = 0.5$, the strongest and sharpest Ortho-II electron diffraction intensity occurs at an oxygen deficiency of only $\delta = 0.4$ (see for example [7.61]). The domain size and the ordering quality are apparently maximized at $\delta = 0.4$.

Dark-field images allow the visualization of the ordered domains in real space. Ortho-II domains will appear bright when selecting a superstructure reflection. Figures 7.6(a) and (b) show images corresponding to the diffraction

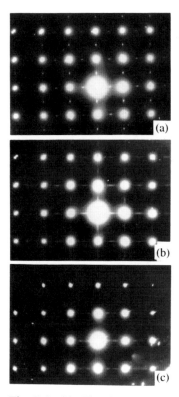

Fig. 7.4. (a) The [001] zone diffraction pattern with sharp superstructure
reflections at positions $h + \frac{1}{2} k \, l$ appearing in well-ordered Ortho-II material.
Since the material is twinned, spots seem to appear along both basic directions;
(b) [001] zone diffraction pattern with streaked superstructure intensity,
corresponding to domains that are short range ordered along a, occurring at
oxygen contents at the slopes above and below the Ortho-II plateau; (c) [001]
zone diffraction pattern with spots at positions $h + \frac{1}{3} k \, l$ and $h + \frac{2}{3} k \, l$ due to an
Ortho-III structure occurring in samples with oxygen contents around 6.8.

patterns of Figs. 7.4(a) and (b). The images show that the Ortho-II domains are
elongated or lenticularly shaped, with the short axis along the a-direction.
Largest Ortho-II domains (40 Å × 200 Å) and a maximal total Ortho-II
volume, occur at oxygen deficiencies $\delta = 0.4$. For the $\delta = 0.4$ sample of ref.
[7.3], from measurements of the Ortho-II peak surface in a [100] electron
densitogram, the Ortho-II volume fraction was an estimated 50%, in accordance with our results. Samples showing elongated Ortho-II diffraction spots,
correspondingly show Ortho-II domains in dark-field images (Fig. 7.6(b)) with
reduced dimensions along a (size : 10 Å × 200 Å).

In high-resolution images along the $[100]_p$ direction, domains of vertical c-
axis stacking (Fig. 7.5) as well as small domains of a staggered stacking of

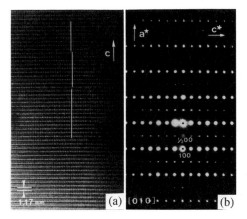

Fig. 7.5. (a) High-resolution image along the [010]-axis of the Ortho-II phase. The cell doubling can be seen by the white dots that appear every $2a_p$. In this view along the chains, chains are stacked vertically along c. Note the presence of anti-phase boundaries with displacement vector $R = \frac{1}{2}[100]$. (b) Diffraction pattern along the [010] zone axis. Note the splitting of the outer spot rows due to twinning.

Fig. 7.6. (a) Dark-field image in which Ortho-II ordered domains are seen bright. The corresponding diffraction pattern is shown in Fig. 7.4(a). (b) Dark field image corresponding to Fig. 7.4(b). Domain size is short along a and large along the CuO chain direction b. The band structure in these images is due to twinning.

successive two-dimensional ordered CuO planes (Fig. 7.7) occur. The diffraction pattern of Fig. 7.5(b) corresponds to the most common, vertically stacked Ortho-IIa variant ($c_{IIa} = 11.7$ Å $\approx c_p$). As can be judged from the alignment of the superstructure spots at positions $h + \frac{1}{2} 0\, l$ with the rows of basic spots $h\, 0\, l$, the two-dimensional arrangements in the CuO plane vertically coincide. This is confirmed by the high-resolution image of Fig. 7.5(a). It also reveals anti-phase boundaries, the material at both sides of which is displaced by $\frac{1}{2}[100]_{OIIa}$.

When the material is heated *in situ* (i.e. inside the microscope vacuum) in a furnace or by the electron beam, disordering reduces the sharp superstructure spots in the [010] diffraction patterns to continuous streaks along the c-direction. In these materials, the alternative staggered stacking variant is encountered. Its correlation length is of the order of five unit-cells along c, too short to be revealed in electron diffraction patterns. However, optical diffraction experiments, using the high-resolution image as a diffracting grating produce the corresponding pattern (Fig. 7.7(c)). The latter structure is A-face centred and thus has doubled a- as well as c-lattice parameters: $c_{IIb} = 2c_{IIa}$. Although the staggered variant Ortho-IIb is less frequently observed and only appears ordered to short range, it is likely to be energetically close to the vertically stacked Ortho-IIa.

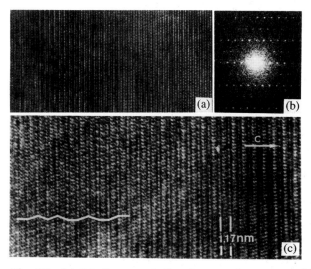

Fig. 7.7. (a) Medium-resolution image of an area exhibiting predominantly the face-centred variant of the Ortho-II phase; (b) optical diffraction pattern of this image, showing spots due to face centering; (c) high magnification of the same area. The cell doubling is again revealed by white dots. Note the presence of stacking faults.

7.2.6 Ortho-III, Ortho-IV, ...

Electron microscopic observations of the Ortho-III phase have been reported far less than of the Ortho-II phase [7.8, 7.34, 7.35]. Clear evidence for the occurrence of this phase is given in the [001] diffraction of Fig. 7.4(d). Due to the short structural coherence length of this phase, the results could not be confirmed either by X-ray, or by neutron diffraction. The Ortho-III phase ideally appears at an oxygen deficiency $\delta = \frac{1}{3}$, and compared to the Ortho-I phase, has one out of three Cu_1O_1 chains depleted of oxygen in an ordered succession. (See Fig. 7.2(c)). Lattice parameters are $a_{III} \approx 3a_I$, $b_{III} \approx b_I$ and $c_{III} \approx c_I$. The tripling of the unit-cell is reflected in the diffraction pattern by the appearance of two superstructure spots at positions $h + \frac{1}{3}k\ l$ and $h + \frac{2}{3}k\ l$. Superstructure spots for the Ortho-III phase are always very broad, due to substantial disorder in the cell-tripling alternation, or to the limited size of the domains.

The high-resolution images of Fig. 7.8, taken along the [001] zone, were

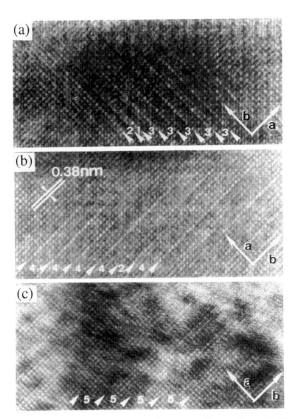

Fig. 7.8. High-resolution images along the [001] zone of specimens which were heat-treated; (a) cell-tripled superstructure; (b) and (c) superstructures with four- and five-fold a-parameters.

recorded in heat-treated samples. Besides a disordering of the oxygen frame-work in the CuO plane, part of the oxygen of the CuO plane is lost during a heat treatment in the electron microscope's vacuum. The phase shown in Fig. 7.8(a), therefore is presumably a variant of the Ortho-III phase, with an alternation of two oxygen-depleted chains, and one filled chain, in the CuO plane (Fig. 7.2(d)). The oxygen content for this phase is $x = 6.33$.

Superstructures with longer periodicities of $4a_l$ and $5a_l$, only appear oc-casionally on a very local scale; they are hardly evidenced by any diffraction technique, but they have been observed in real space by high-resolution electron microscopy (Fig. 7.8). Other periodicities, have been reported in the literature and are also attributed to CuO-chain ordering, but these observations are not well-confirmed [7.8, 7.10, 7.36, 7.37]. Identifying these phases as Cu_1O_1-chain ordered phases, they should probably be considered as metastable phases at oxygen contents intermediate to that of the stable Ortho-I, Ortho-II and Ortho-III phases. Theoretical studies seem to support this conclusion [7.38–7.39].

7.2.7 YBCO at higher temperatures

The phases described until now are induced by the ordering of the oxygen and vacancies in the CuO plane. Due to the weak bonding of oxygen in the CuO plane, *in-situ* heating inside the electron microscope will cause disorder and oxygen desorption. A sample can be heated by focusing the electron beam on the sample; this will heat the sample very locally and unwanted effects due to beam damaging may result. Alternatively, heating can also be effected in the heating holder, usually allowing heating to over 1000 °C at conditions of high mechanical stability.

Starting with a fully oxidized specimen ($YBa_2Cu_3O_7$ i.e. $\delta = 0$) and increas-ing the temperature progressively in a heating holder, avoiding beam damage by using only 100 kV electrons, a short-range ordered Ortho-II superstructure is formed, revealed by broad streaked reflections at $h + \frac{1}{2}k\,l$ positions in diffraction mode. Occasionally broad reinforcements at $h + \frac{1}{3}k\,0$ and $h + \frac{2}{3}k\,0$ are produced indicating the occurrence of ill-defined phases with a mean period of $3a_p$ (Ortho-III). Superstructure reflections fade into streaks as the material is still orthorhombic. The superstructure can be recovered by cooling to somewhat lower temperatures, indicating that the loss of the superstructure is not only due to a loss of oxygen, but also to increasing thermal disorder. Only at a slightly higher temperature the spot splitting due to twinning disappears and promptly a $\sqrt{2}a_p \times \sqrt{2}a_p$ phase diffraction pattern appears ($T > 500$ °C; Fig. 7.9). This phase will be discussed in the next paragraph.

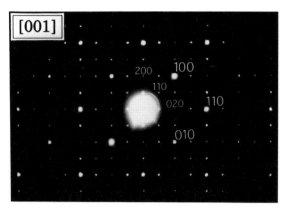

Fig. 7.9. Diffraction pattern of $2\sqrt{2}a_p \times 2\sqrt{2}a_p$ structure in a thin area. Different superstructure spots have different relative intensities. The indexing is based on the $2\sqrt{2}a_p \times 2\sqrt{2}a_p$ unit cell.

Heating to still higher temperatures results in the formation of a new cubic phase with a slightly larger lattice parameter than a_p as concluded from the resulting diffraction patterns. The lattice of this new phase is closely related to the lattice of $YBa_2Cu_3O_{7-\delta}$. In the new phase the tripling of the c-parameter has disappeared suggesting that the new phase is a cation-disordered cubic perovskite. The appearance at high temperature of a cubic phase with a disordered cation sublattice was already suggested earlier on the basis of the observation of three orientation variants of the $YBa_2Cu_3O_{7-\delta}$ structure, in which the c-axes are mutually perpendicular [7.40]. The decomposition suggests that the $2\sqrt{2}a_p \times 2\sqrt{2}a_p$ phase is the limiting phase compatible with the heavy ion framework of the $YBa_2Cu_3O_{7-\delta}$ structure, and that the $2\sqrt{2}a_p \times 2\sqrt{2}a_p$ phase appears at conditions of pressure and temperature at the boundaries of the stability field of the 123-phase.

7.2.8 The $2\sqrt{2}a_p \times 2\sqrt{2}a_p$ phase

The $2\sqrt{2}a_p \times 2\sqrt{2}a_p$ phase was discussed in many studies of the $YBa_2Cu_3O_{7-\delta}$ compound and considered as an oxygen ordered superstructure [7.11, 7.12, 7.22, 7.41–7.46] or as due to partial decomposition [7.47]. In [7.41] it is suggested that Ba has precipitated from the material, before the $2\sqrt{2}a_p \times 2\sqrt{2}a_p$ phase occurs and that this phase would thus have a stoichiometry that deviates from 1:2:3. This conjecture is based on their observation of impurity phases at the surface of the material appearing during the process of *in situ* heating. We have, however, confirmed the existence of this phase without the overall appearance of the secondary phases.

Detailed electron diffraction evidence moreover shows that the hypothesis of oxygen ordering does not hold, and we will propose here evidence for an alternative model. The $2\sqrt{2}a_p \times 2\sqrt{2}a_p$ phase is reproducibly obtained by heating the material *in situ* to temperatures of about 500 °C at a rate of 5 °C/min [7.48]. Only when the tetragonal δ range is reached does the $2\sqrt{2}a_p \times 2\sqrt{2}a_p$ phase appears, as suggested by the disappearance of twin-split reflections prior to the appearance of the new phase. Observations of the $2\sqrt{2}a_p \times 2\sqrt{2}a_p$ phase in the orthorhombic stoichiometry range [7.11, 7.12] are probably erroneous, and/or must be attributed to sample inhomogeneities. An upper limit of the oxygen content was obtained by producing the super-structure in a sample with an initial oxygen deficiency of $\delta = 0.75$. Thermo-dynamic measurements [7.49] yield a value $\delta \approx 0.85$ at the temperatures and pressures of our experiment.

The [001] zone diffraction pattern is shown in Fig. 7.9. The superstructure has a square unit mesh of dimensions $2\sqrt{2}a_p \times 2\sqrt{2}a_p$, diagonal with respect to the perovskite mesh. Characteristic for the [001]-zone diffraction pattern is the sharpness and intensity of the superstructure spots. There is no intermediate transition state [7.28, 7.50] and diffuse or streaked scattering as in the oxygen ordering phases discussed above. The sharp extra reflections are weak at their appearance but intensify rapidly.

The $[1\bar{1}0]$ zone pattern is reproduced in Fig. 7.10. In the $[hk0]$ zone diffraction patterns streaks appear along c^*, indicating that, although the superstructure within a (001) plane is well-ordered, the correlation between the ordered planes in successive unit-cells is weak. The streaks are observed to be intensity modulated, typical of a system in which the scattering elements are highly correlated pairs [7.51]. The period of the modulation was measured to be $3.5c_0{}^*$, corresponding to a real space separation of $0.286c_0$. The pair of layers that fits the distance well within the experimental error is the CuO_2 pair (spaced $0.278c_0$ [7.52]), not the pair of BaO-layers (spaced $0.389c_0$ [7.52]). Thus, rather than the BaO pair (which surrounds the $CuO_{1-\delta}$ layer), the CuO_2 layer pair is involved in the formation of this superstructure.

Tilting experiments with the c^*-axis as tilt-axis have allowed determination of the reflection conditions in the [001] zone diffraction pattern. In Fig. 7.10 streaks at positions $h + \frac{1}{4}$, $h + \frac{1}{4}$ and $h + \frac{1}{4}$, $h + \frac{3}{4}$ are systematically absent. Indexed with respect to the superstructure mesh one obtains: streaks $h0l$ with h odd and $0kl$ with k odd are absent. Reflection conditions in the plane thus become $h0$ with h even and $0k$ with k even. Note that double diffraction fills in the forbidden reflections in the [001] zone diffraction pattern. Knowledge of

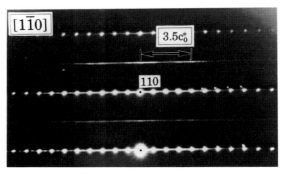

Fig. 7.10. $[1\bar{1}0]_p$ zone diffraction pattern showing the sinusoidal 3.5 $c_p{}^*$ periodic modulation, superimposed on a sinusoidal modulation with period $c_p{}^*$.

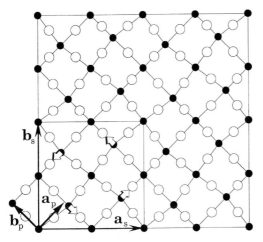

Fig. 7.11. CuO_2 layer of the deformation modulated structure. The displacements of Cu (black dots) are indicated by arrows; they are strongly exaggerated. Perovskite and superstructure meshes are indicated by subscripts 'p' and 's' respectively.

the reflection conditions allows the determination p4gm (No. 12) as the plane group of the superstructure. Since virtually no ordering is present along c, a spacegroup cannot be derived. Space groups compatible with the plane group p4gm are P42₁2 (No. 90), P4bm (No. 100), P4$\bar{b}$2 (No. 117) and P4/mbm (No. 127).

The absence of a transition state and the sharpness of the reflections at the very high oxygen deficiencies are untypical of the other oxygen ordered phases in $YBa_2Cu_3O_{7-\delta}$. However, since most models reported for the $2\sqrt{2}a_p$

$\times 2\sqrt{2}a_p$ phase are based on oxygen-vacancy ordering in the CuO chain plane of the compound [7.22, 7.12, 7.22, 7.41–7.46], we have first tried to construct a model based on an oxygen-vacancy ordering in the $CuO_{1-\delta}$ plane. Some of these models even included vacancies in the Ba layer, in order to reduce the oxygen content of the model structure below $1 - \delta = 0.25$. The models constructed were all consistent with the determined plane group, but diffraction pattern simulations for these models as well as for those proposed in the literature, all produced unsatisfactory matches with the experiment.

Strong evidence against a hypothesis of oxygen ordering for the $2\sqrt{2}a_p \times 2\sqrt{2}a_p$ phase in the $YBa_2Cu_3O_{7-\delta}$ compound comes from the observation of the $2\sqrt{2}a_p \times 2\sqrt{2}a_p$ phase in the compounds $(Y_{0.75}Ce_{0.25})_2(Sr_{0.85}Y_{0.15})_2$ $AlCu_2O_9$ and $Bi_{1.8}Pb_{0.4}Sr_2Ca_2Cu_3O_{10+x}$. Experimental results for the three compounds that produce the $2\sqrt{2}a_p \times 2\sqrt{2}a_p$ phase are very similar. Details can be found in [7.53]. Neither of the latter two compounds contains a $CuO_{1-\delta}$ plane. The only structural element common to the three compounds is the CuO_2 layer; moreover HREM unambiguously located the superstructure within the layers [7.53].

Because of the small likelihood for the presence of oxygen vacancies in the CuO_2 layer, and therefore also of oxygen ordering in this layer, the only reasonable model for the superstructure common to the compounds discussed is one of a displacive modulation in the CuO_2 layers. A plausible deformation is one of the CuO_5 pyramids in between the BaO and the Y layers. A possible mechanism could be of Jahn–Teller type. Simple models were constructed with displacements within the (001) plane.

The symmetry of the displacement configuration within a layer is determined on the basis of the plane group derived above. Only one displacement configuration of the Cu framework, shown in Fig. 7.11, is compatible with the plane group. The Cu atoms at positions $(\pm\frac{1}{4}, \pm\frac{1}{4}, 0)$ are displaced over a distance $\sqrt{2}\Delta a_o$, other Cu positions remain unchanged. The oxygen atoms remain positioned halfway between two Cu atoms. The other layers in the model are left unaffected. Except for a translation, the result does not depend on the choice of either one of the four space groups compatible with the plane group.

A correlation between the two layers of a CuO_2 layer pair is necessarily introduced in order to explain the modulation of the streaks in $[h, k, 0]$ zone diffraction patterns. For the $YBa_2Cu_3O_{7-\delta}$ compound, our model assumes a vertical (along c) coincidence of the displacement configuration. The $CuO_{1-\delta}$ layer is considered to contain only a low fraction of randomly filled oxygen sites, not necessarily critically related to the modulation in the CuO_2 layers.

A more elaborate discussion of the model for the compounds

$(Y_{0.75}Ce_{0.25})_2(Sr_{0.85}Y_{0.15})_2AlCu_2O_9$ and $Bi_{1.8}Pb_{0.4}Sr_2Ca_2Cu_3O_{10+x}$ can be found in [7.53].

7.3 Oxygen ordering and Ba-displacements in the YBCO-247 compound

The compound $Y_2Ba_4Cu_7O_{15-\varepsilon}$, is a regular intergrowth of $YBa_2Cu_3O_{7-\delta}$ (containing only single CuO layers) and $YBa_2Cu_4O_8$ (containing double CuO layers). The oxygen non-stoichiometry is directly related to the oxygen content in the single $CuO_{1-\delta}$ chains of the $YBa_2Cu_3O_{7-\delta}$ blocks. The reversible stoichiometry range was determined as $0.87 \geqslant \varepsilon \geqslant -0.3$ [7.54]. T_c varies from 33 to 95 K over the same interval according to [7.54]. The reported T_c values in ref. [7.55] are much lower, however, (ranging from 14 to 68 K in the interval $0.72 \geqslant \varepsilon \geqslant 0.09$). This situation is very similar to YBCO-123, where T_c not only depends on the oxygen content, but also on the order of oxygen within the $CuO_{1-\delta}$ plane. This observation of different T_c values for different oxygen content however suggests that in the 247-compound oxygen ordering also occurs, and has its importance for T_c. Since single CuO layers exist in the 247-structure in an identical surrounding as in the 123-compound, the oxygen vacancy ordered phases of the 123-compound are likely to occur also in the 247-compound.

Figure 7.12 shows different sections of reciprocal space for oxygen-deficient material; the ordering effects are clearly visible. Patterns (a) and (b) show a cell-tripling superstructure along the [100] direction; pattern (c) shows evidence for a cell doubling superstructure. The presence of streaks in all $[hk0]$ sections suggests a lack of (or a weak) correlation between the ordering in two successive single CuO planes. This observation is not unexpected since the correlation was already weak for YBCO-123, where the distance between successive $CuO_{1-\delta}$ planes is 11.7 Å, while for YBCO-247 the distance between two single $CuO_{1-\varepsilon}$ planes is more than 27 Å.

Analogous to the situation for YBCO-123, the diffuse streaks show a sinusoidal modulation with a period $11.5c^*_{247}$. The distance $d = 0.087c_{247}$ matches, well within the experimental error, the distance between the two BaO layers adjacent to a single CuO layer. We therefore suggest a symmetrical relaxation of the two BaO layers in a BaO—CuO—BaO triplet [7.51]. The occurrence of maxima of the modulations of the streaks at positions $h = 0$, implies that no offset in the plane exists between both BaO layers of the triplet.

A model for the displacements in the BaO layer adjacent to the oxygen vacancy ordered $CuO_{1-\varepsilon}$ layer follows from electrostatic considerations. The ordering in the $CuO_{1-\varepsilon}$ plane in alternating full and empty Cu_1O_1 chains leads

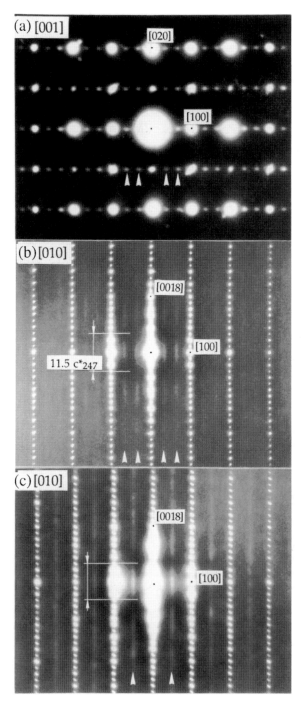

Fig. 7.12. (a) The [001] zone axis diffraction pattern with intense Ortho-III superstructure reflections at positions $h + \frac{1}{3} k \, l$ and $h + \frac{2}{3} k \, l$. Superstructure reflections are marked by short arrows; (b) [010] zone axis diffraction pattern

to local asymmetries in the electrostatic potential; the (100) symmetry plane through the Ba sites of the adjacent planes is now removed and consequently a relaxation of the Ba-O framework is to be expected [7.5, 7.19, 7.56]. The O^{2-} ions will be displaced along the c-direction, in the direction of the missing chain. The Ba^+ anions will be relaxed away from the oxygen vacancy chain. Due to the higher scattering factor of the Ba atoms, the contribution to the superstructure scattering by the Ba layers cannot be neglected. Based on this model, image calculations were performed and compared to high-resolution images along different $[h/0]$ zones (Fig. 7.13). The correspondence confirms the proposed model; however the image-to-simulation comparison technique does not allow the displacements to be accurately determined. The displacements which gave the better fit were: $\Delta x_{Ba} = 0.10$ Å, $\Delta z_{Ba} = 0.15$ Å, along a_p, and c_p respectively. The oxygen atoms of the BaO layer, adjacent to the vacancy rows have been displaced over $\Delta z_O = 0.30$ Å. The same effect has

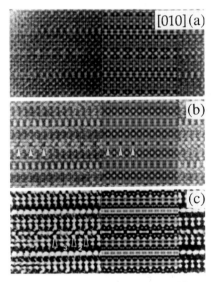

Fig. 7.13. High-resolution [010] zone axis images of Ortho-II material (a-direction horizontal; c-direction vertical). Simulated images based on the deformation model are shown as insets; (a) thickness $d = 4b_o$, defocus $f = -200$ Å; (b) $d = 12b_o$, $f = -200$ Å; (c) $d = 24b_o$, $f = -150$ Å. The $2a_o$ periodicity is indicated in (b) by short arrows. Note that in (c), a mixture of $2a_o$ and $3a_o$ is present.

Fig. 7.12 (*cont.*). with modulated Ortho-III streaks. The period of modulation of the streak is indicated; (c) [010] zone axis diffraction pattern with modulated Ortho-II streaks at positions $h + \frac{1}{2} k l$.

been observed in $YBa_2Cu_3O_{7-\delta}$, by means of X-ray diffraction [7.19]. Since in the $YBa_2Cu_3O_{7-\delta}$ compound no extra periodicities arise because of the Ba relaxation, electron microscopy yields no evidence for the displacements in $YBa_2Cu_3O_{7-\delta}$. In ref. [7.19] the Ba displacement in $YBa_2Cu_3O_{7-\delta}$ was determined by X-ray diffraction to vary between $\Delta x_{Ba} = 0.03$ Å and $\Delta x_{Ba} = 0.1$ Å.

7.4 Oxygen vacancy ordering in $Y_NSr_2MCu_2O_{5+2n\pm x}$ compounds (M = Co, Ga, Al)

The 'pure' Sr-doped 123-compound cannot be synthesized at ambient pressure without replacement of some of the other cations. The presence of new structural blocks when substituting square-planar-coordinated Cu atoms of the CuO chain plane, by octahedrally or tetrahedrally coordinated metal atoms opens the possibility to synthesize new superconductors. In the compounds with substitutions that prefer tetrahedral coordinations, M = Co (YSr_2CoCu_2O) [7.57, 7.58], M = Al (($Y_{0.75}Ce_{0.25})_2(Sr_{0.85}Y_{0.15})_2AlCu_2O_9$) and M = Ga (($Y, Ce)_nSr_2GaCu_2O_{7+2n}$) oxygen-ordered superstructures in the MO plane have been encountered.

7.4.1 Average structure

The tetrahedron of oxygen atoms around the M ion consists of the two oxygen atoms of the SrO layer forming the apices of the CuO_5 pyramid and of two oxygen atoms in the MO layer, one at an O_1 site, the other at an O_5 site. The tetrahedra are linked at the corners and form chains along the $[110]_p$ or $[1\bar{1}0]_p$ directions. A Co—O separation in the plane, larger than the original Cu—O separation is allowed for by a rotation of the tetrahedra along an axis parallel to c. The Co ions concomitantly leave the original Cu positions and move slightly towards the center of the tetrahedron. The corner linking causes successive tetrahedra (along $[110]_p$ or $[1\bar{1}0]_p$) to be rotated in opposite senses. No oxygen vacancies are expected in this framework, as heating in reducing atmospheres has proved not to extract any oxygen from the structure up to temperatures of 800 or 1000 °C.

Two chain variants of opposite 'phase' exist, as represented in Fig. 7.14, called L- and R-type. An L-chain is transformed into an R-chain by a mirror operation perpendicular to the chain direction or by a rotation of 180° about the c-axis. In [7.57, 7.58], the average structure was determined on the basis of X-ray data for the 1212-Ga compound. In the basic undistorted unit cell all chains are parallel, and of the same type. The rectangular quasi-square $(001)_p$ mesh is

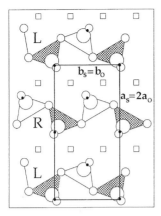

Fig. 7.14. Models for the structure of the chain layer CoO in which L- and R-chains regularly alternate. The unit mesh is indicated (subscript 'o': basic cell, 's': supercell). Small circles represent oxygen atoms, large circles Co atoms, squares oxygen vacancies and black dots are the perovskite Cu positions.

diagonal relative to the perovskite mesh ($a_o \lesssim b_o = \sqrt{2}a_p$). The $a_o[100]$ direction is chosen perpendicular to the chains, the $b_o[010]$ direction along the chains. Cell parameters are $a_o \approx b_o = 5.4$ Å. Along the c-direction, the chains are parallel in two successive CoO layers and are shifted with respect to each other in such a way that they form a staggered arrangement when they are viewed end-on. This shift over $[100]_p$ (or equivalently $[010]_p$) or $\frac{1}{2}[110]_o$ causes a centering of the lattice and a doubling of the c parameter: $c_o = 2c_p = 23.4$ Å. The space group Ic2m (No. 46) is body-centred.

7.4.2 Local structure

With respect to the above average model, electron diffraction and HREM experiments [7.59] indicate the presence of an orthorhombic superstructure. This lowering in symmetry textures the material into domains of two variants. The X-ray technique, being insensitive to local structural variations only allows one to determine a structure averaged over the two variants.

In the superstructure model determined in [7.59] L- and R- chains alternate within the CoO plane, leading to a doubling of the a_o lattice parameter: $a_s = 2a_o$; the other lattice parameters remain unchanged: $b_s = b_o$; $c_s = c_o$.

Mono-domain diffraction patterns along the simple crystallographic zones $[001]_o$ and $[2\bar{1}0]_o$ ($= [1\bar{1}0]_s$) are reproduced in Fig. 7.15 and in Fig. 7.16(b). The intense spots are the basic spots, produced by the perovskite blocks. The reflection condition $h + k + l = $ even is in accordance with the spacegroup

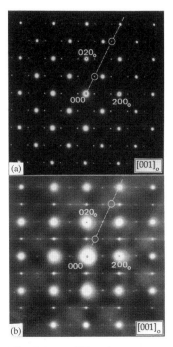

Fig. 7.15. Diffraction pattern of the 1212-Co phase along the [001] zone. The section through the streaks present in the $[2\bar{1}0]_o$ or $[1\bar{1}0]_s$ zone (Fig. 7.16), is indicated.

Ic2m. The weak spots in the $[001]_o$ zone (Fig. 7.15) and the closely separated rows of slightly elongated spots parallel to the c^*-axis, in the x $[2\bar{1}0]_o$ zone (Fig. 7.16(b)) are due to the orthorhombic superstructure.

The doubling of the lattice parameter along a_o ($a_s = 2a_o$), as required by the presence of the weaker spots in the $[001]_s$ and $[1\bar{1}0]_s$ diffraction patterns, is achieved in a natural way by assuming that L- and R-chains alternate in a given CoO-layer as shown in Fig. 7.14.

Successive layers of alternating chain sequences can be stacked with respect to the underlying perovskite block in four different positions. From Fig. 7.16 the c-axis ($c_s = 22.8$ Å) must be doubled as compared to that of the reference 123-compound ($c_p = 11.4$ Å). Since the perovskite blocks are stacked vertically, this must be achieved by a staggered configuration of the chain positions in the successive CoO layers, as observed in the high-resolution image of Fig. 7.16(a). The vertical stacking of chains can be excluded on this basis.

Doubling of the a_o and c_p parameters can be achieved simultaneously if the sublattice of chain positions has a B-centred unit cell and if, moreover, within each layer chains of different type alternate on this sublattice. A characteristic

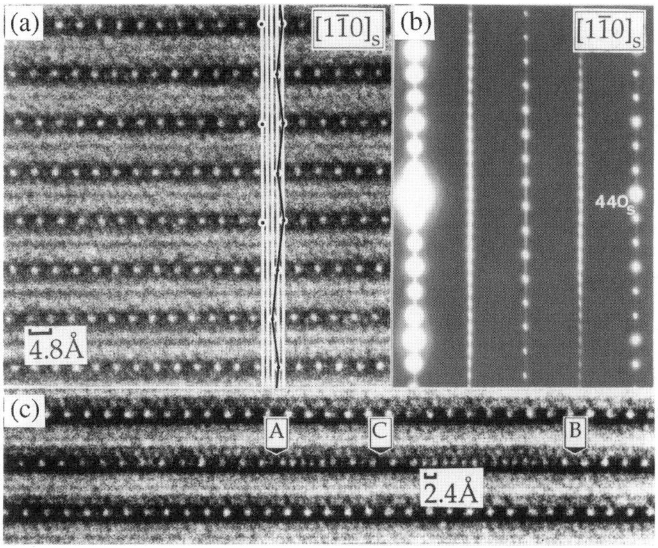

Fig. 7.16. High-resolution image of the Co-1212 phase along the $[2\bar{1}0]_o$ or $[1\bar{1}0]_s$ zone. Most atomic layers are imaged as lines, due to the small intercolumn separation. In the CoO layers the bright dots reveal the super-structure (spacing: 0.48 nm). In (a) the regular stacking can be observed in the upper five dot rows; in the next row a defect appears. In (b) the corresponding diffraction pattern is shown. In certain CoO layers, A to B in (c), the 0.24 nm spacing between bright dots is revealed, indicating the presence of an intralayer line defect. Note that the widely spaced dot arrays in A (or B) and in C are out of phase. The $[001]_s$ direction points upward.

feature of the $[001]_s$ diffraction pattern of Fig. 7.15, produced by a single variant of the Co-superstructure is that the superstructure reflections with $h = 4m + 2$ (m integer) for k even, as well as those with $h = 4m$ for k odd, are systematically absent. In the $[1\bar{1}0]_s$ section of Fig. 7.16 the rows with $h = k =$ odd contain superstructure spots for all l, whereas rows with $h = k = 4m + 2$ contain only spots for l odd. The rows with $h = k = 4m$ coincide with the spots due to the basic structure. It is easily shown that the systematic extinctions of the superstructure reflections are fully accounted for by the model proposed.

The model assumes a perfect $\cdots$ –L–R–L–R– $\cdots$ arrangement of parallel chains, $a_o = \frac{1}{2}a_s$ apart, in the CoO layers, the arrangements in successive CoO layers being related by a translation $\frac{1}{4}[122]_s$. Electron microscopic observations of the compound $(Y, Ce)_n Sr_2 GaCu_2 O_{5+2n}$, that has the CuO-chain plane replaced by a GaO plane, lead to the same superstructure model [7.59]. Due to the presence of a fluorite-like layer $(Y, Ce)_n O_{2n}$ instead of a Y-sheet, the

separation between successive GaO planes is larger than the separation between the equivalent planes in the previous compound. The resulting weaker
correlation between the GaO planes, leads to a highly defected compound,
introducing diffuse streaking in electron diffraction patterns. When the CuO-
chain plane is replaced by an AlO layer, corner-linked chains of AlO_4 tetrahedra meander along the $[100]_p$ direction in an intricate fashion. Correlations
along the c-direction in this compound are absent, leading to streaks along c^*
in $[h, k, 0]$ patterns. Experimental details, similar to those encountered for the
Co-doped compound, are discussed in [7.60].

7.5 New Hg-based superconducting materials

7.5.1 Pure Hg-compounds

Recently several higher-order members of the Hg-based superconducting
family $HgBa_2Ca_{n-1}Cu_nO_{2n+2+\delta}$ have been synthesized. The critical temperature reaches a maximum of 133.5 K for the $n = 3$ member of the family [7.62],
while for larger n values, T_c seems to decrease [7.63]. It was furthermore
shown (not very unexpectedly) that T_c increases steadily with pressure at a rate
of more or less 1 K/GPa. In this way, Chu *et al.* [7.64] and Nunez-Regueiro
et al. [7.65] were able to reach values such as 164 K and 155 K (T_c onsets) for
the Hg-1223 compound respectively. This strong pressure effect suggests that
by the appropriate chemical doping, which would introduce a chemical pressure, T_c values of the order of 150K are possible at normal external pressure.
Several substitutions in the Hg plane as well as at the Ba or the Ca positions
have been tried by the Caen group; they have led to the discovery of a series of
new Hg-based superconducting compounds, which can be prepared under
ambient pressure and which have a T_c up to 110 K [7.66–7.70].

The structure of all members is similar; they contain rocksalt-like slabs
$[(BaO)(HgO_\delta)(BaO)]$, alternating with perovskite slabs of the type
$[(CuO_2)(Ca)]_{n-1}(CuO_2)$. The structure for different n-compounds is represented in Fig. 7.17; for increasing n-values an extra $[(CuO_2)(Ca)]$ slab is
inserted, leaving the rest of the average structure unaltered. Although there is a
strong similarity between the Hg and the Tl series, the occupation of the
oxygen sites in the Hg layers and the corresponding one in the Tl layers, is
quite different.

HREM images of the different members of the family are obtained,
particularly along the [100] zone, which is the most instructive one; the cation
configuration can be readily identified from symmetry considerations. There is
clearly a one to one correspondence between the white dot configuration of the

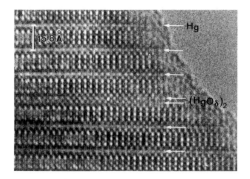

Fig. 7.17. HREM of an isolated defect in a Hg-1223 crystal, where a double layer of HgO$_\delta$ is present, creating locally the 2223 structure. Simulated images of this '2223' structure for a defocus value of 800 Å and thicknesses of 20 Å and 40 Å confirm the interpretation.

HREM image and the cation configuration of Fig. 7.17. These semi-intuitive interpretations have of course to be confirmed by computer simulations [7.76]. They allow us to analyse in detail the atomic structure of planar defects in these compounds. The most common defect encountered, particularly for higher-order members of the family, is the intergrowth of different *n*-members; they occur as isolated defects and are well known from other homologous series. Periodicities up to $n = 8$ have been seen to occur locally, although no single phase material could be produced [7.71]. A most remarkable defect, however, is shown in Fig. 7.18. The image contrast is such that the copper ions are imaged as bright dots, while the mercury plane is imaged as a more or less diffuse line with maxima at the projected atom positions. It is evident that locally at the position indicated, we have now four rocksalt layers rather than the usual sequence of three layers [(BaO)(HgO$_\delta$)(BaO)]. A single Hg layer is now replaced by a double Hg layer of bright dots. The occurrence of such defects with a double (HgO$_\delta$) layer suggested the feasibility of producing '2 2 $n - 1n$' type materials. Such '2 2 $n - 1n$' compounds have indeed been produced by replacing some of the Hg by Pr or Cu [7.72] or some of the Ca by Y [7.73].

   Another important contribution of electron microscopy is to assist chemists in the development of new superconducting compounds by identifying the phases produced and to correlate the results with the synthesis conditions. When attempts were made to produce Hg-containing compounds using different precursors, T_c always tended to be much lower (69 K for the Hg-1223 phase while the Hg-1234 phase does not superconduct) [7.74]. X-ray synchrotron measurements on powder material revealed a large deficiency on the Hg site,

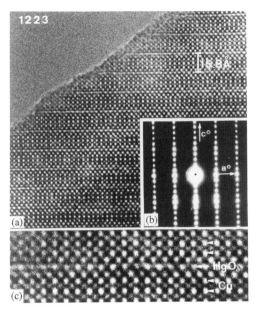

Fig. 7.18. (a) HREM image along [010] of the Hg-1223 compound, showing the individual cations as bright dots. No extended defects are observed. (b) Corresponding electron diffraction patterns. (c) High magnification of part of (a); note that in the HgO_δ layer the intensity of the individual dots is not homogeneous.

the occupancy factor was found to be about 0.7. The electron diffraction patterns only revealed the perfect Hg-1223 structure, without any observable deviation from perfection. Local microanalysis by EDX confirmed the low Hg content of 0.7; this deviation, however, was not compensated by an increase of any of the other cations, such as Cu, which might be expected from observations of Cu-1223 type structures [7.75, 7.76]. On the HREM image, which is particularly sensitive to any local ordering or local substitutions in the (HgO_δ) layer, no such effects can be observed. In the image of Fig. 7.18 all cations are revealed as bright dots and the (HgO_δ) layer can be easily identified as the more diffuse layer; this follows from comparison with computer simulated images. Along this layer, however, we do see variations in the intensity of the different bright dots (Fig. 7.18(c)). We can quantify these variations by making densitometer traces along the (HgO_δ) layer as well as along one of the neighbouring CuO_2 layers. The results are shown in Fig. 7.19, where we have plotted the intensity for different bright dots along a single HgO_δ row and labeled them from 1 to 34; the intensities are corrected for the background. For the (HgO_δ) layer there is a clear pseudo-periodic variation of the intensity

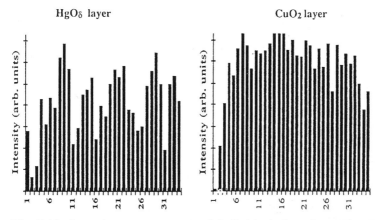

Fig. 7.19. Intensity measurement of individual dots from Fig. 7.18(c); (a) along the HgO$_\delta$ layer; (b) along the neighbouring CuO$_2$ layer.

(Fig. 7.19(a)). Such a variation is not observed in the corresponding measurement for the surrounding BaO or CuO$_2$ layer (Fig. 7.19(b)). These intensity variations are to be related to variations in the Hg sublattice. If the variations were not random but periodic, they would give rise to weak superstructure reflections, visible in the diffraction pattern. If complete columns, or even a single column, were occupied by carbonate groups, this would be detected by HREM [7.77].

7.5.2 Derived Hg-compounds

Pure mercury superconducting materials can only be obtained under high pressure. However, based on the compounds described above, a number of derived structures can be synthesized at ambient pressure when part of the Hg is replaced by other elements. They can be subdivided into three subclasses.

1. Superconducting materials based on the 1201 structure

Attempts to substitute Ba for Sr into the 1201 structure and to stabilize a HgSr$_2$CuO$_{4+\delta}$ type structure are only successful if part of the Sr is replaced by La and some of the Hg is replaced by other elements such as Pb or Pr [7.78, 7.79]. The 1201 structure is very flexible and all these substitutions do not change the structure or the symmetry. Only when Pr (or Ce) is substituted on the Hg sublattice in a ratio close to 1:1, ordering occurs between Hg and Pr along the [100] direction. The diffraction patterns as well as the HREM images strongly resemble those corresponding to the Ortho-II structure observed for the oxygen-deficient YBa$_2$Cu$_3$O$_{7-\delta}$ material. HREM, however, allows us to deduce unambiguously that the ordering takes place in the Hg-O layer; the combination of EDX, HREM and ED allows the superstructure formation to be identified as an ordering between Hg and Pr on the Hg sublattice [7.80].

2. Superconducting materials based on the 1212 structure

Structurally the same results hold as for the 1201 materials; the 1212 structure allows a large number of substitutions without drastically changing its proper-

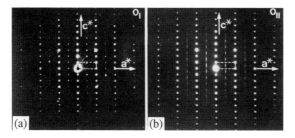

Fig. 7.20. The [010] electron diffraction patterns of the Ortho-I superstructure (a) and the Ortho-II superstructure (b) in $Hg_{0.4}Pr_{0.6}Sr_{1-x}Pr_xCu_2O_{6+\delta}$.

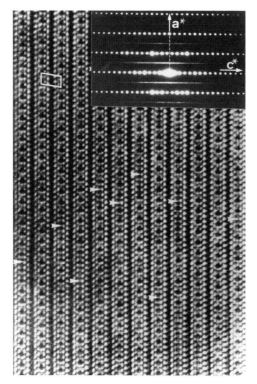

Fig. 7.21. HREM image of a Hg-Pr 1222 compound, showing ordering between Hg and Pr, but a large number of 'antiphase boundaries', indicated by arrowheads, are present.

ties. The highest T_c reached for the Ba-based compounds is 110 K [7.81], while for the Sr-based compounds it is worthwhile noting that, for example, the compound $Hg_{0.4}Pr_{0.6}Sr_{1-x}Pr_xCu_2O_{6+\delta}$ is superconducting below 85 K and has excellent magnetic properties. Again ordering takes place between Hg and Pr along the [100] direction and two different superstructures may be formed, as can be seen from the [010] patterns of Fig. 7.20. HREM evidence of the ordering in the (Hg–Pr) plane is provided in Fig. 7.21. Simulated images, based on an ordering model where only Hg is replaced by Pr, confirm that the ordering takes place in the (Hg–Pr) plane but fail to reproduce exactly the experimental images, proving that non-negligible displacements of cations in neighbouring layers are associated with this ordering.

3. Intergrowths of Hg-1201 with other superconducting materials

Tl^{+III} and Hg^{+II} are very similar ions; structurally they can be intimately mixed so as to form an intergrowth of Hg-1201 and Tl-2201. The resulting material is superconducting though at temperatures below the transition temperature of the individual compounds. More details and other examples can be found in [7.82].

7.5.3 Hg-based oxycarbonates

A wide variety of novel superconducting materials have recently been prepared by incorporating complex ions such as carbonate, phosphate or sulphate groups in the perovskite-like structures of existing high T_c cuprates. For an overview and further references we refer to [7.77, 7.82, 7.83]. Carbonate groups can be substituted into the mercury or thallium based cuprates of the type $Hg_{1-x}Tl_xSr_{4-y}Ba_yCu_2CO_3O_{7-\delta}$. The incorporation of these groups produces interesting modulated structures. Related compounds of the type $ASr_{4-y}Ba_yCu_2CO_3O_{7-\delta}$ where $A = Tl$, Hg or $Hg_{0.5}Pb_{0.5}$ were also found to exhibit remarkable long-period interface-modulated structures. The modulation vector is along the [010] perovskite direction in the samples with $A = Tl$ and along the $[110]_p$ direction in the samples with $A = Hg$ or $A = Hg_{0.5}Pb_{0.5}$. The wavelength is mostly incommensurate but varies between six and eight times the basic perovskite unit.

High-resolution electron microscopy of the compound

$$Hg_{1-x}Tl_xSr_{4-y}Ba_yCu_2CO_3O_{7-\delta}$$

is able to image directly the modulation (Fig. 7.22). Together with the electron diffraction information (inset of Fig. 7.22), one can also deduce the origin of the modulation and the building blocks of the complex structure. The prominently bright dot sequences image columns of carbon and oxygen atoms which are the

Fig. 7.22. HREM image of $Hg_{0.3}Tl_{0.7}Sr_{1.5}Ba_{2.5}Cu_2CO_3O_{7-\delta}$ along [100]; the carbonate groups are imaged as particularly white features. The corresponding diffraction pattern is shown as an inset. An enlarged image and comparison to calculated images allows us to deduce the atomic structure and to locate the carbonate groups.

lightest atoms in the compound. The carbonate groups clearly substitute on the (Hg–Tl) sublattice, in agreement with previous findings. The carbonate sequences contain either three or four dots. Bright and dark dot sequences alternate in anti-phase in successive layers, leading to a centred rectangular arrangement. The corresponding [100] zone diffraction pattern (inset of Fig. 7.22) consists of main reflections, which can be indexed on a tetragonal lattice, and weaker satellite reflections, which decrease in intensity with distance from the main reflections. The pattern looks commensurate; however, careful measurements show that it is in fact incommensurate, the period being slightly smaller than $8b_p$. Actually the periodicity of the modulation can vary between $6b_p$ and $8b_p$, depending on a number of parameters, which will not be discussed here.

The high-resolution images of these compounds not only suggest a model for the stacking along the *c*-direction:

$$[CO + (Tl, Hg)O] - (Sr, Ba)O - CuO_2 - (Sr, Ba)O - [(Tl, Hg)O + CO]$$

but also for the modulated structure within the $[CO + (Tl, Hg)O]$ plane. Incommensurate diffraction patterns originate from a structure in which commensurate carbonate strips of two different widths (three and four times the perovskite unit) are uniformly mixed. The high-resolution images (e.g. Fig. 7.17) give evidence for sequences such as 3434... or 334334....

As to the near future of HREM and the imaging of light elements in electron microscopy, we can announce that we have just installed a new CM30-FEG-Ultra twin microscope at the EMAT laboratory in Antwerp, which is able to produce structural information down to the 1.1 Å level. The interpretation of such images in terms of the projected crystal potential is certainly not straightforward, but with the help of CCD recording and computer treatment of the data, we are starting real quantitative HREM [7.84].

Acknowledgments

We are most grateful to S. Amelinckx (Antwerp) and C. Greaves (Birmingham) for common research on the YBCO-based compounds and to M. Huvé (now in Lille), M. Hervieu, C. Michel and B. Raveau (Caen), C. Chaillout, M. Marezio (Grenoble), E. V. Antipov (Moscow) for valuable discussions and use of common results on the Hg-based compounds and the oxycarbonates.

References

[7.1] R. J. Cava *et al.*, *Physica C* **165**, 419 (1989).
[7.2] A. Ourmazd & J. C. H. Spence, *Nature* **329**, 425 (1987).
[7.3] Y. P. Lin *et al.*, *J. Sol. Stat. Chem.* **84**, 226 (1990).
[7.4] C. Chaillout *et al.*, *Phys. Rev. B* **36**, 7118 (1987).
[7.5] Y. Zhu *et al.*, *Physica C* **167**, 363 (1990).
[7.6] C. H. Chen *et al.*, *Phys. Rev. B* **38**, 2888 (1988).
[7.7] C. N. R. Rao *et al.*, *Phys. Rev. B* **42**, 6765 (1990).
[7.8] C. P. Burmester *et al.*, Mat. Res. Soc. Symp. Proc. **183**, 369 (1990).
[7.9] J. L. Hodeau *et al.*, *Physica C* **153-155**, 582 (1988).
[7.10] R. Beyers *et al.*, *Nature* **340**, 619 (1989).
[7.11] M. A. Alario-Franco *et al.*, *Mat. Res. Bull.* **22**, 1685 (1987).
[7.12] M. A. Alario-Franco *et al.*, *Physica C* **156**, 455 (1988).
[7.13] C. J. Hou *et al.*, *J. Mater. Res.* **5**, 9 (1990).

[7.14] C. Chaillout *et al.*, *Sol. Stat. Comm.* **65**, 283 (1987).

[7.15] M. Hervieu *et al.*, *Mat. Lett.* **8**, 73 (1989).

[7.16] G. Van Tendeloo, H. W. Zandbergen & S. Amelinckx, *Sol. Stat. Comm.* **63**, 603 (1987).

[7.17] R. M. Fleming *et al.*, *Phys. Rev.* B **37**, 7920 (1988).

[7.18] H. You *et al.*, *Phys. Rev.* B **38**, 9213 (1988).

[7.19] T. Zeiske *et al.*, *Physica C* **194**, 1 (1992).

[7.20] Y. Nakazawa & M. Ishikawa, *Physica C* **158**, 381 (1989).

[7.21] Y. Nakazawa & M. Ishikawa, *Physica C* **162-164**, 83 (1989).

[7.22] Y. Ueda & K. Kosuge, *Physica C* **156**, 281 (1988).

[7.23] R. J. Cava *et al.*, *Phys. Rev. B* **36**, 5719 (1987).

[7.24] H. Strauven *et al.*, *Sol. Stat. Comm.* **65**, 293 (1987).

[7.25] P. Gerdanian, C. Picard & B. Touzelin, *Physica C* **182**, 11 (1991).

[7.26] C. Nobili *et al.*, *Physica C* **168**, 549 (1990).

[7.27] D. J. Werder *et al.*, *Phys. Rev. B* **37**, 2317 (1988).

[7.28] J. Reyes-Gasga *et al.*, *Physica C* **159**, 831 (1989).

[7.29] G. Van Tendeloo & S. Amelinckx, *J. El. Mic. Technique* **8**, 285 (1988).

[7.30] E. A. Hewat *et al.*, *Nature* **327**, 400 (1987).

[7.31] C. N .R. Rao & J. Gopalakrishnan, *New directions in solid state chemistry,* Cambridge Solid State Science Series, (Cambridge University Press, 1986).

[7.32] D. Van Dyck, C. Condé & S. Amelinckx, *Phys. Stat. Sol.* (a) **56**, 327 (1979).

[7.33] A. G. Khachaturyan & J. W. Morris, Jr., *Phys. Rev. Lett.* **64**, 76 (1990).

[7.34] S. Yang *et al.*, H, *Physica C* **193**, 243 (1992).

[7.35] S. Rusiecki *et al.*, *J. Less Common Metals,* **164–165**, 31 (1990).

[7.36] R. Cloots *et al.*, *J. Cryst. Growth* **129**, 394 (1993).

[7.37] D. J. Werder *et al.*, *Phys. Rev. B* **38**, 5130 (1988).

[7.38] A. G. Khachaturyan & J. W. Morris, *Phys. Rev. Lett.* **61**, 215 (1988).

[7.39] D. de Fontaine, G. Ceder & M. Asta, *Nature,* **343**, 544 (1990).

[7.40] G. Van Tendeloo *et al.*, *Sol. Stat. Comm.* **63**, 969 (1987).

[7.41] R. Sonntag *et al.*, *Phys. Rev. Lett.* **66**, 1497 (1990).

[7.42] R. Sonntag, Th. Zeiske & D. Hohlwein, *Phys. B* **180–181**, 374 (1992).

[7.43] A. A. Aligia, *Europhys. Lett.* **18**, 181 (1992).

[7.44] S. Semenovskaya & A. G. Gotchaturyan, *Phys. Rev. B* **46**, 6511 (1992).

[7.45] A. A. Aligia, J. Garces & H. Bonadeo, *Physica C* **190**, 234 (1992).

[7.46] S. Semenovskaya & A. G. Khachaturyan, *Phil. Mag. Lett.* **66**, 105 (1992).

[7.47] D. J. Werder, C. H. Chen & G. P. Espinoza, *Physica C* **173**, 285 (1990).

[7.48] T. Krekels, *et al.*, *Physica C* **167**, 677 (1990).

[7.49] M. Tetenbaum *et al.*, *HTS-physics and materials science*, NATO Advanced Study Institute, 13–26, August, Bad Windsheim, Germany (1989).

[7.50] D. Van Dyck *et al.*, *Phys. Stat. Sol* (a) **43**, 541 (1977).

[7.51] M. Verwerft *et al.*, *Appl. Phys. A* **51**, 332 (1990).

[7.52] J. D. Jorgensen *et al.*, *Phys. Rev. B* **36**, 3608 (1987).

[7.53] T. Krekels, S. Kaesche & G. Van Tendeloo, *Physica C* **248**, 317 (1995).

[7.54] J. Y. Genoud *et al.*, *Physica C* **192**, 192 (1992).

[7.55] J. Karpinski *et al..*, *Physica C* **161**, 618 (1989).

[7.56] T. Zeiske *et al.*, *Physica C* **207**, 333 (1993).

[7.57] G. Roth *et al.*, *J. de Physique* **1**, 721 (1991).

[7.58] Q. Huang *et al.*, *Physica C* **193**, 196 (1992).

[7.59] T. Krekels *et al.*, *J. Sol. Stat. Chem.* **105**, 313 (1993).

[7.60] O. Milat *et al.*, *Physica C* **217**, 444 (1993).

[7.61] T. Krekels *et al.*, *Physica C* **196**, 363 (1992)
[7.62] A. Schilling *et al.*, *Nature* **363**, 56 (1993).
[7.63] E. V. Antipov *et al.*, *Physica C* **215**, 1 (1993)
[7.64] C. W. Chu *et al.*, *Nature* **365**, 323 (1993).
[7.65] M. Nunez-Regueiro *et al.* *Science* **262**, 97 (1993).
[7.66] A. Maignan *et al.*, *Physica C* **212**, 239 (1993).
[7.67] M. Hervieu *et al.*, *Physica C* **216**, 264 (1993).
[7.68] D. Pelloquin *et al.*, *Physica C* **216**, 257 (1993).
[7.69] C. Martin *et al.*, *Physica C* **212**, 274 (1993).
[7.70] A. Maignan *et al.* *Physica C* **216**, 1 (1993).
[7.71] G. Van Tendeloo *et al.*, *Physica C* **223**, 219 (1994).
[7.72] C. Martin *et al.*, *Solid State Comm.* **93**, 53 (1995).
[7.73] P. G. Radaelli *et al.*. *Physica C* **235**, 925 (1994).
[7.74] E. M. Kopnin *et al.*, *Physica C* **243**, 222 (1995).
[7.75] M. Alario-Franco *et al.*, *Physica C* **222**, 52 (1994).
[7.76] X. J. Wu *et al.*, *Physica C* **224**, 69 (1994).
[7.77] M. Huvé *et al.*, *Physica C* **231**, 15 (1994).
[7.78] F. Goutenoire *et al.*, *Physica C* **216**, 243 (1993)
[7.79] D. Pelloquin *et al.*, *Physica C* **216**, 257 (1993).
[7.80] G. Van Tendeloo *et al.*, *J. Solid State Chemistry,* **114**, 369 (1995).
[7.81] A. Maignan *et al.*, *Physica C* **216**, 1 (1993).
[7.82] B. Raveau *et al.*, *J. Superconductivity,* **7**, 9 (1994).
[7.83] M. Hervieu *et al.*, *Physica C* **235–240**, 25 (1994).
[7.84] D. Van Dyck, M. Op de Beeck & W. Coene, *Optik* **93**, 103 (1994).

8

Valence band electron energy loss spectroscopy (EELS) of oxide superconductors

Y. Y. WANG and V. P. DRAVID

8.1 Introduction

Transmission electron energy loss spectroscopy (EELS) consists of measuring energy loss dispersion of inelastically scattered high-energy electrons transmitted through a thin film. The high-energy electrons, which interact with the electrons in the solid, lose a certain amount of energy and transfer momentum to the solid. Because of the energy and momentum conservation rules, energy loss and the corresponding momentum-transfer (q) of the probed electron represents the energy and the momentum of the electronic excitations in solids.

Although some optical techniques, such as soft X-ray absorption and optical reflectance measurements, provide comparative information about solids with higher energy resolution, EELS enjoys several unique advantages over optical spectroscopies. First of all, unlike optical reflectance measurements which are sensitive to the surface condition of the sample, the transmitted EELS represents the bulk properties of the material. Secondly, EELS spectra can be measured with q along specific controllable directions and thus, can be used to study the dispersion of plasmons, excitons, and other excitations [8.1–8.5]. Such experiments offer both dynamics as well as symmetry information about the electronic excitations in solids. In addition, the capability to probe the electronic structure at finite momentum-transfer also allows one to investigate the excited monopole or quadrupole transitions, which cannot be directly observed by conventional optical techniques limited by the dipole selection rule.

Because of the significant energy spread of conventional TEM electron sources (e.g. LaB$_6$, W-hairpin filaments with $\Delta E \sim 1$–2 eV), EELS measurements to investigate the electronic structure of solids have been generally limited to dedicated electron energy loss spectrometers with energy resolutions ~ 0.1 eV. These require self-supporting thin films with areas on the scale of square millimeters and ~ 1000 Å thickness; hence requiring difficult sample preparation procedures [8.3, 8.5, 8.6]. With the advent of commercial cold field emission gun (cFEG) transmission electron microscope (TEM) and scanning TEM (STEM), the energy resolution that can be achieved with EELS in a transmission electron microscope is ~0.5 eV or better. In most cases, this energy resolution is high enough to provide meaningful information on the electronic structure in solids. In addition, because of the small probe size (ranging from millimeters to < 1 nm) offered by these microscopes, the EELS measurement of a 'single crystal' becomes more practical for materials which may not be available in large single crystal sizes. Furthermore, the ability of TEM to manipulate the diffraction pattern at high momentum resolution offers a distinct advantage over a dedicated STEM equipped with similar bright FEG sources [8.7, 8.8], especially for low loss EELS.

There have been several articles which review EELS studies of solids [8.1–8.5]. For the EELS study of oxide superconductors, most of the research has been concentrated on the oxygen core level excitations, where the information about the density of hole carriers and their crystallographic confinement can be obtained [8.9]. Unlike the core level EELS studies, only a few low loss EELS studies have been reported for oxide superconductors [8.6, 8.10, 8.11].

While the core level EELS studies of oxide superconductors have provided rich information on the electronic structure and the evolution of charge carriers [8.12, 8.13], we believe that valence band EELS of these compounds also provides equally important information, if not more. First of all, it is now understood that the symmetry of the electronic structure may be critical in understanding the pairing mechanism, due to the recent discovery of the symmetry of the superconductivity gap [8.14], which leads to the proposal of the d-wave pairing in high T_c superconductors. Secondly, because it is understood that the electronic structure of the oxide material is a very strongly correlated system, it will be very interesting to learn the dynamics of the exciton as well as free-carrier plasmon. By angle-resolved photoemission spectroscopy, the unusual dynamics of a single hole in the oxide superconductor as well as in the undoped cuprate has been reported [8.14, 8.15]. However, unlike photoemission, which probes the dynamics of a single hole, electron energy loss spectroscopy can be used to probe the dynamics of electron-hole excitations (excitons), where Coulombic interactions play an important role. These issues can be well addressed by low loss EELS, as outlined in this chapter. Specifically, the anisotropy of bonding states (see Fig. 8.1) can be addressed with momentum-transfer resolved EELS.

Some of the low loss EELS work on high T_c superconductors using a dedicated EELS spectrometer has been reported earlier by several groups [8.6, 8.10, 8.16, 8.17]. However, here, we will only report the recent low loss EELS experiments on high T_c oxide materials using a cFEG TEM.

σ **Bond**   **In-plane** π **Bond**   **Out-of-plane** π **Bond**

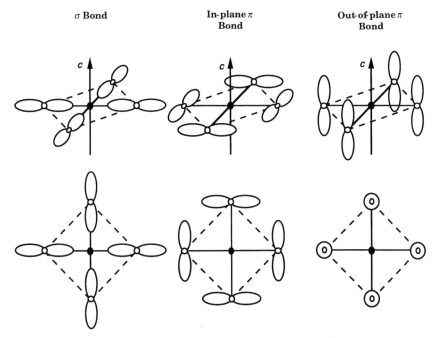

Fig. 8.1. The oxygen σ state, in-plane π state and out-of-plane π state.

8.2 Experimental

Self-supported TEM samples for the oxide superconductors were prepared by mechanical polishing followed by ion beam thinning (IBT) at liquid nitrogen temperature to reduce the ion beam damage of the sample. Some of the results were also confirmed again by using crushed powder dispersed on holey carbon film to assess the effect of ion beam thinning. EELS experiments were deliberately performed in the thinnest region of the specimen to minimize multiple scattering. If found necessary, the spectra were corrected for multiple scattering using the approach described elsewhere [8.18].

EELS experiments were carried out using a 200 kV cold field emission TEM (Hitachi HF-2000), equipped with a Gatan 666 parallel EELS spectrometer interfaced via the Gatan EL/P software to a PowerMac. The microscope also has an Oxford Systems' ultrathin window (UTW) energy dispersive X-ray (EDX) detector for X-ray microchemical analysis. Local chemistry was always confirmed by X-ray microanalysis prior to the EELS measurements. All the data presented in this chapter correspond to defect-free regions of appropriate composition. Experiments were carried out by selecting a smaller collection angle with a controlled momentum-transfer as described later.

Considerable care was taken to minimize radiation dose by employing beam spreading and minimum dose techniques. EELS data were collected using the energy dispersion of 0.05 eV/channel, with the zero loss beam resolution (FWHM) of ~0.5 eV. The EELS spectra were corrected for dark current and readout noise. The channel-to-channel gain variation was eliminated by normalizing the experimental spectra with independently obtained gain spectra of the spectrometer or by adding several channel-shifted spectra during the measurements to average the gain variation. The momentum resolution for valence plasmon measurement is selected as ~0.04 Å^{-1} by an EELS spectrometer aperture of ~1 mm in diameter, and a camera length of 1.6 m.

8.3 Anisotropic dielectric function of cuprates

Anisotropic dielectric properties of materials have been known for quite a while. Most high T_c materials have an anisotropic electronic structure owing to the anisotropy of their crystal structures (e.g. flat CuO_2 planes). However, because the usual optical measurement techniques require large single crystals with flat faces (millimeter size), only a few anisotropic dielectric functions have been measured for cuprate superconductors, which are often difficult to grow to larger sizes with chemical homogeneity.

Earlier attempts to study anisotropic dielectric functions using electron energy loss spectroscopy relied mostly on studying the core level excitations, where the advantage of the perpendicular relation between q_e and $q_\perp$ could be used in the measurements. One example of such studies is the investigation of the O K-edge core level excitations in and out of the CuO_2 plane in $Bi_2Sr_2CaCu_2O_8$ [8.19].

For valence band excitations, however, a different method has to be used to study the anisotropic dielectric properties due to the smaller q_e (~0.02 Å^{-1} for 20 eV energy loss with 100 kV incoming beam energy), which means that the momentum transfer in the low loss region is mostly due to $q_\perp$. This requires the sample to be oriented in a certain direction relative to the direction of the electron beam to study the dielectric response in a particular orientation. Because of the small beam size available in cFEG TEM, a specific orientation of a thin section with millimeter size area can be easily obtained by ion beam thinning techniques to directly measure the anisotropic properties of the material.

8.3.1 Free carrier plasmon in the CuO$_2$ plane

It is well known that for all Cu—O high T_c compounds, the electrical conductivity is much higher within the *ab* plane than out of it. For example, by measuring the infrared reflectance on single crystals (millimeter size) of $La_{1-x}Sr_xCuO_4$, Collins *et al.* [8.20] showed that the free carrier plasmon is confined to the *ab* plane. $YBa_2Cu_4O_8$ (Y124) has a similar CuO_2 plane as that of $La_{1-x}Sr_xCuO_4$, in addition to Cu—O chains which are believed to supply the hole carriers to the CuO_2 plane.

For anisotropy measurements, the $YBa_2Cu_4O_8$ (Y124) crystal was oriented in such a way that either the *a*- or *b*-axis was parallel to the electron beam [8.21]. The resultant diffraction pattern in reciprocal space thus contains b^* (or a^*) and c^* directions. The EELS data were obtained by varying the relative position of the EELS spectrometer and the diffraction pattern. Figure 8.2(a) shows comparative EELS spectra of Y124 with momentum transfer (q) confined to the *c*-direction and within the *ab*-plane. Two low energy excitations are observed in the EELS spectra for momentum transfer confined within the *ab*-plane: one at ~1.4 eV which is the free carrier plasmon and another excitation at about 3.2 eV. The free carrier plasmon almost disappears when the momentum transfer is along the *c*-direction. This clearly implies that the free charge

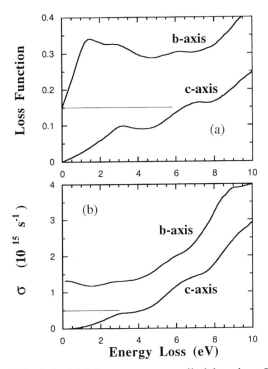

Fig. 8.2. (a) Momentum-controlled low loss EELS spectra of Y124, within the *ab*-plane versus the *c*-direction; (b) a.c. conductivity spectra in the *ab*-plane versus the *c*-direction.

carriers in Y124 are confined largely to the *ab*-plane. This result, though not surprising, confirms the earlier O K-edge X-ray absorption spectroscopy evidence [8.22] for the anisotropy of the charge carriers in Y124. The anisotropy is well reflected in the a.c. conductivity spectra of Y124, along the *c*-direction versus within the *ab* plane, as can be seen in Fig. 8.2(b).

8.3.2 Confinement of the Hubbard transition in the CuO_2 plane in the infinite-layer compound

The compound, $ACuO_2$ (A = Sr, Ca), is the simplest of all the copper-oxide superconductor structures with only CuO_2 planes separated by A-cation planes (A=Sr/Ca cations) and it is a semiconductor [8.23–8.25]. Because this compound can be considered as the limiting compound in the $Ba_2Sr_2Ca_{n-1}Cu_nO_y$ ($n = 1, 2, 3$) and $Tl_2Ba_2Ca_{n-1}Cu_nO_y$ ($n = 1, 2, 3$) family of compounds as the number of CuO_2 layers goes to infinity, it is also referred to as the infinite-layer compound. It has been reported that the infinite-layer compound exhibits both *n*- and *p*-type superconductivity: donor doping with La (for Sr) creates *n*-type superconductivity with a T_c of ~42 K [8.25], while oxygenation under high pressure is claimed to give rise to *p*-type superconductivity for non-stoichiometric $(Sr_{1-x}Ca_x)_yCuO_2$ ($y = 0.90$, $x = 0.30$) with a T_c of ~110 K [8.24]. By atomic-resolution TEM imaging, simulation and nano-probe energy dispersive X-ray analysis, the detailed defect structure of *p*-type doping superconductor material was recently identified as composed of three layer Sr—O blocks, which may provide hole carriers into the Cu—O planes [8.26].

Unlike other copper oxide superconductors which have higher c/a ratio, the c/a ratio in the non-defective infinite-layer compound is close to 1. However, because of a lack of apical oxygen, the electronic structure is expected to be more anisotropic than all other high T_c oxides, despite the similar *c* and *a* lattice constants of the unit cell. This supposition was recently confirmed by X-ray absorption spectroscopy (XAS) measurements [8.27], in which an anisotropic upper Hubbard band was reported. In this section, we will discuss the question of whether the electronic transitions near the Fermi surface behave anisotropically with respect to the *ab*-plane and the *c*-axis.

It is now known that the excitation at ~3 eV in cuprate superconductors is due to a strong hybridization between Cu 3*d* and O 2*p* states (σ), where the σ state is parallel to the Cu—O bond as shown in Fig. 8.1. For O 2*p* state, there are two 2*p* π states, which are perpendicular to the Cu—O bond direction and which are weakly hybridized with Cu 3*d*. One of the π states is perpendicular to the Cu—O plane and the other is in the Cu—O plane, but perpendicular to the Cu—O bond. Because of the lack of apical oxygen in the defect-free infinite-layer compound, it is an ideal sample for studying the anisotropic dielectric function in this energy region.

Figure 8.3 shows low loss EELS spectra obtained with the momentum-transfer (q) confined within the *ab*-plane and along the *c*-axis [8.28]. Significant differences in these spectra are visible below 10 eV. The spectrum from the *ab*-plane has two primary excitations, one at ~2.4 eV and another at 6 eV, while the higher energy portions of both the spectra are dominated by volume plasmon and other collective excitations. The EELS spectra recorded along the *c*-direction are devoid of any excitations at ~2.4 eV. Dielectric functions were calculated through Kramers–Kronig analysis. The real part of the dielectric function at zero energy (ε_0) was used to normalize the energy loss spectra. The value for ε_0 in the *ab*-plane was estimated to be 4.3, consistent with the Cu—O bond length ($a = 3.99$ Å) based on previous work on Cu—O compounds [8.29]. The value for ε_0 along the *c*-axis was calculated by fitting ε_0 in the *ab*-plane and along the *c*-axis with the Lorentzian oscillator. Self-consistent calculation for ε_0 along the *c*-axis was used, and the value obtained was 3.3. Energy loss spectra along the *c*-axis and within the *ab*-plane for these ε_0 values satisfy the *f*-sum rule, which indicates that the normalized constants were properly chosen.

Because the imaginary part of the dielectric function can be directly associated with the interband transitions, only ε_2 spectra along the *ab*-plane and the *c*-axis are shown in Fig. 8.4. Significant differences in ε_2 values between the *ab*-plane and the *c*-axis can be seen below 8 eV, but above 8 eV they appear to be quite similar. The O K-edge is also shown in Fig. 8.4. The

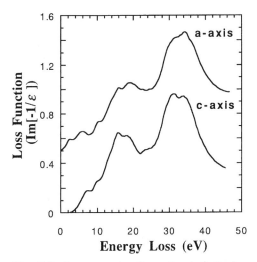

Fig. 8.3. Loss spectra from the infinite-layer compound, along the *ab*-plane (top line) and *c*-axis (bottom line). Note the clear differences between the spectra below 10 eV.

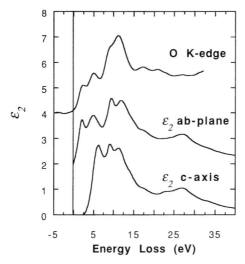

Fig. 8.4. Imaginary part of the dielectric function for the infinite-layer compound calculated from the loss spectra. The O K-edge loss spectrum is also given for comparison.

energy scale for the O K-edge was aligned using the second peak in the K-edge spectra and the ε_2 spectrum for the *ab*-plane. It is worth noting that the energy of the excitation in the ε_2 spectrum is different from the peak position in the loss function. This is a clear example that we cannot use the energy loss spectrum alone to decide the energy of the excitations.

The first excitation in Figure 8.4 along the *ab*-plane is associated with the CuO_2. In all of the undoped cuprates, there is a sharp excitation below the broad ~3 eV excitation. In the present case, due to the energy resolution (0.5 eV) of the spectrometer, the ε_2 spectrum for the *ab*-plane represents a mixture of the sharp excitation and the broad excitation. The anisotropic dielectric function indicates that the Hubbard band lies mostly in the *ab*-plane. This is consistent with the very anisotropic O K-pre-edge data reported from XAS measurements, which indicate that only $1 \pm 1\%$ of the upper Hubbard band is out of the *ab*-plane for the infinite-layer compound [8.27]. The reason for this significant anisotropy in the Hubbard band for the infinite-layer compound is clearly associated with the lack of apical oxygen in the structure.

8.3.3 Hybridization between O σ and Cu 3d states in the Cu—O chains

As indicated in the previous section, the excitation at 3 eV for cuprates is very anisotropic, i.e. stronger in the Cu—O plane, but weaker out of the Cu—O

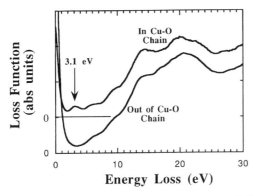

Fig. 8.5. Electron energy loss spectra of Ca_2CuO_3 along the Cu—O chain and out of it.

plane. However, such experiments certainly cannot differentiate whether it is the O σ or in-plane π state which is hybridized with the Cu $3d$ state, resulting in the strong 3 eV excitation in the Cu—O plane. In order to pinpoint the direction of the excitation, pure Cu—O chain material can be used to carry out this experiment. One such material is the low-pressure phase of Ca_2CuO_3 (Sr_2CuO_3). In these materials, only Cu—O chains are present. Figure 8.5 shows energy loss spectra in the Cu—O chain and out of the Cu—O chain. The profile of the high-energy portion of the spectra is very similar. In the spectrum along the Cu—O chain direction, there is a strong excitation at $\sim$3.1 eV which is much weaker in the spectrum away from the Cu—O chain direction. It can be interpreted as a transition which involves hybridization between Cu $3d$ and O $2p\,\sigma$ states.

8.4 Momentum-transfer (q) resolved electron energy loss spectroscopy

In the above sections, we reported the use of electron energy loss spectroscopy to study the anisotropic properties of the dielectric function in high T_c materials. These studies are all limited to the small angle scattering regime, which means that energy loss data are comparable to the optical data. However, large momentum-transfer EELS can provide additional information, such as dispersion of the free carrier plasmon, effective mass of the exciton, and optically forbidden transitions, all of which are considered to be crucial to the electronic structure of solids. By mapping dispersion properties as well as optically forbidden transitions in momentum space, we can deduce symmetry information about the electronic structure. In this section, we will report on the use of momentum-transfer resolved electron energy loss spectroscopy to obtain the symmetry of the electronic excitations in $BaBiO_3$ as well as in $Sr_2CuO_2Cl_2$.

Nucker *et al.* [8.16] first reported anisotropic dispersion of the free carrier plasmon in $Bi_2Sr_2CaCu_2O_8$. Here, we will only discuss the anisotropic dispersion of the insulator case not only because of its simplicity but also for the subsequent understanding of the insulator-to-metal transition in the normal state of these compounds that occurs with doping.

8.4.1 *Symmetry of the electronic excitations in BaBiO₃*

$BaBiO_3$ is the parent compound of $Ba_{1-x}K_xBiO_3$ and $Ba_{1-x}Pb_xBiO_3$ oxide superconductors [8.30, 8.31]. With an odd number of electrons in the valence band, $BaBiO_3$ is an insulator despite the metallic prediction of the band structure calculations [8.32]. Owing to the different Bi—O bond lengths, a model involving charge density wave (CDW) instability was proposed to explain the insulating nature of $BaBiO_3$ [8.33]. The optical spectroscopy investigations have assigned the observed excitation near 2 eV as a transition across the CDW gap [8.34–8.36].

In this study, we have employed momentum-transfer (q) resolved EELS to probe the momentum-transfer dispersion of the valence band excitations in $BaBiO_3$. We report an optically forbidden transition (~4 eV) and the effective mass of the optical gap (~2 eV), both of which exhibit crystalline anisotropy.

Figures 8.6(a) and (b) are the energy loss spectra at different q for $BaBiO_3$ along [100] and [110], respectively [8.37]. As q approaches zero, a pronounced excitation at ~2.5 eV is seen in both spectra, which corresponds to the ~2 eV excitation observed in optical spectroscopy [8.34–8.36, 8.38]. With increasing q, the energy position of the 2.5 eV excitation along the [100] direction remains almost stationary; whereas the energy position of the same excitation along the [110] direction disperses towards higher energy. The dispersion relationships with q of the excitation along [100] and [110] directions are shown in Fig. 8.7(a), which clearly indicates that the dispersion of the dipole-allowed transition is very strongly anisotropic.

In addition to the threshold excitation, another broad excitation appears at about 4.5 eV with increasing q, as can be seen in Fig. 8.6. The oscillator strength of this feature is stronger along [100] than along [110]. The oscillator strengths of the forbidden transition for different momentum-transfers are plotted against q^2 in Fig. 8.7(b), for both [100] and [110] directions. The linear relationship between the oscillator strength and q^2 is consistent with the major characteristic of either a monopole or quadrupole transition. This dipole-forbidden transition is strongly anisotropic as can be seen in Fig. 8.6. The ratio

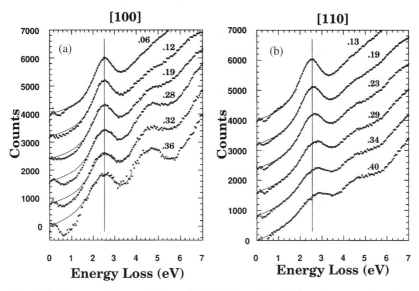

Fig. 8.6. Energy loss spectra of $BaBiO_3$ with different momentum-transfer q (in units of Å^{-1}) in [100] (a), and [110] (b) directions. The zero loss is stripped by fitting as an asymmetric Lorentzian function. The energy region below 1.0 eV is extended by a Drude model.

of the oscillator strength along [100] and [110] is estimated from the slopes of the q^2 dependent plots in Fig. 8.7(b), and is ~3.1. In an earlier paper [8.37], we proposed this excitation to be a transition from the non-bonding oxygen $2p\sigma$ orbital of the valence band to the empty Bi $6s$. These wavefunctions are illustrated in Fig. 8.8. By using symmetry principles, the ratio of the transition strength between [100] and [110] has been calculated to be 4.0, which compares favorably with the experimental result of ~3.1.

The excitation at ~2 eV has been assigned as the transition across the optical gap induced by a charge density wave (CDW). Although the realistic band structure for a CDW does not give an optical gap as large as ~2 eV, we used the assumption that the 2 eV excitation is associated with the CDW gap and theoretically estimated anisotropic dispersion for the optical gap. However, for perfect nesting, the numerical calculation of the dielectric function, $\varepsilon_2(q, w)$ indicates that the energy band gap is isotropic and dispersionless in terms of momentum transfer, q. This calculated result disagrees with the strong anisotropic dispersion observed by angle-resolved EELS. In order to interpret the anisotropic dispersion, p-wave symmetry of a small exciton picture was introduced.

The symmetry of the p-wave function is shown in Fig. 8.9. Due to the effective hopping integral t' of the quasiparticle, the excitons can move

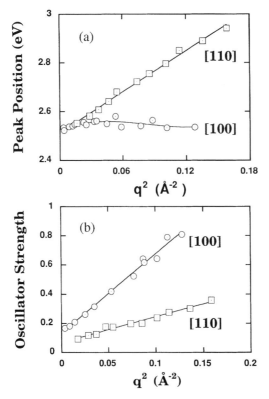

Fig. 8.7. (a) The position of the 2.5 eV peak vs. q^2; (b) the oscillator strength of the 4.5 eV excitation vs. q^2, for BaBiO$_3$.

between the next nearest neighbor sites which belong to the same sublattice. Neglecting the kinetic energy, the optically active gap is given by the energy of the local excitons in Fig. 8.9. Detailed calculation shows that the dipole-allowed transition can be expressed as:

$$2t' \sin(q_x)\sin(q_y) \approx t'q^2 \sin(2q) \quad \text{for } P_1,$$
$$-2t' \sin(q_x)\sin(q_y) \approx -t'q^2 \sin(2q) \quad \text{for } P_2,$$

(8.1)

where q is the angle between $\mathbf{q}$ and the x-axis and t' is the hopping integral of the quasiparticle. By symmetry, calculation indicates that $|I_1|^2 = C(1 + \sin(2q))$ for the P_1 exciton, and $|I_2|^2 = C(1 - \sin(2q))$ for the P_2 exciton, where C is a constant. The dispersion in this model shows a strong anisotropy; dispersionless along [100] and dispersive along [110]. The oscillator strength of the model shows that along [110] only the exciton P_1 is optically active, while along [1$\bar{1}$0] only the exciton P_2 is optically active. The exciton

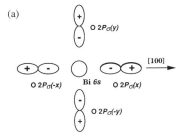

(a)

Bonding and Antibonding States

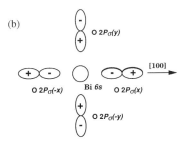

(b)

Non-Bonding State: $d(x^2-y^2)$ Symmetry

Fig. 8.8. Illustration of bonding–antibonding wave function ϕ_0, which has s-symmetry and the non-bonding wave function ϕ_1, which has $d(x^2 - y^2)$ symmetry. The amplitude of this wave function is zero along the [110] direction.

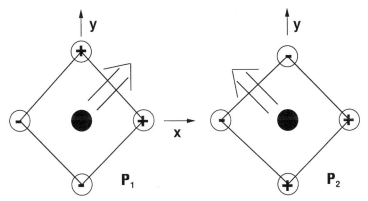

Fig. 8.9. Two small excitons in the $x - y$ plane with p-symmetry. The exciton consists of a quasiparticle (solid circle) at sublattice A and a state formed from a linear combination of four nearest neighbor quasiholes (open circles) at sublattice B. The signs indicate the relative phases of the quasiholes. The arrows indicate the directions of the exciton motion with the strongest oscillator strength.

energy with a stronger oscillator strength is always shifted towards higher energy. All these are qualitatively consistent with our experimental results.

It should be pointed out that although these models can qualitatively explain the physical properties of these excitations, detailed calculation is needed for quantitative comparison. However, we believe that the symmetry information would not change significantly even in a more sophisticated model. As we will show in the next section, a more sophisticated model involving similar symmetry ideas is developed to explain the similar electron energy loss spectra for $Sr_2CuO_2Cl_2$ [8.39].

8.4.2 Symmetry of the charge transfer exciton in $Sr_2CuO_2Cl_2$

$Sr_2CuO_2Cl_2$ is an ideal insulating isomorph to the superconducting cuprates. In this system, the apical oxygen site is replaced by Cl, resulting in a two-dimensional electronic structure in the CuO_2 plane due to the small radius of the Cl p-wavefunction. In addition, because of the Cl replacement at the apical site, the excitations below the 6 eV energy range occur mainly in the CuO_2 plane.

In our earlier study [8.39], we have employed momentum-transfer resolved EELS to investigate the dispersion of the valence band excitations in $Sr_2CuO_2Cl_2$. We report an optically forbidden transition (at $\sim$4.5 eV) and the dispersion of the optically allowed threshold transition, both of which are strongly anisotropic. The optically allowed excitation is found to have a large dispersion range of $\sim$1.5 eV. This is remarkable in view of the recent photoemission experiment [8.15], which reported a narrow bandwidth of $\sim$0.35 eV for the single hole state in the same compound. Our experiment strongly supports the notion that although single hole motion disturbs the antiferromagnetic spin background, motion of the electron–hole pair in the insulating compound does not.

Figures 8.10(a) and (b) show electron energy loss functions at different momentum transfers for $Sr_2CuO_2Cl_2$, along [100] and [110], respectively. There is an optically allowed transition in both directions, whose position is at 2.8 eV for small q, corresponding to the 2 eV excitation observed in optical spectroscopy [8.40, 8.41]. With increasing q along [100], a broad transition at about 4.5 eV gradually appears. As q increases, the position of the optically allowed transition along [110] disperses systematically towards higher energy by about 1.5 eV, while the energy position along [100] is less dispersive. This indicates that the dipole-allowed transition becomes anisotropic at finite q in the CuO_2 plane and that the bandwidth of the excitation is very large. To further illustrate the anisotropic properties of the transition, we measured the

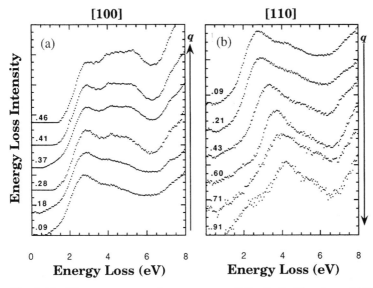

Fig. 8.10. Electron energy loss spectra of $Sr_2CuO_2Cl_2$ along [100] (a), and [1$\bar{1}$0] (b) with different momentum transfers, given by the inserted numbers ($\mathring{A}^{-1}$). All intensities are in arbitrary units.

peak positions in the energy loss function along different orientations for a fixed value of $q = 0.38 \, \mathring{A}^{-1}$. The peak position of the optically allowed transition is plotted against the angle of the orientation in Fig. 8.11. The periodicity in the direction of momentum transfer is due to the C_4 symmetry of the CuO_2 plane.

We have also measured the intensity of the optically allowed transition relative to the other valence excitations in the loss function. As q increases, the intensity decreases rapidly along [100], while the intensity along [1$\bar{1}$0] changes little. This indicates that even the oscillator strength of the dipole-allowed transition is anisotropic.

Another important feature is the optically forbidden transition located at 4.5 eV, at finite q. The transition is strongly anisotropic, appearing along [100], but not along [1$\bar{1}$0]. This implies a certain symmetry in the excitation.

To gain more insight, we carried out a Kramers–Kronig analysis to extract the excitation spectrum, or $\varepsilon_2(q, \omega)$, from the EELS data. We used $\varepsilon(q, \omega = 0) \approx \varepsilon_0 \equiv \varepsilon(q = 0, \omega = 0)$ where $\varepsilon_0 = 4.83$ was obtained from previous optical measurements on this material [8.40, 8.41]. Figure 8.12 shows the resultant dispersion and the peak intensity of the optically allowed excitation, or the peak position and the weight of ε_2, respectively, as functions of the scattering angle. The transition peak position at 2.8 eV for small q in the

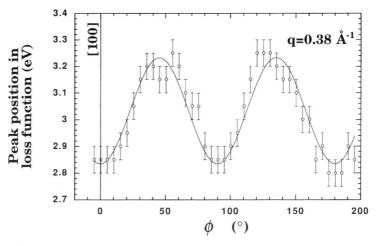

Fig. 8.11. Measured peak position of the dipole-allowed transition in the loss function versus the orientation angle of q relative to [100], for a fixed q of 0.38 Å^{-1} in the $x - y$ plane. The solid line is a fit of the data to $-\cos(2f)$ for the purpose of illustration.

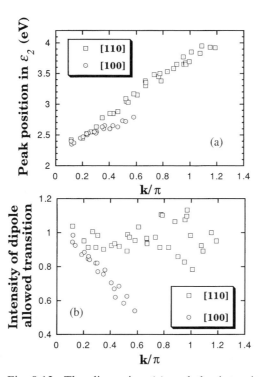

Fig. 8.12. The dispersion (a) and the intensity (b) of the optically allowed excitation along the [100] and [110] directions in ε_2 obtained by Kramers–Kronig analysis.

energy loss function is shifted to 2.4 eV in ε_2. The optically forbidden transition centered at ~4.5 eV in the loss function remains at the same energy position in ε_2. The dispersion of the 2.4 eV excitation in the dielectric function along [110] is similar to that in the energy loss function, but along [100] the dispersion is substantially larger than in the loss function. However, the excitation dispersion remains anisotropic along the different orientations. The oscillator strength of the optically allowed excitation declines quickly along [100], while it remains almost unchanged along [110].

It is interesting to note that the oscillator strength of the optically forbidden excitation at 4.5 eV compensates for the strength of the optically allowed excitation. As the scattering angle increases along [100], the strength of the forbidden excitation grows, and the strength of the allowed one decreases. However, along [110] there is no forbidden excitation, and the intensity of the optical branch remains unchanged. This compensation between the forbidden and the allowed excitations indicates that these two excitations are closely related to each other.

The most striking result derived above is the large energy dispersion of ~1.5 eV for the dipole-allowed transition along [110]. This is especially interesting because of the narrow dispersion of ~ 0.35 eV observed in the angle-resolved photoemission experiments in the same material [8.15]. It has been established that the insulating CuO_2 plane is antiferromagnetic and can be well described by the spin-$1/2$ Heisenberg model for a square lattice [8.42]. In the photoemission experiment, the electron is removed by a photon, leaving a single hole in the antiferromagnetic background. The motion of this single hole disturbs the background, resulting in a large effective mass. The narrow bandwidth observed in the angle-resolved photoemission experiment is attributed to this strong correlation, and it is related to the magnetic interaction J, of the order of 0.1 eV [8.15, 8.42]. In inelastic electron scattering experiments a fast electron passes through a thin film and generates an electron–hole pair in the solid. Normally, in a semiconductor or other band-like insulating material, the effective mass of an exciton is larger than either the hole mass or the electron mass [8.43]. However, in these materials the effective mass is smaller by a factor of about four than the single hole mass, based on calculation that the effective mass is roughly inversely proportional to the bandwidth. We argue that this mass anomaly provides further evidence for the strong electron correlation in cuprates. The general idea underlying the larger dispersion bandwidth of the exciton is that, contrary to a single hole, the electron–hole pair in a CuO_2 plane moves freely in the lattice without disturbing the antiferromagnetic spin background and the exciton mass can be much smaller than the single hole mass.

A microscopic model for the exciton in the undoped Cu-oxides has been proposed earlier [8.39] in which a quantitative agreement between theoretical calculations and experimental EELS results is presented. Here, we shall only briefly sketch some basic aspects of the theory. The exciton model is illustrated in Fig. 8.13. The ground state of the undoped Cu-oxides is antiferromagnetic [8.42] with the configuration of Cu-$3d^9$ and O-$2p^6$. Using the hole notation, there is one hole of spin-$1/2$ at each Cu site. The low energy charge excitation is a hole transferred from a Cu $3d(x^2 - y^2)$ state to its neighboring O $2p\sigma$ state. This charge transfer excitation model, assuming the additional hole primarily sits on the oxygen site, is supported by earlier experiments [8.12, 8.13]. In the simplest picture, the O holes are due to transitions between O $2p\sigma$ and Cu $3d(x^2 - y^2)$. The O hole (or the quasihole) is bounded to the Cu-$3d^{10}$ (the missing Cu hole, or the quasi-electron) due to the Coulomb attraction. Because of the strong hybridization between the Cu and O holes, the O hole is also coupled to its neighboring Cu hole (Cu $3d^9$), forming a spin singlet which is similar to the spin singlet in hole doped cuprates [8.44]. Therefore, the exciton in cuprates consists of a Cu-$3d^{10}$ and its nearest neighbor Cu—O singlet. As far as the spins are concerned, the exciton is similar to a pair of bound holes. When the exciton moves through the lattice, the antiferromagnetic spin configuration remains unchanged. The spin-up holes remain on one

Fig. 8.13. Illustration of the spin singlet exciton model in the undoped Cu—O plane. The exciton consists of a vacancy at a Cu site and a Cu—O spin singlet at the neighboring site. The latter is represented by a solid line with two anti-parallel spins.

sublattice, and the spin-down holes remain on another sublattice. This is equivalent to free exciton motion with essentially no coupling to the antiferromagnetic background, which results in the smaller exciton mass observed in our experiment.

The observed optically allowed and forbidden transitions may be related to the different symmetries of the excitons. Because of the four-fold symmetry of the Cu-oxide plane, four excitons for a spatially fixed Cu-$3d^{10}$ can be constructed according to the relative phases of the four neighboring O $2p\sigma$ holes. They have s-, $d(x^2 - y^2)$- and two different types of p-wave symmetries. Detailed theoretical calculation shows that the different symmetry excitons have different energy levels [8.39] and a local molecular orbital model can be used to illustrate the symmetries of the excitons [8.45]. According to the calculation, the s-wave has the highest energy level, two p-waves are in the middle, and the d-wave has the lowest energy level. The importance of symmetries in cuprates has been recognized previously [8.44, 8.46]. Since the p-wave state is dipole allowed in the transition from Cu-$3d(x^2 - y^2)$, it is natural to relate it to the optically allowed transition at small q. Both s-wave and d-wave excitons are dipole inactive at $q = 0$ and are optically forbidden. As discussed in detail in ref. [8.21], at a finite q along [100] the s-wave exciton is coupled to the p-wave, which explains the gradual appearance of the broad forbidden excitation at 4.5 eV and the decrease in oscillator strength of the dipole-allowed transition in this direction; whereas along [110], the s-wave exciton is essentially decoupled from the p-wave, which explains the absence of the optically forbidden transition in this direction. In our experiments, we did not observe the lower energy, optically forbidden transitions with d-symmetry due to the nature of the excitation [8.44] and the poor energy resolution of the EELS spectrometer. The d-exciton is expected to be located at the mid-infrared region, and because it is optically forbidden for $q = 0$, it is expected to be very weak or unobservable by reflectance or absorption optical spectroscopy.

Thus, q-resolved electron energy loss spectra analysis for $Sr_2CuO_2Cl_2$, indicates a large energy dispersion ($\sim$1.5 eV) of the optically allowed transition, and the discovery of an optically forbidden transition which is at a higher energy position and is very anisotropic. The large dispersion implies a mass anomaly in the exciton and provides further evidence for the breakdown of the one electron picture in cuprates. In addition to the anisotropy of the forbidden transition, it is observed that the intensity of the forbidden transition compensates the intensity of the allowed transition, indicating that these two are closely related to each other based on the f-sum rule. A spin singlet exciton model, in which the exciton moves freely in an antiferromagnetic spin background, is briefly described to account for the smaller mass and the symmetries of the excitations.

8.5 Conclusions

Transmission EELS in a cold FEG TEM is a useful analytical technique which complements optical spectroscopy techniques with the added advantages of q-resolved studies, capable of detecting optically forbidden transitions and obtaining dynamic information about excitations in solids. In addition, the cFEG TEM parallel electron energy loss spectroscopy with small probe size (nanometer scale) and better energy resolution ($\sim$0.5 eV) has provided a powerful way to determine the dielectric properties of single crystal oxide superconductor materials without requiring a large size single crystal, making the EELS measurement much more easy to perform.

The anisotropic dielectric properties can be easily obtained using electron energy loss spectroscopy. In hole-doped cuprate material, it is noted that the free carrier plasmon is strong in the CuO_2 plane, but weak out of the CuO_2 plane, which is consistent with previous optical measurements. Similarly, it is also discovered that in the infinite-layer compound, the excitation at $\sim$2–3 eV, presumed to be a Hubbard transition, is strong in the Cu—O plane, but weak out of the plane, which is again consistent with the notion that there is a strong hybridization between O $p\sigma$ and Cu-$3d(x^2 - y^2)$. Furthermore, in pure Cu—O chain compounds it is discovered that there is a strong excitation at 3.1 eV in the Cu—O chain and weak excitation out of the chain. Such information has already proved very useful in understanding the electronic structure of cuprates.

Momentum-transfer resolved EELS has opened a unique avenue for investigation of the symmetries of the electronic structure in solid state materials. By studying the insulating compound $BaBiO_3$ as well as $Sr_2CuO_2Cl_2$ with this technique, we have been able to extract symmetry information on the electronic excitations. Due to its correlation with high-temperature superconductivity, we believe that this information has already given us significant insight into the electronic structure of this strongly correlated system. Further studies of such compounds are expected to provide significant clues as to what kind of mediated boson (phonon, exciton, or some other excitation) is needed to form a Cooper pair in these exotic oxide superconductors.

These experiments further illustrate that electron energy loss spectroscopy with cFEG TEM is a powerful method to study the electronic structure of solids. The drawback of this technique is the poor energy resolution compared with that of the optical techniques. With the advance of spectroscopy technology, it is our hope that the energy resolution for electron energy loss spectroscopy could be improved significantly without losing the practical value of this technique.

Acknowledgments

The authors would like to thank F. C. Zhang, N. Bulut, A. L. Ritter, S. E. Schnatterly, and M. V. Klein for valuable discussions. This work is supported by the National Science Foundation through the Science and Technology Center for Superconductivity (STCS, NSF Cooperative Agreement No. NSF-DMR-91-20000).

References

[8.1] H. Raether, *Springer tracts in modern physics* ed. H. Hohler (Springer, Berlin, 1965).
[8.2] J. Danials *et al.*, *Springer tracts in modern physics* **Vol. 54** (Springer, Berlin, 1970).
[8.3] S. E. Schnatterly, *In solid state physics* **Vol. 14**, 275 (Academic Press, New York, 1979).
[8.4] H. Raether, *Excitation of plasmons and interband transitions by electrons, Springer tracts in modern physics* **Vol. 88** (Springer-Verlag, Berlin, Heidelberg, New York, 1980).
[8.5] J. Fink, *Adv. Electron Electr. Phys.* **75**, 121 (1989).
[8.6] Y. Y. Wang & A. L. Ritter, *Phys. Rev. B.* **43**, 1241 (1991).
[8.7] R. D. Leapman & J. Silcox, *Phys. Rev. Lett.* **42**, 1362 (1979).
[8.8] R. D. Leapman, P. L. Fejes, & J. Silcox, *Phys. Rev. B.* **28**, 2361 (1983).
[8.9] J. Fink *et al.*, *Physica C* **185–189**, 45 (1991); J. Fink *et al.*, *J. Electron Spectrosc. and Rel. Phenom.* **66**, 395 (1994).
[8.10] N. Nucker *et al.*, *Phys. Rev. B* **39**, 12 379 (1989).
[8.11] Y. Y. Wang, H. Zhang & V. P. Dravid, *Microscopy Research and Technique* **30**, 208 (1995).
[8.12] H. Romerg *et al.*, *Phys. Rev. B* **42**, 8768 (1990).
[8.13] C. T. Chen *et al.*, *Phys. Rev. Lett.* **66**, 104 (1991).
[8.14] D. S. Marshall *et al.*, *Phys. Rev. B.* **52**, 12 548 (1995).
[8.15] B. O. Wells *et al..*, *Phys. Rev. Lett.* **74**, 964 (1995).
[8.16] N. Nucker *et al.*, *Phys. Rev. B.* **44**, 7155 (1991).
[8.17] C. Tarrio & S. E. Schnatterly, *Phys. Rev. B* **38**, 927 (1988).
[8.18] Y. Y. Wang, *Ultramicroscopy* **33**, 151 (1990).
[8.19] N. Nucker *et al.*, *Phys. Rev. B* **39**, 6619 (1989).
[8.20] R. T. Collins *et al.*, *Phys. Rev. B* **39**, 2251 (1989).
[8.21] H. Zhang *et al.*, *Phyica C* **208**, 231 (1993).
[8.22] A. Krol *et al.*, *Phys. Rev. B* **45**, 2581 (1992).
[8.23] T. Seigrist *et al.*, *Nature* **334**, 231 (1988).
[8.24] M. Azuma *et al.*, *Nature* **365**, 775 (1992); Z. Hiroi, M. Azuma, M. Takano, & Y. Takeda, *Physica C* **208**, 286 (1993).
[8.25] J. D. Jorgensen *et al.*, *Phys. Rev. B* **47**, 14 654 (1993).

[8.26] H. Zhang *et al.*, *Nature* **370**, 352 (1994).

[8.27] E. Pellegrin *et al.*, *Phys. Rev. B* **47**, 3354 (1993).

[8.28] Y. Y. Wang *et al.*, *Phys. Rev. B* **48**, 9810 (1993).

[8.29] S. L. Cooper *et al.*, *Phys. Rev. B* **42**, 10785 (1990).

[8.30] D. G. Hinks, *MRS Bulletin*, **15**, 55 (1990).

[8.31] H. Sato *et al.*, *Nature* **338**, 241 (1989).

[8.32] L. F. Mattheiss & D. R. Hamann, *Phys. Rev. Lett.* **60**, 2681 (1988).

[8.33] D. E. Cox & A. W. Sleight, *Solid State Commun.* **19**, 969 (1976).

[8.34] H. Sato *et al.*, *Nature* **338**, 241 (1989).

[8.35] S. H. Blanton *et al.*, *Phys. Rev. B* **47**, 996 (1993).

[8.36] M. A. Karlow *et al.*, *Phys. Rev. B* **48**, 6499 (1993).

[8.37] Y. Y. Wang *et al.*, *Phys. Rev. Lett.* **75**, 2546 (1995).

[8.38] Y. Y. Wang *et al.*, *Ultramicroscopy* **59**, 109 (1995).

[8.39] Y. Y. Wang *et al.*, *Phys. Rev. Lett.* **77**, 1809 (1996).

[8.40] P. Abbemonte *et al.*, private communication.

[8.41] S. Tajima *et al.*, *Physica C* **168**, 117 (1990).

[8.42] D. Vaknin *et al.*, *Phys. Rev. B* **41**, 1926 (1990).

[8.43] D. C. Mattis & J.-P. Gallinar, *Phys. Rev. Lett.* **53**, 1391 (1984).

[8.44] F. C. Zhang & T. M. Rice, *Phys. Rev. B* **37**, 3759 (1988).

[8.45] For simplicity, a local molecular orbital model (see J. P. Lowe, *Quantum chemistry* (Academic Press, Inc., San Diego, 1993)), which involves one Cu surrounded by four O, can be used to illustrate the different energy levels for different symmetries. According to the simple matrix calculation, the intra-molecular energies for the exciton states are $2t_1$ for the *s*-wave, $-2t_1$ for the *d*-wave, and 0 for two *p*-waves, where t_1 is the hole hopping integral between two nearest O atoms, and $t_1 > 0$. This indicates that the *s*-wave has the highest energy level, *d*-wave has the lowest energy level, and two *p*-wave states are in the middle.

[8.46] M. J. Rice & Y. R. Wang, *Phys. Rev. B* **36**, 8794 (1987).

9

Investigation of charge distribution in Bi$_2$Sr$_2$CaCu$_2$O$_8$ and YBa$_2$Cu$_3$O$_7$

Y. ZHU

9.1 Introduction

The charge carriers in high temperature superconductors are the electron holes confined to the CuO$_2$-plane [9.1, 9.2], and thus, the distribution of charge plays a key role in determining their superconducting properties. Several groups of researchers have calculated the electronic structure of different superconducting oxides [9.3–9.5], and core-level spectroscopic studies are plentiful, both emission spectroscopy, and absorption spectroscopy with incident electrons and incident X-rays. In absorption spectroscopy, attention has focused on the near-edge structure of the K- and L-edge of copper, and, in particular, the K-edge of oxygen which exhibits clear signatures of the electron holes that are responsible for superconductivity [9.6, 9.7]. On the other hand, there have been few experimental studies of the spatial distribution of the electron charge in these superconductors.

In high-temperature superconductors, the density of electron holes is typically considerably less than 1% of the total density of electrons. However, the electron diffraction patterns and images of these superconductors, with their high local concentration of charge, is expected to be strongly influenced by the charge distribution. One reason for this expectation is the large crystal unit cell, resulting in reflections at small angles which are very sensitive to the charge. We realize this from the classical picture of the scattering of fast electrons by an atom. Charged particles interact with the electrostatic potential, and thus, for small scattering angles, which correspond to large impact parameters, the incident particle sees a nucleus that is screened by the electron cloud. Thus, the scattering amplitude is mainly determined by the net charge of the ion at small scattering angles, q. In mathematical terms the scattering amplitude of incident electrons, f_e, is given by the Mott formula:

$$f_e(\theta) = \frac{8\pi^2 m_0 e^2}{h^2} \left(\frac{\lambda}{\sin\theta}\right)^2 (Z - f_x), \qquad (9.1)$$

where Z is the charge of the nucleus, λ the wavelength of the incident electrons, m_0 the rest mass of the electron, h Planck's constant, and e the charge on the electron. The scattering amplitude for incident X-rays, f_x, is determined by the spatial distribution of electrons around the atom. Near the forward direction, where f_x is close to Z, small changes in f_x may drastically alter f_e. For example, for a Ba atom $Z^{Ba} = 56$, and the X-ray scattering amplitude at $\theta = 0°$ for an ionized Ba^+ and Ba^{++} atom is $f_x^{Ba^+} = 55$ and $f_x^{Ba^{++}} = 54$, respectively. The difference in their scattering amplitude is $\Delta f_x = (f_x^{Ba^{++}} - f_x^{Ba^+})/f_x^{Ba^+} = 1.8\%$. In contrast, for electrons, $\Delta f_e = \{(Z - f_x^{Ba^{++}}) - (Z - f_x^{Ba^+})\}/(Z - f_x^{Ba^+}) = 100\%$. Thus, electron diffraction has a greater sensitivity than X-ray diffraction in addressing the valence electron distribution in crystals with Bragg reflections present at small reciprocal distances i.e. at small scattering angles [9.8–9.10].

Owing to the screening at low angles, an electron hole and a bonding electron, after reversal of sign, both have the scattering amplitude of a stripped hydrogen atom, H^+. This scattering amplitude is shown in Fig. 9.1 where it is compared with the scattering amplitude of a neutral Bi atom. Note that the scattering amplitude of the hole is larger than that of the high Z neutral Bi

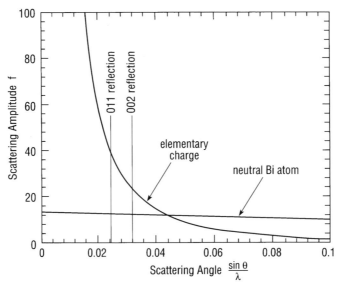

Fig. 9.1. Electron scattering amplitude of an electron hole and a Bi atom at small scattering angles. Also indicated is the angle at which the innermost reflections, the superstructure reflections {110}, and the fundamental reflections {002} of $Bi_2Sr_2CaCu_2O_8$ are located.

atom at small angles. It also follows from these considerations that the electron scattering amplitude of any charged ion approaches that of H^+ multiplied by the excess charge, or valency, when the scattering angle becomes small.

In the following sections, we will discuss our studies on charge distribution in $Bi_2Sr_2CaCu_2O_8$ and $YBa_2Cu_3O_7$ high temperature superconductors by using advanced transmission electron microscopy. In the former, we consider structural deviation due to the charge modulation based on computer simulations of experimental images and diffraction patterns; in the latter we use a more quantitative approach to address the charge distribution by determining the 001 structure factor using a novel diffraction technique.

9.2 $Bi_2Sr_2CaCu_2O_8$

9.2.1 A general nanoscale description: difference structure

The major structural feature in $Bi_2Sr_2CaCu_2O_8$ (hereafter denoted as Bi/2212) is the structural modulation which results in superlattice reflections. To address such modulation, we use a standard method of crystallography, adding and subtracting matter, to describe the deviation from the average structure [9.11]. Let us start with an average structure with total scattering amplitude $\overline{\psi}$ and let the actual structure y deviate from the average structure by a perturbation Δy. The scattering amplitude of the crystal in the kinematical case at position s in reciprocal space is thus

$$\psi(s) = \overline{\psi} + \Delta\psi, \tag{9.2}$$

and the intensity

$$I = \psi\psi^* = (\overline{\psi} + \Delta\psi)(\overline{\psi} + \Delta\psi)^* = \overline{\psi}\overline{\psi}^* + \overline{\psi}\Delta\psi^* + \overline{\psi}^*\Delta\psi + \Delta\psi\Delta\psi^*. \tag{9.3}$$

Because $\overline{\psi}$ is periodic, it has values other than zero only at the Bragg position of what we define here as the fundamental reflections, and thus only the last term contributes to the intensity outside the fundamental reflections. Thus, it suffices to consider Δy to assess what is outside the fundamental reflections, i.e. the diffuse scattering and superstructure reflections.

The deviation from the average structure that gives rise to Δy, and thus, the superstructure reflections, can be described in terms of the removal and addition of ions in the average structure in such a way that charge balance is maintained. Displacement of an atom, for example, is to remove it from its position in the average structure and add it at its position in the actual structure. Sometimes it is convenient to distinguish between modifications of the average structure that leave the composition of the material unchanged, and those that do not. To the first category belong displacement, interchange of atoms between two sites, and charge transfer. We restrict ourselves to the situation where the deviation from the average structure can be described by identical clusters [9.12] of scattering amplitude ΔF at the position r_j. Thus

$$\Delta\psi = \sum_j \Delta F \exp(2\pi i s \cdot r_j). \tag{9.4}$$

Here $\Delta F = F_I - F_R$, where we have removed scattering matter of scattering amplitude F_R and inserted scattering matter of amplitude F_I at the position r_j. Moving a single atom or electron A from position r_j to $r_j + \Delta r$ gives

$$\Delta F = f_A \sum_j [\exp(2\pi i s \cdot (r_j + \Delta r)) - \exp(2\pi i s \cdot r_j)]$$

$$\approx f_A 2\pi i s \cdot \Delta r \sum_j \exp(2\pi i s \cdot r_j). \tag{9.5}$$

This approximation applies at small $s.\Delta r$. For modifications that change the composition,

$$\Delta F = (f_A - f_B) \sum_j \exp(2\pi i s \cdot r_j). \tag{9.6}$$

Here f_A and f_B are the scattering amplitudes of the species added and removed from position r_j. In the case of introducing a vacancy or interstitial, either f_A or f_B is zero.

We conclude that displacements over small distances contribute negligibly to the electron diffraction pattern at small angles (Fig. 9.2(a)) because of the short s-vectors. Displacement of atoms over larger distances, of the order of half the unit cell, and substitution may contribute equally in all parts of reciprocal space apart from the general fall-off of the scattering amplitude with angle. Charge transfer gives major contributions only close to the origin of reciprocal space, as is apparent from the curve of Fig. 9.1 and also from Fig. 9.2(b) which shows the scattering amplitude from the superlattice in Bi/2212 due to the charge. We also note here the complimentary of the displacement of atoms over short distances and charge transfer over larger distances, a feature unique to electron diffraction. Charge transfer shows up at reflections at small angles while displacement of atoms occurs at large angles.

9.2.2 Direct imaging of charge modulation

Experimental electron diffraction patterns and electron microscope images are usually compared with calculations made under the assumption that the atoms

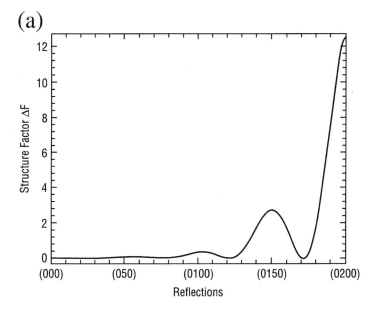

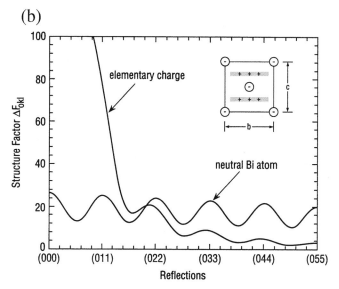

Fig. 9.2. Structure factor ΔF describing the deviation from an average structure in $Bi_2Sr_2CaCu_2O_8$; (a) deviation due to the displacement of atoms, (b) due to the charge. Note, the charge transfer contributes to reflections at small angles while the lattice displacement contributes at large angles.

are neutral, and for small crystal unit cells with interplanar spacings less than 5 Å this is a good approximation in most cases. Decades ago, Cowley [9.8] pointed out that electron diffraction should be able to reveal ionicity in crystals, and effects of deviation from neutral atoms have been retrieved from con-

220   *Y. Zhu*

vergent beam electron diffraction and used to study charge distribution [9.9, 9.13, 9.14]. These studies required interpretation procedures involving many parameters to extract interesting information about the charge distribution, while the system we study enables us to directly image the charge distribution with a resolution of about 1 nm [9.10].

Bi/2212 has the idealized crystal unit cell shown in Fig. 9.3(b), i.e. when the

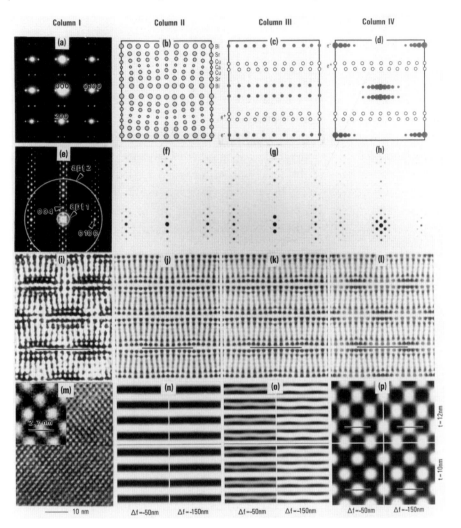

Fig. 9.3. Electron diffraction patterns and images. Experimental images are shown in Column I and calculated images in Columns II, III and IV, based on three different models of the charge distribution. (a) refers to the (001) projection, all others to the (100) projection. The bar in the experimental images (Column I and inset in (m)), and in the calculated images (j), (k), (l) and (p), indicates the size of the crystal unit cell along the *b*-axis. Column II is based on a model with neutral atoms with atomic positions from Ref.

mixing of superstructure periods that leads to incommensurability is disregarded. The unit cell can be described as an average structure with structure factor, F_g, and a deviation from this, ΔF_g. It is useful to think of the average structure as consisting of five identical units along the b-axis with lattice parameters $a = 0.538$ nm, $b = 5a = 2.69$ nm and $c = 3.06$ nm, giving rise to the fundamental reflections. The large crystal unit cell results in reflections at small angles, and such reflections may change their intensities by orders of magnitude by charge transfer over distances that are a considerable fraction of the crystal unit cell.

To further address the question of displacement of atoms versus charge transfer, we performed dynamical calculations of electron diffraction and images using a standard multislice program. We modified the program to include the scattering amplitudes of holes and electrons, and performed calculations in the (100) projection of Bi/2212 for three different models of the crystal. In Fig. 9.3(b) and Column II, the model is based on neutral atoms using the displacement parameters reported by Horiuchi *et al.* [9.17]; similar parameters were reported in other work [9.15, 9.16]. In the model of Column III, we introduced charged atoms and used the charge distribution from the electron structure calculations of Gupta & Gupta [9.18]. They found a hole concentration in the CuO_2 layer of 0.37 per Cu (or Bi) atom in the material and an excess electron concentration of the same magnitude in the BiO layer. Their charge distribution is relative to atoms of the following valance: O^{2-}, Ca^{2+}, Sr^{2+}, Cu^{2+}, and Bi^{3+}. Thus, relative to neutral atoms, which are the frame of reference in our calculations, the excess charge is that shown in Fig. 9.4. We added these charges to the model which was based on neutral atoms, and assumed the charge is evenly distributed in the CuO_2 and BiO planes to arrive

---

Fig. 9.3 (*cont.*). [9.17], as indicated in (b). Column III is based on a model the same as in (b) except for the addition of charge modulation along the *c*-direction as calculated in Ref. [9.18], but with an evenly distributed charge in the *b*-direction, as indicated in (c). Column IV is the same as (b) but with additional modulation of the charge in the *b*-direction as well as in the *c*-direction, as indicated in (d). (e) Experimental diffraction pattern showing the size of the objective apertures, apt. 2 for the calculated high-resolution electron microscope image in the same row, and apt. 1 for the lower-resolution image in the row below. (f), (g), and (h) are calculations for different charge distributions. (i) Experimental HREM image with corresponding HREM image calculations in (j), (k), and (l) for a crystal thickness of 27 Å; other parameters used are also the same as Ref. [9.17]. (m) Experimental lower-resolution image with corresponding calculations for the different models at different thickness and different defocus values in (n), (o), and (p). The inset in (m) is magnified for comparison with the calculated images (n), (o) and (p).

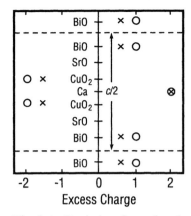

Fig. 9.4. Deviation from the charge of neutral atoms on the different planes along the *c*-direction; o from formal valence, and x after creation of the holes as calculated in Ref. [9.18].

at the calculated results of Column III. In the last model (Fig. 9.3(d)) and column IV, we maintained the average concentration of charge in each CuO_2 plane and BiO plane at the same level as in the previous model, but we modulated the charge in the BiO layer along the *b*-axis. With reference to the formal valence, this model shows a pileup of electron charge in the region of the BiO double layer where the atomic planes expand along the *b*-direction; this agrees with the notion that electron doping increases the Bi—O distance [9.19]. It was established by X-ray [9.15], neutron [9.16], and electron diffraction, and by high resolution electron microscopy (HREM) [9.17] that the deviation from the average structure of Bi/2212 is reasonably well described by the introduction of a displacement field. However, in electron diffraction superstructure reflections are seen also near the forward direction, Fig. 9.3(a), and they should not be there in a kinematical diffraction pattern unless the displacements are extremely large. Unfortunately electron diffraction is prone to multiple scattering. Thus, we cannot rule out the possibility that the superstructure reflections near the center of these diffraction patterns are caused by the other reflections further out in reciprocal space; these may act as new incident beams bringing superstructure spots caused by displacement into the central region of the diffraction pattern.

In our comparisons, the diffraction pattern of the last model with the pronounced {011} reflections (Fig. 9.3(h)) shows the best agreement with observations, Fig. 9.3(e). However, these calculations are for very thin crystals, while the maximum thickness of the illuminated region under our parallel beam diffraction experiments could be much larger. A notable difference in the calculated HREM images is the dark, fairly large elliptic regions for the model with charge modulation in the BiO layer. Very similar features are seen in the high resolution images, Fig. 9.3(i). Similar HREM images were observed by others previously [9.20, 9.21], which were pointed out as being a striking feature [9.21]. In a subsequent calculation [9.17] these workers substituted 40% of the atoms of some of the Bi sites by Sr, thereby improving the agreement with experiments. We obtain excellent agreement with our experimental images (Fig. 9.5) without interchanging atoms. Nevertheless, we tested the possibility that substitution could be the source of the inner reflections by

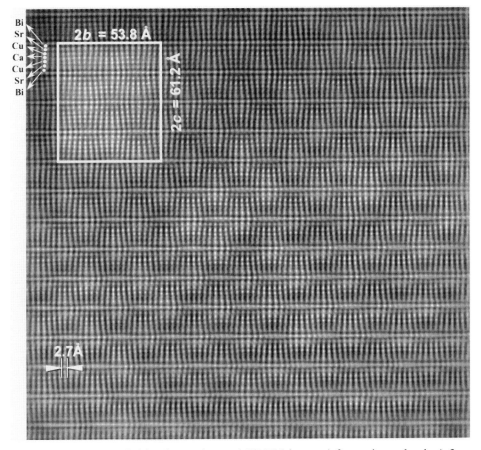

Fig. 9.5. A digitized experimental HREM image (after noise reduction) from an area with a slight increase in crystal thickness from top to bottom. The embedded image consisting of four unit cells is calculated using the displacive and charge modulation shown in Fig. 9.3(b) and Fig. 9.3(d). Parameters used for the simulation are the same as for Fig. 9.3(l). Good agreement between the experiment and calculation confirms the validity of the model.

making calculations with a gradual change in the occupancy across the unit cell. The image of Fig. 9.3(m) was obtained using a much smaller objective aperture so that, in addition to the direct beam, only the four innermost superstructure spots and the two fundamental reflections (00 ± 2) were inside the aperture (apt. 1), as seen in Fig. 9.3(e). Primarily, modulation along the c-direction is seen in the calculated images for different thicknesses and defocus values using the models without charge modulation in the b-direction, Figs. 9.3(n) and 9.3 (o). This change is what we expect since the dominating beams now are {002} and (000). When charge modulation along the b-direction is introduced, good agreement with the observed image is achieved. A noteworthy point is that contrary to high resolution images obtained using a large objective

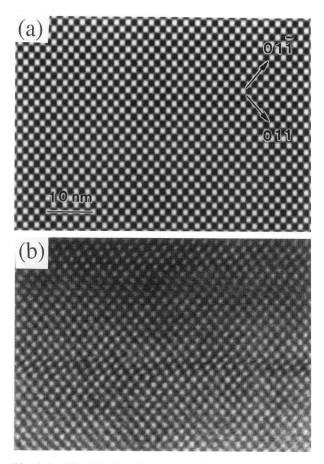

Fig. 9.6. (a) Calculated image with only the incident beam and the four superstructure spots of type {011} inside the objective aperture; (b) a similar experimental image where disorder of the image features caused by charge modulation is present, both incommensurability and out of phase boundaries.

aperture (apt. 2, Fig. 9.3(i)–(l)), these images (Fig. 9.3(p)) are very robust, meaning that they are relatively insensitive to small changes in thickness, and extremely insensitive to changes in imaging conditions.

By further decreasing the objective aperture to exclude the fundamental reflections (00 ± 2), the image (Fig. 9.6(a)) is almost exclusively caused by the charge modulation, though admittedly, the spatial resolution is poor. An aperture that includes one or two fundamental reflections is probably the optimum if this approach is to be developed into a quantitative technique. Then, if the structure as determined by X-ray or neutron diffraction, for example, is well known, the fundamental reflection may be used as a reference beam in an electron interferometric approach to study charge distribution. In Fig. 9.6(b), we show that images obtained with a very small objective aperture are very powerful for mapping the long-range disorder of the charge modulation and this also demonstrates the incommensurability of the charge distribution. Presently, the technique is not quantitative and we cannot make a clear choice between the rather simplistic picture of charge transfer presented here and more detailed models based on electronic-structure calculations of the spatial distribution of valence electrons, such as those we used for $YBa_2Cu_3O_7$ (see section 9.3.2). What appears clear, however, is that there is a considerable modulation of the valence electron density in the BiO double layer, and this may be caused, at least in part, by additional oxygen in this layer.

9.3 $YBa_2Cu_3O_7$

9.3.1 A novel diffraction method: parallel recording of electron diffraction intensity as a function of thickness

Since $YBa_2Cu_3O_7$ does not have structure modulation which generates super-lattice reflections, the method we used for Bi/2212 cannot be directly applied to this material. However, as $YBa_2Cu_3O_7$ also has a relatively large lattice parameter along the c-axis, we can study the charge distribution in the crystal by determining the amplitude and phase of Bragg reflections at low angles. As discussed above, the spatial distribution of electrons in crystals can be addressed by X-ray and by electron diffraction. When the amplitudes and phases of the X-ray structure factors are known, the spatial distribution of electrons in the crystal, $r(r)$, can be obtained by Fourier transform. Similarly, the electrostatic potential in the crystal, $U(r)$, can be extracted by Fourier transform of ideal electron diffraction data. When isolated atoms are brought together to form a solid, it is mainly the outer valence electrons that are rearranged; this necessitates very accurate X-ray diffraction experiments to register such minute differences, $r(\mathbf{r}) - r_a(\mathbf{r})$. Here, $r_a(\mathbf{r})$ is the electron distribution to be expected in the crystal when the electron clouds around the atoms are unperturbed by the fact that the atoms are in the vicinity of other atoms, i. e. using the scattering amplitudes of isolated atoms to calculate the structure factors of the crystal. In electron diffraction, since the interaction takes place between the incident electrons and the electrostatic potential of the crystal, the difference between $U(\mathbf{r})$ and $U_a(\mathbf{r})$ can be very large.

Figure 9.7 illustrates the great potential of low-angle reflections in addressing the charge distribution in a crystal, where the Fourier components of the electrostatic potential for inner reflections of $YBa_2Cu_3O_7$ are calculated by assuming ionic bonding and different distributions of electron holes in the crystal unit cell. We start with an ionic model for the charge distribution by delocalizing the transferred electrons around the atoms so that the scattering amplitudes of Y^{3+}, Ba^{2+} and Cu^{2+} [9.22], and O^{2-} [9.23] are in agreement with tabulated values converted from X-ray scattering amplitudes. We used the

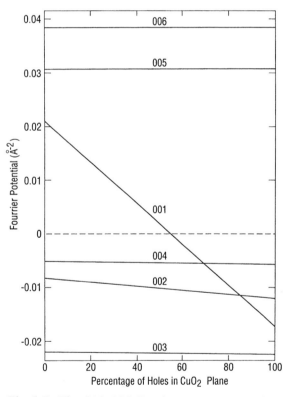

Fig. 9.7. The 001–006 Fourier components of the electrostatic potential in $YBa_2Cu_3O_7$ assuming ionic bonding, for different distributions of holes in the CuO_2 plane.

atomic positions determined by neutron diffraction [9.24] to calculate these structure factors of the orthorhombic unit cell of size $a = 0.382$ nm, $b = 0.388$ nm and $c = 1.169$ nm. We note, in Fig. 9.7, the huge variation in the value of the 001 Fourier component. With the origin chosen at the Cu atom in the CuO chain, its value changes from a large positive to a large negative one when the holes are moved from the CuO chains to the CuO_2 planes. Thus, if the kinematical intensities and the phases of these reflections could be retrieved from electron diffraction experiments, we would be well armed to determine the charge distribution in large unit cell crystals.

However, a major obstacle in electron diffraction has been the difficulty of accurately measuring intensity. In special cases, this problem was circumvented by observing in relatively thick regions of crystal, typically 100 nm, features in the diffraction patterns associated with the strong dynamical coupling between the Bragg beams [9.25, 9.26] and, more recently, by many-beam simulations of the whole convergent beam electron diffraction (CBED) pattern [9.27, 9.28] until there is a good fit with the experimental data. These procedures were quite successful in determining structure factors in small unit cell crystals [9.29], but are not suited for the reflections at small scattering angles from crystals with large unit cells.

A somewhat different approach is to use a small convergent beam angle to study large unit cell crystals, and very thin regions of the crystal to reduce the dynamical effects [9.30]. This was the technique we applied at first in attempting to quantify charge distributions in high-temperature superconductors by operating our transmission electron microscope at 200 keV. As was pointed out by Anstis *et al.* [9.9], we found that strong reflections at small angles cause strong coupling between the Bragg beams and thus, very rapid oscillations of the beam intensities with thickness, so that the diffraction pattern was greatly altered by small changes in thickness, as shown in Figs. 9.8(a) and (b). We could not control the thickness to within the accuracy required, about 1 nm, because our crystals were wedge-shaped at the edge, rather than being platelets a few tens of nanometers thick. During the work, we realized that by focusing the electron probe above (or below) the specimen, as shown in Fig. 9.9, we could obtain thickness profiles, or Pendellosung plots, of many reflections simultaneously, starting from zero thickness up to a maximum that could range from 10 nm to several hundred nanometers, depending on the distance from the specimen to the cross-over [9.31]. Figs. 9.10(a) and (b) show such patterns of the (00*l*) row of $YBa_2Cu_3O_7$ at a low and high magnification, i.e. with the probe farther from, and closer to, the specimen.

The technique used in this study does not significantly differ in principle from using the thickness fringes in dark-field images that were previously used

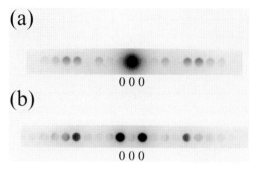

Fig. 9.8. Electron diffraction patterns of the (001) reciprocal row in $YBa_2Cu_3O_7$; (a) and (b) show diffraction patterns at estimated thicknesses of 10 nm and 45 nm, respectively.

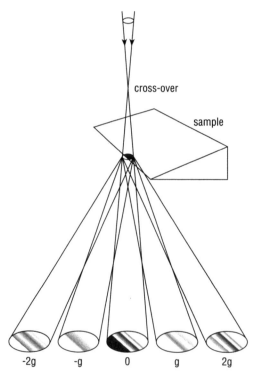

Fig. 9.9. Schematic of the experimental configuration used to obtain the diffraction patterns with parallel recording of diffraction intensities as a function of thickness.

to determine structure factors in small unit cell crystals [9.32]. The great advantage of our approach is that many reflections in large unit cell crystals can be recorded simultaneously, thereby ensuring that the exposure and the crystallographic direction of the incident beam are exactly the same for all

(a)

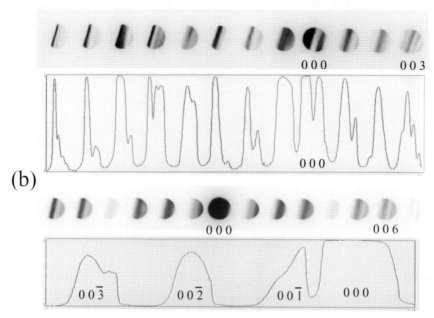

(b)

Fig. 9.10. Examples of diffraction patterns and line scans of the intensity profiles for different reflections as a function of thickness using the diffraction condition shown in Fig. 9.9. The probe is farther from the specimen in (a) than in (b) resulting in lower magnification of the thickness oscillations. The line scan in (b) is from 003 to 000 only. Note the onset of each intensity profile.

reflections. A further advantage compared with other techniques of convergent beam electron diffraction, is that the increase in intensity can be followed at very small thickness where the relationship between the structure factors and the intensities is simple, and where the problems of normal and anomalous absorption are minimized. This technique can be improved further by preparing a perfect crystal wedge, thus obtaining a linear relationship between image and thickness. With imaging plates or a CCD (charge couple device) camera, the intensity as a function of thickness of many reflections can be recorded digitally. We expect that this will improve the spatial resolution of the charge distribution by allowing accurate determinations to be made of the structure factors of several reflections.

9.3.2 Distribution of electron holes

To determine the amplitude of the inner reflections, we focus attention on the onset of increase in intensity near zero thickness where the kinematical approach holds, and the relationship between the intensities, I_g, of the different reflections, g, the Fourier components of the potential, U_g, and the thickness, z, is

$$I_g = |\psi_g \psi_g^*|^2 \propto |U_g|^2 z^2. \qquad (9.7)$$

Thus, in the very thin region, less than 10 nm, the amplitudes of the Fourier components can be extracted. From Fig. 9.7, we note that the 003 reflection is hardly influenced by the charge distribution, and therefore can be used as reference to determine the absolute value of the 001 and 002 Fourier components. By analyzing several diffraction patterns using relatively high spatial resolution, i.e. with a small probe close to the specimen as is the case in Fig. 9.10(b), we found that the potentials of the 001 and 002 reflections are both close to half the value of the 003. Assuming fully ionic bonding and holes only in the CuO chains and CuO_2 planes, this estimated absolute value of the 001 structure factor corresponds to 80% of the holes being in the CuO_2 plane for the negative sign, and 28% for the positive one (see Fig. 9.7). In the thicker regions of the specimen, where the diffraction is dynamical due to coupling between many beams, there is phase information. We used a multislice program to perform the dynamical calculations and included 47 beams along the 001 reciprocal row. With 28% of the holes in the CuO_2 planes, there was poor agreement with the experimental thickness profiles from thicker regions, while the agreement was good with 80% of the holes in the CuO_2 planes. Calculations for hole fractions of 76% and 56% in the CuO_2 plane, which are shown in Figs. 9.11(a) and (b), demonstrate the great sensitivity of electron diffraction to charge distribution. By refinement, the best fit was with a hole fraction of $76 \pm 8\%$ in the CuO_2 plane. The diffraction pattern shown in Fig. 9.8(b), with the electron probe focused on the specimen, is consistent with this charge distribution if the thickness is 45 nm. This remarkable pattern, with virtually no intensity in the direct beam 000 and the 002, 003, and 004 reflections, could not be reproduced in calculations in which the hole concentration in the CuO_2 plane deviated by more than 8% from 76%. From bond valence calculations, using the bond distance determined by Jorgensen *et al.* [9.24], Brown [9.33] concluded that 2/3 of the holes were in the CuO_2 planes. Hole distributions also have been calculated from first principles [9.4] and determined by core-level absorption spectroscopy [9.7]. In those studies, some holes were found in the BaO planes. We tried models with holes in the BaO plane in addition to the CuO chain and the CuO_2 plane, but achieved reasonable agreement only with the majority of the holes in the CuO_2 planes. Thus, whether the remaining small fraction of the holes were in the BaO or CuO planes had little influence on the calculated diffraction pattern.

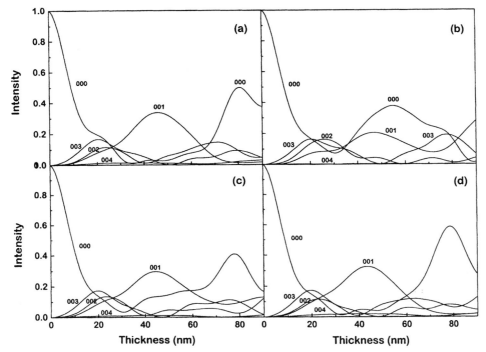

Fig. 9.11. Dynamical calculations of the thickness profiles of electron diffraction intensities for different hole distributions assuming fully ionic bonding in (a) and (b) and with 76% and 56% of the holes in the CuO_2 plane, respectively. In (c) there is partly covalent bonding assigning the charge $+3$, $+2$, $+1.62$, and -1.69 to the Y, Ba, Cu, and O ions, respectively (from Ref. [9.3]), and with an additional charge transfer of 0.08 holes per unit cell from the chains to the planes in (d).

In these calculations, we assumed fully ionic bonding, but other charge distributions also may be consistent with our observations. Introducing some covalent character by assigning the valencies $+3$, $+2$, $+1.62$, and -1.69 to the Y, Ba, Cu, and O atoms, respectively, as suggested from electronic structure calculations using the local density approximation [9.9], did not give a good fit to the observations; see Fig. 9.11(c). However, with the small correction of moving 0.08 holes per unit cell from the CuO chain to the CuO_2 plane, the agreement became good; see Fig. 9.11(d). This demonstrates the limitation of knowing only a few structure factors with sufficient accuracy for assigning a charge distribution. At the same time, it once again demonstrates the great sensitivity of electron diffraction to charge transfer in crystals with large unit cells.

In this case we determined only the 001 structure factor with sufficient accuracy to give valuable information about charge transfer. The value of this structure factor is -0.45 ± 0.15 nm, which may be more fully appreciated by converting it into the X-ray structure factor which then becomes -8.55 ± 0.11 electrons. The determined value of the structure factor of the (001) reflection, which is very sensitive to the distribution of valence electrons in the crystal unit cell, corresponds in the purely ionic model to $76 \pm 8\%$ of the electron holes being located in the CuO_2 planes of the *fully oxygenated samples*. Note that the effect of moving the Ba atoms only 0.0005 nm would influence the 001 *X-ray structure factor* more than an inaccuracy of this magnitude. Also, in a previous electron diffraction study that relied on the coupling between low- and high-angle reflections, the positions of the atoms and their Debye-Waller factors had to be known accurately [9.34]. However, with our approach, the reflections at low scattering angles of the thin areas are much more dominated by charge distribution than changes in the positions of the atoms.

9.4 Conclusions

While it has been long known that, in principle, electron diffraction can provide useful information on ionicity in crystals, we have experimentally verified this, and demonstrated that electron microscope images on a nanometer scale and electron diffraction patterns (amplitudes and phases of Bragg reflections at low angles) are very sensitive to the distribution of valence electrons and charge transfer over a few tenths of a nanometer. Although in this chapter we discussed only the charge distribution in $YBa_2Cu_3O_7$ and $Bi_2Sr_2Ca_{1-n}Cu_nO_{2n+4+d}$ $(n = 2)$ cuprates, the methods described can be applied to the other structurally complicated Bi-based superconductors, $n = 1$ (Bi/2201) and $n = 3$ (Bi/2223), and most likely to the closely related Tl-based superconductors, as well as other crystals with large unit cells.

Acknowledgments

The author would like to thank J. Tafto for his great contribution to the work during his summer visits at Brookhaven. This research was supported by the U.S. Dept. of Energy, Division of Materials Sciences, Office of Basic Energy Science, under contract No. DE-AC02-76CH00016.

References

[9.1] W. E. Pickett, *Rev. Mod. Phys.* **6**, 433 (1989).
[9.2] N. Nücker *et al.*, *Phys. Rev. B* **39**, 6619 (1989).

[9.3] H. Krakauer, W. E. Pickett & R. E. Cohen, *J. Supercond.* **1**, 111 (1988).

[9.4] A. M. Oles & W. Grzelka, *Phys. Rev. B* **44**, 9531 (1991).

[9.5] W. E. Pickett, *Rev. Mod. Phys.* **61**, 433 (1989).

[9.6] J. Fink *et al.*, *J. Electron Spectrosc. Rel. Phenom.* **66**, 395 (1994).

[9.7] N. Nücker *et al.*, *Phys. Rev. B* **51**, 8529 (1995).

[9.8] J. M. Cowley, *Acta Cryst.* **6**, 516 (1953).

[9.9] G. R. Anstis *et al.*, *Acta Cryst.* **A29**, 138 (1973).

[9.10] Y. Zhu & J. Tafto, *Phys. Rev. Lett.* **76**, 443 (1996).

[9.11] See e.g. J. M. Cowley, *Acta Cryst.* **A32**, 83 (1976).

[9.12] Y. Zhu & J. Tafto, *Philo. Mag. A* **74**, 307 (1996).

[9.13] J. M. Zuo, J. C. H. Spence & R. Hoier, *Phys. Rev. Lett.* **62**, 547 (1989).

[9.14] K. Gjonnes, N. Boe & J. Tafto, *Ultramicroscopy* **48**, 37 (1993).

[9.15] K. Imai *et al.*, *Jpn. J. Appl. Phys.* **27**, L1661 (1988).

[9.16] Y. Gao *et al.*, *Acta Cryst.* **A49**, 141 (1993).

[9.17] S. Horiuchi *et al.*, *Jpn. J. Appl. Phys.* **27**, L1172 (1988).

[9.18] R. P. Gupta & M. Gupta, *Phys. Rev. B* **49**, 13154 (1994).

[9.19] A. Q. Pham *et al.*, *Phys. Rev. B* **48**, 1249 (1993).

[9.20] Y. Matsui *et al.*, *Jpn. J. Appl. Phys.* **27**, L372 (1988).

[9.21] Y. Matsui *et al.*, *Jpn. J. Appl. Phys.* **27**, L827 (1988).

[9.22] P. A. Doyle & P. S. Turner, *Acta Cryst.* **A24**, 390 (1968).

[9.23] D. Rez, P. Rez & I. Grant, *Acta Cryst.* **A50**, 481 (1994).

[9.24] J. D. Jorgensen *et al.*, *Phys. Rev. B* **41**, 1863 (1990).

[9.25] O. Terasaki & D. Watanabe, *Acta Cryst.* **A35**, 895 (1979).

[9.26] J. Gjonnes & R. Hoier, *Acta Cryst.* **A27**, 313 (1971).

[9.27] K. Tsuda & M. Tanaka, *Acta Cryst.* **A51**, 7 (1995).

[9.28] R. Vincent, D. M. Bird & J. W. Steeds, *Philo. Mag. A* **50**, 765 (1984).

[9.29] J. M. Zuo, J. C. H. Spence & R. Hoier, *Phys. Rev. Lett.* **62**, 547 (1989).

[9.30] A. Olsen, P. Goodman & H. J. Whitfield, *J. Solid State Chem.* **60**, 305 (1985).

[9.31] Y. Zhu & J. Tafto, *Philo. Mag. B* **75**, 785 (1997).

[9.32] A. Ichimiya & R. Uyeda, *Z. Naturforsch.* **32a**, 750 (1977).

[9.33] I. D. Brown, *J. Solid State Chem.* **90**, 155 (1991).

[9.34] K. Gjonnes, N. Boe & J. Tafto, *Ultramicroscopy* **48**, 37 (1993).

10

Grain boundaries in high T_c materials: transport properties and structure

K. L. MERKLE, Y. GAO and B. V. VUCHIC

10.1 Introduction

The electric transport properties of a high temperature superconductor are largely determined by the absence or presence of high-angle grain boundaries and their arrangement within the material. Ultimately, the grain boundary properties are governed by the grain boundary structure and composition at the atomic level. An important goal of electron microscopy investigations is to establish correlations between the electrical transport behavior and the structure and chemical composition of grain boundaries. In this chapter we examine, via the example of $YBa_2Cu_3O_{7-x}$ (YBCO) grain boundaries, grain boundary structures in high T_c materials and their influence on grain boundary transport properties. The weak-link behavior of high-angle grain boundaries will be discussed in view of results from structural investigations at different length scales ranging from the macroscopic to the mesoscopic and microscopic down to the atomic scale.

Polycrystals of $YBa_2Cu_3O_{7-x}$ typically can carry critical currents that are two orders of magnitude lower than the critical current densities in corresponding single crystals. This reflects the average reduction of the critical current due to high-angle grain boundaries. Thus, high-angle grain boundaries present a considerable impediment to high current applications of high T_c materials. Conversely, the weak-link nature of grain boundaries can be of considerable value when applied to the design of microelectronic devices, such as superconducting quantum interference devices (SQUIDS). In fact, one of the first commercial applications of high T_c materials was based on the function of grain boundaries as Josephson junctions [10.1].

Following Chaudhari *et al.* [10.2], who measured the superconducting properties of individual grain boundaries in thin films, Dimos *et al.* [10.3, 10.4] demonstrated a strong correlation between the critical current densities across the grain boundaries and the grain boundary misorientations for several grain boundary geometries, which included tilt, twist, and general grain boundaries. The ratio of the critical current density across the grain boundaries to the average value of the critical current densities in the adjacent grains is found to decrease dramatically as the misorientation angle θ is increased and reaches a value of 10 to 15°, while beyond this range a saturation value is approached, almost two orders of magnitude smaller than that for $\theta = 0°$. Later work suggested that certain special boundary geometries in the high-angle range may, however, be associated with strong coupling characteristics [10.5, 10.6]. Further work on the misorientation dependence by Ivanov and coworkers indicated as a function of θ an exponential decrease in the critical current across the interface in [001] tilt grain boundaries in YBCO films [10.7]. The critical current density [according to refs. 10.4, 10.7, 10.51, 10.59, and 10.60] for grain boundaries with misorientation about [001] is shown in Fig. 10.1.

Obviously, the dramatic decrease in critical current is connected with the increasing structural disorder at the grain boundary as the high-angle grain boundary regime is approached, which also may be associated with possible changes in the elemental composition of the grain boundary relative to the bulk

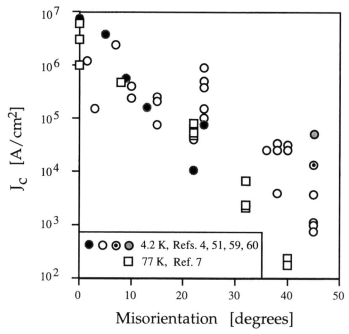

Fig. 10.1. Critical current densities J_c for grain boundaries in YBCO films grown by epitaxy on suitably treated single crystal or bicrystal substrates, plotted against the [001] misorientation angle [10.4, 10.7, 10.51, 10.59, 10.60].

material. On the other hand, the considerable scatter in the data is also an indication that other factors, such as the dependence on the exact synthesis conditions for the preparation of the films can introduce considerable variation in the grain boundary properties for a given boundary geometry.

Because of these difficulties associated with systematic grain boundary studies we have at present a very limited understanding of the correlation of grain boundary electric transport properties to the microstructure. Clearly, it is well recognized that detailed information on the structure and composition of grain boundaries is necessary as a basis for an understanding of the correlation between the structure of grain boundaries and their electric properties.

In this chapter we shall consider several basic issues that need to be addressed concerning the connection between structure and properties of grain boundaries in high T_c materials. First, we shall examine issues connected with the geometry of grain boundaries. Grain boundary models need to be established at different size scales, including a description of the macroscopic, mesoscopic, and atomic scale grain boundary structure. Next, we turn to simple oxide materials, their grain boundary structure and grain boundary models to examine what can be learned from general structural features in these materials concerning the complex structures of grain boundaries in high-temperature superconductors. Thirdly, possible connections between grain boundary structure and electric transport properties will be pointed out. Then we will give examples of recent advances in obtaining direct correlations between structural observations using electron-optical methods and electric transport measurements in thin films. Finally, we briefly examine the potential contributions from high-resolution techniques to our understanding of grain boundary transport properties in high-temperature superconductors. Throughout the investigations of grain boundaries in high-temperature superconductors there has been a need for structural information at all levels, from the macroscopic to the atomic scale structure of grain boundaries. It is obvious that electron-optical methods play a crucial role in establishing an understanding of the complex interrelations between grain boundary structure and properties in high T_c materials.

10.2 Grain boundary structure

For the purpose of this paper 'global grain boundary structure' refers to the interface between two grains over a finite boundary area, typically of a dimension commensurate with the size of sample used for electric transport measurements. The 'local grain boundary structure' refers to the positions of atoms within the grain boundary as well as their chemical identity. Thus grain boundary structure has to be considered over many size scales, from the macroscopic to the atomic level.

10.2.1 Grain boundary geometry

The geometry of a planar interface between two grains is characterized by five macroscopic and three microscopic parameters (rigid-body shifts). In the usual convention the five macroscopic degrees of freedom (DOFs) involve a mis-orientation axis (two DOFs) and a misorientation angle (one DOF) plus the grain boundary plane (two DOFs). This description is particularly well suited for a bicrystal in which two grains of a given misorientation are joined. This description is also the one on which coincidence site lattice (CSL) models are based. An alternative way to define the macroscopic geometry of a planar grain boundary is given by the interface-plane scheme, which uses the two crystal-lographic planes joined at the interface (four DOFs) plus a twist angle ψ between the two planes (one DOF) [10.8]. The latter description is generally used for heterophase interfaces, but can also offer some advantages when discussing grain boundaries, since the physical properties of grain boundaries often strongly depend on the planes joined at the interface [10.9].

10.2.2 Grain boundary models

The atomic level structure of grain boundaries has been an important issue for the past several decades. In cubic materials geometrical constructs of periodic grain boundaries can be obtained for certain misorientation axis-angle combi-nations that are associated with coincident site lattices (CSLs). The CSLs are formed by the coincident sites of two hypothetically interpenetrating crystal lattices, where Σ is the reciprocal density of CSL sites. Much of the discussion of grain boundary structure and properties has revolved around the description of grain boundary structures in terms of the CSL, the displacement-shift complete (DSC) and the O-lattice [10.10, 10.11].

The CSL is a useful tool to illustrate the atomic level geometry of interfaces. Based on the unrelaxed atom positions, rigid models of grain boundaries can be constructed by this scheme. Figure 10.2 illustrates the construction of a grain boundary from the CSL model. The rigid model of a grain boundary resulting from this procedure typically has some atoms located in much closer proximity than the interatomic distances. Thus considerable relaxations invol-ving rearrangements of atomic columns and rigid-body displacements are necessary to arrive at a realistic model of a grain boundary.

CSL models are only possible under very strict geometrical conditions. For

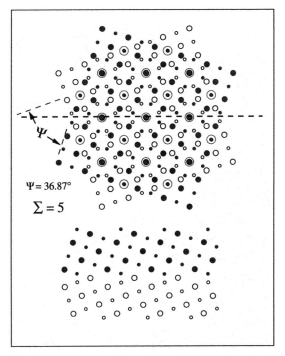

Fig. 10.2. Construction of a grain boundary by the CSL scheme. A projection along [001] of two interpenetrating fcc lattices is shown for $\Sigma = 5$. A rigid model of a grain boundary is obtained by removing full and open atoms from opposite sides of the grain boundary (dashed line) respectively.

example, even in cubic materials exact CSLs are only possible for certain misorientations. However, for cubic systems much of the misorientation space is at least close to some exact CSL lattice. In contrast to this, crystals with lower symmetry generally do not allow the formation of CSL lattices. For rotations about the *c*-axis, hexagonal and tetragonal crystals can still form coincident sites, whereas the orthorhombic crystal structure does not allow exact CSLs. Nevertheless, as illustrated in Fig. 10.3 for YBCO, rotations about the *c*-axis can bring atoms almost into coincidence (near coincidence) between atoms of the interpenetrating lattices.

Another possibility for obtaining CSLs exists by approximating the crystal structure by 'pseudocubic' or tetragonal unit cells. This approach of applying the CSL to non-cubic systems has been discussed in the literature by means of the constrained coincident site lattice [10.12] which has, among others, also been applied to YBCO grain boundaries.

The rigid atomic models of grain boundaries constructed by these CSL schemes can serve a useful purpose in providing a first approximation for

(a)
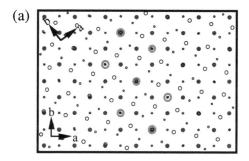

(b)
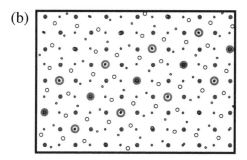

Fig. 10.3. (a) Interpenetrating YBCO lattices at $\theta = 36.87°/[001]$ misorientation ($\Sigma = 5$). Except for the origin there are no exact coincidences, but six columns are within $0.15\,\text{Å}$ of each other (ringed atoms). (b) $\theta = 37.43°/[001]$, exact coincidence along one plane, near coincidences in neighboring CSL planes.

periodic high-angle grain boundaries, especially when applied to short period, symmetric grain boundaries and in cases where the relaxations are small. However, many discussions of grain boundary structure have unfortunately implicitly or directly related certain geometrical features of the CSL boundaries ('special boundaries') to grain boundary properties. There is, however, no evidence that physical properties can be predicted on the basis of CSL analysis alone [10.13].

For large, complex unit cells that contain more than one or two atomic species, a further complication arises when a boundary model is to be constructed according to the scheme illustrated in Fig. 10.2. After selecting a plane normal appropriate for the grain boundary to be modeled, the position of the plane chosen within the CSL unit cell strongly affects the resulting grain boundary core structure and in particular its chemical composition. Thus, the boundary may have strong deviations from stoichiometry or may even have a higher density than the bulk [10.14]. While some of these unphysical features of a rigid model could be resolved by applying additional boundary conditions to the selection of the grain boundary plane, such as charge neutrality and density appropriate for bulk, it is clear that such models are of very limited value. Rigid models need in any case to be modified to take into account atomic relaxations at the grain boundary core. Models derived by computer simulations can account for relaxations and have been applied extensively to metal and simple oxide grain boundaries [10.15, 10.16], but are at present generally not available for grain boundaries in high-temperature superconductors. Therefore, considerable efforts have been extended to derive realistic atomic-scale models of grain boundaries by analysis of high-resolution electron microscopy (HREM) experiments as will be discussed below in Section 10.4.

10.2.3 *Global grain boundary structure*

An electric transport measurement of a grain boundary is necessarily performed across a finite grain boundary area. Thus, when considering structure-property correlations, the global structure of the boundary may involve considerable variations of the local grain boundary structure along this finite length. When the grain boundary is free to adjust its local geometry the grain boundary plane may be different along different sections of the boundary. Moreover, grain boundary precipitates may completely interrupt the current flow across the grain boundary. Thus, characterization of such a grain boundary involves macroscopic, mesoscopic, and atomic-scale observation. This can be accomplished by electron-optical observations at different levels of magnification, an example of which is shown in Fig. 10.4.

A scanning electron microscope (SEM) was used to record the secondary electron image in Fig. 10.4(a). The YBCO grain boundary runs horizontally across the center of the image and appears completely straight at this level of magnification. The misorientation across the grain is 45°/[001], i.e. a 45° rotation about the [001] axis, which can be verified by the recorded electron backscattered patterns shown in the insets. A boundary is imaged in the transmission electron microscope (TEM) in Fig. 10.4(b) under dark field illumination, using a strong reflection from one of the grains. In this manner only one of the grains is imaged, showing the demarcation between the grains very clearly. At this level of magnification it is obvious that the grain boundary strongly deviates from a straight line. In fact the grain boundary makes excursions to either side of the average line given by the template onto which the YBCO film was deposited. This meandering which is often found for YBCO films grown by epitaxy [10.17, 10.18] has its origin in the nucleation and growth mode and the much higher growth rate along the *a*–*b*-directions compared with the *c*-direction. In this manner a grain nucleated on one side

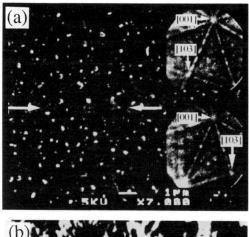

Fig. 10.4. Electron-optical observation of YBCO grain boundary junction at different length scales [10.17, 10.19]. Thin films were grown by pulsed metallorganic beam epitaxy (POMBE) using sputter-induced epitaxy modification to form $\theta = 45°/[001]$ grain boundaries. (a) SEM secondary electron image of grain boundary shows a straight interface horizontally across the image. Two insets of electron backscattering patterns from the two grains indicate the orientation of the grains. (b) Dark-field TEM of meandering grain boundary. (c) Asymmetric facets. (d) Lattice fringe image of stepped grain boundary consisting entirely of (100)(110) asymmetric facets, i.e. the steps connecting the long facets have the (110)(100) grain boundary configuration.

can grow across the template boundary without changing its orientation. A lattice fringe image in Fig. 10.4(c) shows that the grain boundary is well structured at the atomic scale and displays well developed asymmetric facets. The HREM image in Fig. 10.4(d) of a section in which the average grain boundary plane deviates from the asymmetric orientation reveals that the (110)(100) asymmetric facets are indeed connected at the steps between the facets by the same type of grain boundary, namely short sections of (100)(110) asymmetric facets [10.19, 10.20].

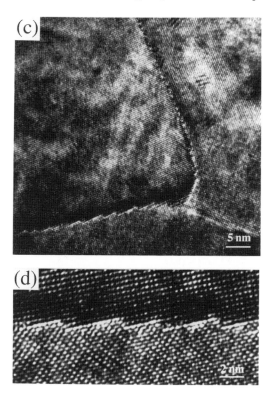

Fig. 10.4 (*cont.*).

10.3 Oxide grain boundaries

Because of the small coherence length in high T_c superconductors, typically only a few nanometers at most, it is obvious that the electric transport properties of grain boundaries in these materials ultimately are governed by their atomic-scale interfacial structure. Within the past decade HREM investigations have contributed a considerable amount of information regarding the atomic structure of grain boundaries in ceramic oxides. In this section we shall examine some important results from this work on simple oxides in order to see what can be learned from these observations and to what extent these results may be relevant and applicable to the microstructure of high-temperature superconductors.

10.3.1 Grain boundary phases

The HREM observations revealed that well-structured grain boundaries in which the lattice structure is being maintained right up to the boundary are the rule in pure oxides [10.21–10.23]. Many technically important ceramics often contain, however, thin amorphous phases at the grain boundary [10.24–10.26]. Such impurity phases of course can be very detrimental to the properties of high-temperature superconductors. Therefore, early efforts in preparing high T_c materials were directed at improving the purity and stoichiometry to avoid the deposition of undesirable impurities at the grain boundary and to prepare grain boundaries that were fully oxygenated. It was also found that during the preparation of YBCO materials which involved anneals in atmospheres such as air, small concentrations of CO_2 can have a devastating effect on transport properties across grains. Although the grains in a bulk sample were fully superconducting, a thin boundary layer, which was at the smallest concentration not wide enough to be detected by conventional TEM, was sufficient to prevent the supercurrent transport across the grains [10.27].

10.3.2 Symmetric high-angle grain boundaries

Turning back to pure oxides, bicrystals of NiO have been extensively studied by HREM [10.21, 10.28, 10.29]. The bicrystals were prepared to form [001] tilt grain boundaries at a variety of misorientation angles. This grain boundary geometry can be directly compared to the [001] tilt grain boundaries in YBCO. The latter are particularly important in the study of grain boundary current transport since the Cu—O_2 plane normal of YBCO is parallel to the tilt axis. Thus the current carrying Cu—O_2 planes remain parallel to each other in going across the grain boundary. The weak-link properties can be tuned to a certain extent by adjusting the tilt angle. It is also relatively straightforward to grow *c*-oriented YBCO films on suitable substrates, which can be prepared to form a bicrystal junction. Hence, this geometry is used for the manufacture of grain boundary junctions.

An HREM image of the $\Sigma = 5$, (310) symmetric tilt grain boundary in NiO at a misorientation of $\theta = 37°$ is shown in Fig. 10.5. Note the relatively dense core structure and the change in structure as the grain boundary translates across a small step in Fig. 10.5. This observation of multiple structures for the same macroscopic grain boundary geometry indicates that the grain boundary energies of the two structures are quite close. The incorporation into the boundary of a step which does not have dislocation character and is less than the interplanar distance of the CSL in height is equivalent to saying that the two grain boundaries have a different rigid-body translation parallel to the grain boundary, as indicated in the schematic CSL model in Fig. 10.5 [10.30]. In contrast to the observation, lattice statics calculations by Duffy & Tasker which simulate the grain boundary at 0 K [10.15] had suggested a much more

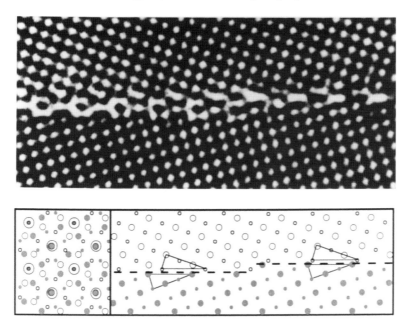

Fig. 10.5. HREM image of [001] tilt grain boundary in NiO. The $\Sigma = 5/(310)$ symmetric tilt grain boundary shows two variants, separated by a small step. The schematic CSL drawing (below) illustrates that the step introduces an effective rigid-body translation parallel to the grain boundary [10.30].

open structure for this grain boundary. Moreover, the calculated structure had mirror symmetry of atomic columns, in contrast to the observation. A much closer agreement between observed and calculated structures is found, however, for other [001] tilt grain boundaries in NiO, such as the (320) symmetric tilt grain boundary [10.31].

The (210) and (310) symmetric tilt grain boundaries have the smallest structural repeat units for [001] tilt grain boundaries and for this reason do not show any strong connection between the grain boundary planes crossing the grain boundary. A more typical case of a [001] tilt grain boundary is depicted in Fig. 10.6 where the (510) symmetric tilt grain boundary ($\theta = 22.6°$) clearly shows good structural coherence in between the localized regions of misfit. Note that the structural units are quite distinct, but are by no means completely identical. The topology of the arrangement of atomic columns also suggests that relaxations parallel to the tilt axis are present. Moreover there are indications that some columns at the grain boundary core are not fully occupied [10.28]. Similar conclusions have recently been drawn by Browning *et al.* from Z-contrast observations of $SrTiO_3$, using a dedicated STEM [10.32].

The presence of relatively large open spaces in high-angle grain boundaries in ionic oxides of the NaCl structure have been suggested to be due to the fact that like charges in close proximity to each other are energetically prohibited [10.15]. HREM observations (see Figs 10.5 and 10.6) show, however, that relatively dense-packed structures are possible in NiO grain boundaries. In metals, extensive grain boundary simulations for a wide range of grain boundary geometries have indicated that the grain boundary excess volume (or volume expansion, i.e. rigid-body translation normal to the boundary) is directly related to grain boundary energy [10.9, 10.33]. Low-energy grain boundaries are relatively dense-packed, while the excess volume increases roughly linearly with grain boundary energy. In oxide grain boundaries no similar connection has been established as yet. However, from the general structural similarities that exist one would expect a similar trend for the oxides. In general the grain boundaries are comprised of regions of good atomic match followed by regions of mismatch in such a manner that the average local atomic environment in the grain boundary is as close as possible to the bulk coordination.

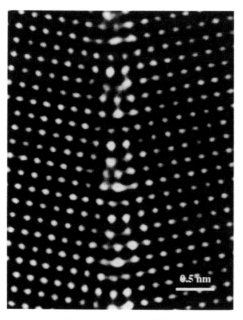

Fig. 10.6. HREM image of the (510) symmetric tilt grain boundary in NiO ($\theta = 22.6°$). Note the well-connected (200) planes across the grain boundary between the regions of misfit [10.28].

10.3.3 Asymmetrical grain boundaries

Symmetrical tilt grain boundaries are very special boundaries that can have small planar unit cells and short structural repeat units. However, for a given bicrystal, there are many more possibilities for forming asymmetrical grain boundaries. Indeed, asymmetric grain boundaries have been observed by HREM in numerous investigations [10.34–10.40]. Asymmetric grain boundaries often have lower energy than the corresponding symmetric ones. Also, in many bicrystal experiments a variety of asymmetric grain boundary facets are found, even in samples designed to study symmetric grain boundaries. In many instances, one of the interface borders is formed by a low-index (dense-packed) plane. This suggested that such configurations are associated with low grain boundary energy [10.35, 10.37, 10.38, 10.41].

Coexistence of symmetrical and asymmetrical grain boundaries is found in HREM observations in practically all grain boundary studies [10.9, 10.21]. Symmetrical grain boundaries may have small-period grain boundary structures, but the energetics of grain boundaries appears to be determined more by the packing density of atomic planes joined at the boundary than the size of the structural repeat units. It has been amply documented that grain boundary geometry alone is not a criterion of low grain boundary energy in itself [10.13]; nevertheless the possibility of lowering the energy of an interface by its choice of interface plane within a given bicrystal is an important relaxation mechanism and often determines the structures present at the atomic scale. This can also lead to phenomena such as the formation of microfacets at the atomic scale. The structures depicted in Fig. 10.4 give a good example of this behavior for a high-temperature superconductor.

10.4 Grain boundaries in YBCO

10.4.1 High-angle grain boundaries

Compared with cubic systems, the reduced symmetry of the crystal structure in high-temperature superconductors will allow, on the one hand, a greater multitude of possible grain boundary geometries just because of the larger number of crystallographic combinations that allow the formation of a grain boundary. On the other hand, the large crystal anisotropy should make a great number of grain boundary configurations energetically rather unfavorable. For a given misorientation this will result in a strong selection of the grain boundary plane provided the grain boundary can relax to an energetically favorable orientation. A typical example is shown in Fig. 10.7, where a grain boundary from a

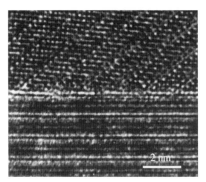

Fig. 10.7. HREM image of grain boundary in polycrystalline YBCO. The grain boundary appears free of impurity phases and is formed on the basal plane of one crystal, which borders a high-index plane in the second grain.

polycrystal compact is shown to form a grain boundary that includes the basal plane on one side of the grain boundary, while the opposing border is formed by a high-index plane. Therefore, asymmetric grain boundaries that include the basal plane are often observed in polycrystalline high temperature superconductor materials.

Special grain boundaries are formed by misorientations of 90° about the orthogonal crystal axes in the orthorhombic structure. These misorientations, of course, would not form grain boundaries in cubic systems. Examples of grain boundaries in YBCO formed by $\theta = 90°$, about [100], [010], and [001], are shown in Fig. 10.8. Invariably the grain boundary plane assumes either the symmetrical configuration or forms an asymmetrical grain boundary which is made up of two low-index planes that are joined at the interface.

The 90° grain boundaries in YBCO are of special interest, since some of these have been found to show weak-link free behavior [10.6, 10.42]. The 90°/[001] tilt grain boundaries typically appear in the form of symmetrical (110) twins on transformation from the tetragonal to the orthorhombic phase. For the same misorientation (100)(010), incommensurate grain boundaries have been observed in a few instances in YBCO films grown by metal-organic vapor deposition. Figure 10.8(a) shows an example of a section of a (100)(010) grain boundary from a small island grain that is, due to the small misfit between a and b lattice parameters, coherently connected to the matrix. More interesting are the 90°/[100] grain boundaries, because those were not expected to show weak-link free behavior. Basically only two distinct grain boundary inclinations have been found for the 90°/[100] tilt grain boundaries in YBCO [10.43]. These are pictured in Fig. 10.8(b), which shows facets of the (013) symmetric grain boundary and the (010)(001) asymmetric boundary. Boundaries of the

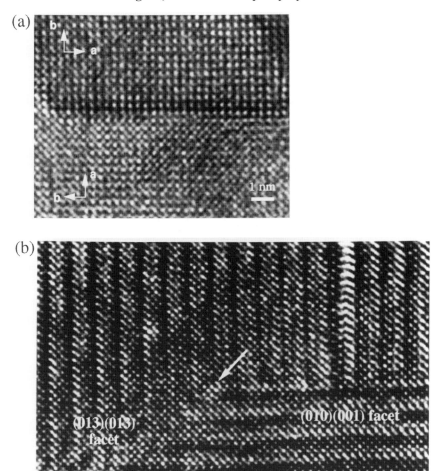

Fig. 10.8. HREM images of various 90° grain boundaries in YBCO; (a) [100] tilt, *a* and *b* interchanged across boundary; (b) [100] tilt grain boundary showing two distinct facets, the (013) symmetric boundary with the CuO_2 planes joining at the interface and the (010)(001) asymmetric grain boundary; (c) same as the asymmetric boundary in (b), but viewed at 90°, with one grain in the [001] projection; (d) tilt and twist facets combined (*continues overleaf*).

same macroscopic geometry are also found in step-edge junctions; their structure is discussed in detail in Chapter 13.

Here we just point out that the (010)(001) tilt boundary in Fig. 10.8(b) is atomically abrupt, while the slight lateral lattice mismatch across the boundary

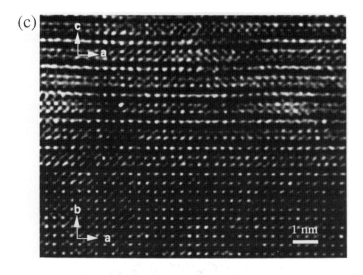

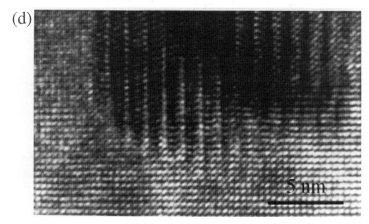

Fig. 10.8 (*cont.*).

is accommodated by a stacking fault. The atomic structure of this grain boundary has been investigated using HREM image simulations based on several atomic structure models. The results indicated that the grain boundary in Fig. 10.8(b) is well represented by either of two models in which the boundary plane is either the (010) lattice plane or the (001) plane. This ambiguity is due to the fact that the two models differ only by the occupation of oxygen sites in the grain boundary plane, and the electron scattering of oxygen atoms is too weak to affect the HREM images [10.43]. However, different terminations have also been found in studies of step-edge junctions [10.44]. Fig. 10.8(c) shows another HREM image of a (010)(001) boundary, but viewed in a direction perpendicular to the projection in Fig. 10.8(b). This image also gives a clear picture of the terminating layers and makes it evident that the boundary plane for this interface can be either a (010) or a (001) lattice plane.

Clearly, since the CuO_2 planes, which play an essential role in the superconducting transport, are disrupted across the (010)(001) boundaries, they are expected to show weak-link behavior. However, since the CuO_2 planes are continuous across (013)($0\bar{1}3$) boundaries, the latter may provide good supercurrent conducting paths across the 90° [100] tilt grain boundaries, resulting possibly in weak-link free behavior [10.43]. Measurements across individual (013)($0\bar{1}3$) boundaries would have to be performed to prove this conjecture. Measurements by Eom *et al.* on thin textured YBCO films that contained 90° twist grain boundaries have indicated that the weak-link free behavior of 90°/[100] bicrystals may be due to facets that consist of twist grain boundaries. From measurements in two different directions it was concluded that the critical current density across the 90° [100] tilt grain boundaries is lower by a factor of ∼30 than the value across the 90° [010] twist grain boundaries [10.6].

The HREM image in Fig. 10.8(d) shows a curved boundary between the grain on the top and the grain on the bottom. The boundary has a 90°/[100] misorientation and, like the 90° [100] tilt boundaries, this boundary is structurally coherent and free of any second phase. A grain boundary along a horizontal plane in this image would be a pure twist grain boundary.

However, in contrast to the previously described tilt grain boundaries the boundary is not planar and as well defined as in the (010)(001) boundary. Thus, although the grain boundary has mainly twist character, the grain boundary is of mixed type and it is evident that the crystalline structure does not abruptly change across the boundary. It appears possible that the irregular shapes of the boundary plane could increase the contact area of the CuO_2 planes across the boundary.

10.4.2 Low-angle grain boundaries

The strong decrease in critical current (J_c) as a function of misorientation angle as shown in Fig. 10.1 naturally begs for an explanation in terms of the microstructure of low-angle grain boundaries, since the main drop in J_c takes place within the region that is normally associated with low-angle grain boundaries. It is well known that low-angle grain boundaries are well described by the Read-Shockley model. Thus, tilt grain boundaries consist of a wall of edge dislocations. The dislocation spacing D is given by the Read–Shockley formula $D = |\boldsymbol{b}|/\sin\theta$, where $\boldsymbol{b}$ is the dislocation Burgers vector and θ is the boundary tilt angle. This structure was confirmed for low-angle [001] tilt grain boundaries in YBCO, where the Burgers vector **b** was found to be [100] or [010]; see [10.45]. The edge dislocations are associated with long-range strain fields, that can be imaged by TEM, and whose range is commensurate to the dislocation spacing. It has been suggested that the superconducting order parameter is suppressed in the vicinity of the dislocation core since the local strains are expected to affect the oxygen deployment in the immediate neighborhood of the dislocation core [10.46].

At low misorientations ($<10°$) the lattices between the grain boundary dislocations are well connected from one grain to the other and one therefore expects sufficient intergrain coupling to maintain high critical currents. As the dislocation spacing is decreased at larger misorientation angles the effect of the dislocation cores on the electric transport will be felt, resulting in a decreased critical current across the grain boundary. The strong drop-off in J_c with θ should be explainable in terms of the structural changes that occur. Since this involves, to first approximation, just a reduction in dislocation spacing, one should be able to describe this behavior in terms of the reduced supercurrent that can be carried across and near the grain boundary cores.

HREM images of dislocation cores in a low-angle [001] tilt grain boundary ($\theta = 3.5°$) of YBCO are shown in Fig. 10.9. The atomic columns of the metal ions appear white in this image. Two sets of lattice fringes are visible in both grains and show the smooth continuation of lattice planes across the grain boundary in the region between dislocation cores. At the latter, strongly localized distortions are visible. By drawing a Burgers circuit around a dislocation the Burgers vector, **b**, of the dislocations was determined to be

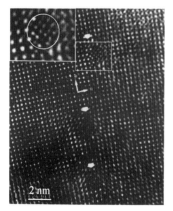

Fig. 10.9. Low-angle [001] tilt grain boundary in YBCO with extended and reconstructed dislocation cores (arrows) [10.66].

equal to [100]. In contrast to observation in most pure metals and oxides, the dislocation cores are considerably larger in size ($\sim$2 nm) than their Burgers vector (0.38 nm).

HREM image simulations of model dislocation core structures were performed as shown in Fig. 10.10. Since the core region appears quite different from that of the adjacent grains, two atomic models based on different compositional configurations at the core were used for the image simulations. Model 1 maintained the $YBa_2Cu_3O_{7-x}$ composition at the dislocation core, whereas in model 2 it was assumed that only Cu and O are present within the core, i.e. the Y–Ba columns at the dislocation core of the first model are replaced by Cu-rich columns ($\oplus$ in Fig. 10.10(b)) in the second model. It was found that model 2 gives a reasonably good match to the experimental image [10.45]. Therefore, we must assume that the core regions are not superconducting.

Consequently, as the dislocation cores get more closely spaced with increasing tilt angle, the well-coupled regions get smaller and smaller. Indeed, using the Read–Shockley formula together with the observed size of the cores, one

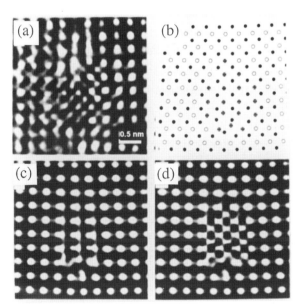

Fig. 10.10. (a) HREM image of one of the dislocation cores in Fig. 10.9. (b) Dislocation core model for [001] tilt grain boundary in YBCO. HREM simulations were performed for two models. (c) Model 1 HREM simulation assumes that the grain boundary core chemical composition is identical to bulk. (d) For the model 2 simulations the Y–Ba atomic columns were replaced by Cu. This model image closely resembles the observed HREM image (after [10.65]).

can estimate that the dislocation cores will overlap at $\theta = 11°$, which is, according to Fig. 10.1, close to the observed transition from the strong coupling to the weak coupling regime [10.3, 10.4]. We note that the reconstruction of dislocation cores and the associated enrichment in Cu at the grain boundary, which also has been reported for high-angle grain boundaries, using high spatial resolution X-ray energy dispersive spectroscopy [10.47, 10.48], is expected to be sensitive to the processing conditions used. In any case, the tendency towards copper enrichment at the grain boundaries in $YBa_2Cu_3O_{7-x}$ will increase the degree of the weak-link behavior of the grain boundaries in addition to the effects of structural disorder and the effects of strains associated with grain boundary dislocations.

10.5 Direct correlation between grain boundary structure and electric transport properties

In the previous section some correlations between certain grain boundary geometries and the grain boundary transport behavior were pointed out. However, the connection between structure and properties is rather indirect, since the measurements were not performed on the same sample. Numerous studies of the transport properties of many types of grain boundaries have been reported in the literature, including studies of the microstructure and chemistry of bulk and thin-film grain boundaries [10.42, 10.44, 10.48–10.57]. However, a direct comparison of the structure of a particular sample to its transport properties is seldom made. Since there are many factors that can affect the superconductivity behavior of a grain boundary, it is essential for further progress in this field that the connection between structure and property be made as directly as possible. Aside from the misorientation (which was the only variable considered in Fig. 10.1), many factors, such as the detailed grain boundary atomic-structure, orientation of the grain boundary plane, grain boundary segregation, grain boundary phases, oxygen deficiency etc., can strongly impact on grain boundary transport properties.

In order to provide the opportunity for a direct analysis of structure–property correlations, measurements of structure and properties need to be performed on the same sample. Recent work in our laboratory has focused on the study of $45°/[001]$ tilt grain boundaries in YBCO [10.17, 10.19, 10.58–10.60]. Thin films were deposited epitaxially, using several techniques and deposition conditions. In each case grain boundaries were introduced into predetermined patterns suitable for electrical characterization. The individual thin-film bi-epitaxial grain boundary junctions then were electrically characterized by using d.c. four-probe measurements. Subsequently, the structure of these same grain boundaries was examined using transmission electron microscopy. Films of widely different grain boundary transport characteristics, including resistive and superconducting boundaries, and their respective microstructures, were investigated.

An optical micrograph of a thin-film sample, patterned for resistivity measurements is shown in Fig. 10.11 during the thinning process used to obtain electron-transparent sections of the grain boundary [10.19]. The sample contains two grain boundaries, indicated by arrows, and the probe configuration was such that electrical characterization of both grain boundaries as well as the grain was possible. Details of the experimental procedures and results can be found elsewhere [10.19, 10.20, 10.61]. Examples of resistance–temperature and current–voltage characteristics of grain boundaries and a grain are shown in Figs. 10.12 and 10.13. The electrical characterizations were typically made at junction widths of about 40 μm (see Fig. 10.11).

As shown in Fig. 10.4, the grain boundaries are not made up of a single, well

Fig. 10.11. A sample suitably shaped for four-point resistivity measurements is prepared for transmission electron microscopy. The hole formed by dimpling from the substrate side followed by ion milling is positioned such that both grain boundaries (arrows) can be observed by TEM [10.19].

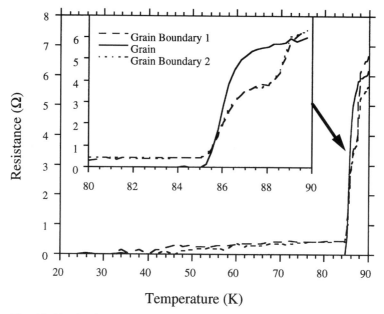

Fig. 10.12. Resistance–temperature relationship for a 45°/[001] tilt grain boundary in YBCO. In contrast to the grain characteristics, the grain boundary shows a foot structure, which is due to thermally activated phase slippage [10.17, 10.59].

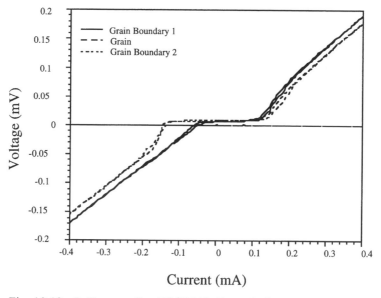

Fig. 10.13. *I–V* curve for 45°/[001] tilt grain boundary in YBCO, showing Josephson junction type behavior [10.58].

defined grain boundary facet, but display a meandering geometry. Thus, the macroscopic geometry of the grain boundary plane is not necessarily the one that is present at the microscopic scale. The superconductive coupling across a meandering grain boundary is not well understood at present. Since the phase relations between the Cooper pairs are important, the supercurrent coupling across such a convoluted interface must be extremely complex. Consequently, no simple Fraunhofer pattern has been observed for these grain boundaries when the critical current was measured as a function of the magnetic field [10.19]. To obtain a more uniform orientation of the grain boundary plane at the microscopic level it would be desirable to study considerably smaller junction widths in samples that contain extended facets of uniform geometry.

As mentioned above, widely different structures and properties were obtained in films prepared under different synthesis conditions. Although they all showed the meandering structure, the most important structural features affecting the transport behavior were non-superconducting precipitates at the grain boundary and structural disorder at the grain boundary. The grain boundary that was resistive had a considerable fraction of the grain boundary decorated by precipitates; however, the sections which were free of precipitates indicated within a width of about 1 nm a region of structural disorder, as shown in Fig. 10.14 [10.17, 10.19]. The bright contrast at this grain boundary also suggests a reduced density at the grain boundary. Thus, such a limited perturbance of the grain boundary structure is sufficient to strongly inhibit the current transport across the grain boundary. Nevertheless, detailed normal state transport measurements on such a grain boundary, including $1/f$ noise measurements, revealed certain similarities to the transport behavior of superconducting grain boundaries [10.62]. This appears to be largely due to microscopic regions that maintain a limited coupling across the grain boundary.

The superconducting grain boundary shown in Figs 10.4(c) and (d) is atomically very well structured and also basically consists of only one type of

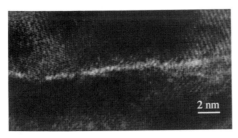

2 nm

Fig. 10.14. This HREM image of a resistive grain boundary shows structural disorder within an approximately 10 Å wide region at the grain boundary [10.17].

grain boundary. Although the grain boundary meanders, practically all interfaces consist of (100)(110) asymmetric facets. Transport measurements on this grain boundary show distinct differences in the magnetic field dependence compared to other superconducting 45°/[001] tilt grain boundaries. The weak-link behavior of this grain boundary can be understood by a model that assumes that at low fields, flux penetration along the boundary is essentially limited to normal regions, for example precipitates, at the grain boundary. In modeling the magnetic field dependence, it was found that the size of the well connected regions was very important [10.19].

This work clearly demonstrated that it is important to establish direct correlations between microstructure and transport properties of grain boundary junctions. At this point, both the complex and varied structures of grain boundaries that have been found and the extremely complicated transport behavior, present considerable challenges concerning the development of models for the connection between structure and transport properties. However, it is clear that progress in this field is directly dependent on our ability to develop such models.

10.6 Discussion

An understanding of the influence of grain boundary microstructure on the transport behavior requires consideration of the grain boundary structures and composition at the macroscopic, mesoscopic and microscopic levels. To date very few investigations have been able to establish direct connections between microstructures and transport properties, and those have been limited to rather special interface geometries. Moreover, complete characterization at the different levels is not always possible. In addition, grain boundary geometries often are not well controlled, even in thin-film experiments. However, the best promise for making progress in this field is the study of geometrically well defined grain boundaries of well characterized atomic-level structures, which could be accomplished for a limited range of well chosen macroscopic geometries.

The role of the grain boundary plane needs to be established, especially in view of anisotropic superconducting order parameters. In thin films, considerably smaller grain boundary junctions than those used in our work would be feasible and it should thus be possible to prepare samples with reasonably well defined facets. Of course, at inclinations such as for the 45°/[001] grain boundaries, where asymmetric facets dominate supposedly because of their much lower energy than other inclinations (see Fig. 10.4), symmetric facets may not form, at least not exclusively. Whenever there is coexistence of different facets it will generally not be possible to separate the effects of one type of facet from the other.

Similar to the meandering grain boundaries in YBCO films, the grain boundary plane is often not well defined in bicrystals and may include twist as well as tilt or general grain boundaries [10.42, 10.63]. Therefore, the role of the grain boundary plane inclination is difficult to assess at present. Comparisons between different grain boundary geometries have been performed in a number of investigations, using EELS studies, but typically no characterization of the transport behavior and structure was available for those grain boundaries [10.56, 10.57, 10.64, 10.65]. Characterizations of the microstructure alone without electromagnetic characterizations are clearly not sufficient to provide the necessary insights into the grain boundary transport behavior.

On the other hand certain features of the atomic-level grain boundary structure directly impact on transport behavior. For example, the well-connected lattice planes in low-angle grain boundaries, which are found in between the grain boundary dislocations, suggest good coupling within these regions, whereas the supercurrent transport would be largely excluded at grain boundary dislocations. Thus, the grain boundary transport may be described by a strong channel and weak channel in parallel. The weak channel may be caused by either a reconstructed dislocation core structure and/or the severely strained lattice at the dislocation cores. Considerable coherence between the lattices can be observed in simple oxides (for example in the $\theta = 22.6°$ grain boundary in Fig. 10.6) up to angles much higher than the transition from the strong to weak coupling ($\sim 10°$). In analogy to this, [001] tilt grain boundaries in YBCO are expected to be similarly structured. Such structures, although weak-linked, may possess more uniform and stronger coupling than incoherent sections of high-angle grain boundaries (e. g. (310) symmetric grain boundaries, Fig. 10.5).

Indeed, there have been considerable improvements recently in the noise characteristics of SQUIDS produced from YBCO films grown on $SrTiO_3$ bicrystals at misorientations near 24°. It appears possible that the weak-link character of such grain boundaries would be tunable to a certain extent by adjusting the misorientation angle. Although coherence may play an important role in determining grain boundary transport characteristics, chemical disorder, or introduction of point defects at grain boundary cores could have an even stronger effect on grain boundary properties.

Recent investigations of 45°/[001] tilt grain boundaries have found considerably higher J_c values than had previously been reported for grain boundaries with similar macroscopic geometry (see Fig. 10.1). The former were produced by the sputter-induced epitaxy modification technique and were extremely well structured atomically (see Fig. 10.4(d)). Thus, the particular processing conditions used can have a profound influence on the quality of the resultant grain boundaries.

For junctions of finite width, the transport behavior can be extremely complex, since each section of the grain boundary may behave differently. Moreover, information on the magnetic flux distribution is needed for modeling the transport behavior. It would be desirable to directly relate flux line behavior to microstructural features. Magneto-optic techniques may be used for information on the flux penetration at grain boundaries, albeit at relatively low resolution ($\sim$1 μm). Although considerable progress has been made recently in establishing direct correlations between grain boundary transport and structure, our knowledge of the connection between structure and properties is rather elementary at present.

10.7 Summary and conclusions

Electron microscopy techniques are essential tools needed for the investigation of the local structure and composition of grain boundaries in high-temperature superconductors. While we have emphasized structural aspects here, analytical characterizations (see Chapters 8 and 11) are extremely important and must be part of establishing direct connections to transport properties.

In order to establish correlations between structure and transport properties for grain boundaries in high T_c superconductors it is necessary both to perform investigation on the same sample and to characterize the grain boundary structure over the whole range of length scales associated with a grain boundary junction, down to the atomic level. Further progress in our understanding of these complicated interrelations is expected from experiments on extended, well developed grain boundary facets that are distinguished by a single inclination of the grain boundary plane.

Investigations of the structure of grain boundaries in high-temperature superconductors have given a rather complex picture. Improvements in obtaining more uniform grain boundaries, structurally and compositionally, will have considerable impact not only on grain boundary structure/property research, but also on the technologically important ability to achieve highly reproducible junction behavior.

Acknowledgments

This work was supported by the U.S. Department of Energy, Basic Energy Sciences–Materials Science under Contract No. W-31-109-ENG-38 (KLM) and the National Science Foundation (DMR 91-20000) through the Science and Technology Center for Superconductivity (YG and BVV).

References

[10.1] K. Char *et al.*, *Appl. Phys. Lett.* **59**, 733 (1991).
[10.2] P. Chaudhari *et al.*, *Phys. Rev Lett.* **60**, 1653 (1988).
[10.3] D. Dimos *et al.*, *Phys. Rev. Lett.* **61**, 219 (1988).
[10.4] D. Dimos, P. Chaudhari & J. Mannhart, *Phys. Rev. B* **41**, 4038 (1990).
[10.5] S. E. Babcock *et al.*, *Nature* **347**, 167 (1990).
[10.6] C. B. Eom *et al.*, *Nature* **353**, 544 (1991).
[10.7] Z. G. Ivanov *et al.*, *Supercond. Sci. Technol.* **4**, 439 (1991).
[10.8] D. Wolf, in *Materials interfaces, Atomic-level structure and properties*, eds. D. Wolf & S. Yip (Chapman & Hall London 1992) p. 1.
[10.9] D. Wolf & K. L. Merkle, in *Materials interfaces, Atomic-level structure and properties*, eds. D. Wolf & S. Yip (Chapman & Hall, London, 1992), p. 87.
[10.10] D. A. Smith & R. C. Pond, *Int. Met. Rev.* **205**, 61 (1976).
[10.11] R. W. Balluffi, A. Brokman & A. H. King, *Acta metall.* **30**, 1453 (1982).
[10.12] A. H. King, A. Singh & J. Y. Wang, *Interface Science* **1**, 347 (1993).
[10.13] A. P. Sutton & R. W. Balluffi, *Acta Metall.* **35**, 2177 (1987).
[10.14] K. L. Merkle, *Scripta Met.* **23**, 1487 (1989).
[10.15] D. M. Duffy & P. W. Tasker, *Phil. Mag. A* **47**, 817 (1983).
[10.16] D. Wolf, *Physica B* **131**, 53 (1985).
[10.17] B. V. Vuchic *et al.*, *Mat. Res. Soc. Proc.* **357**, 419 (1995).
[10.18] D. J. Miller *et al.*, *Appl. Phys. Lett.* **66**, 2561 (1995).
[10.19] B. V. Vuchic, Thesis, Northwestern University (1995).
[10.20] B. V. Vuchic *et al.*, *J. Mat. Res.* **11**, 2429 (1996).
[10.21] K. L. Merkle & D. J. Smith, *Ultramicroscopy* **22**, 57 (1987).
[10.22] K. L. Merkle, *Mat. Res. Soc. Proc.* **82**, 383 (1987).
[10.23] M. G. Norton & C. B. Carter, in *Materials interfaces, Atomic-level structure and properties*, eds. D. Wolf & S. Yip (Chapman & Hall London 1992) p. 151.
[10.24] D. R. Clarke, *J. Amer. Ceram. Soc.* **70**, 15 (1987).
[10.25] D. R. Clarke, *J. Amer. Ceram. Soc.* **72**, 1604 (1989).
[10.26] H. Kleebe *et al.*, *J. Amer. Ceram. Soc.* **76**, 1969 (1993).
[10.27] U. Balachandran *et al.*, *Ceramic Transactions* **18**, 341 (1991).
[10.28] K. L. Merkle & D. J. Smith, *Phys. Rev. Lett.* **59**, 2887 (1987).
[10.29] K. L. Merkle & D. J. Smith, *Mat. Res. Soc. Symp. Proc.* **122**, 15 (1988).
[10.30] K. L. Merkle, *J. Phys. Chem. Solids* **55**, 991 (1994).
[10.31] K. L. Merkle, *Interface Science* **2**, 311 (1995).
[10.32] N. D. Browning *et al.*, *Interface Science* **2**, 397 (1995).
[10.33] D. Wolf, *Scripta Met.* **23**, 1913 (1989).
[10.34] H. Ichinose & Y. Ishida, *Phil. Mag. A* **43**, 1253 (1981).
[10.35] H. Ichinose & Y. Ishida, *J. de Phys.* **46 C4**, 39 (1985).
[10.36] W. Krakow & D. A. Smith, *J. Mater. Res.* **1**, 47 (1986).
[10.37] K. L. Merkle *et al.*, in *Ceramic microstructures '86*, eds. J. A. Pask & A. G. Evans (Plenum, New York, 1987) p. 241.
[10.38] K. L. Merkle & D. Wolf, *Phil. Mag. A* **65**, 513 (1992).
[10.39] U. Dahmen & N. Thangaraj, *Mat. Sci. Forum* **126–128**, 45 (1993).

[10.40] L. K. Fionova & A. V. Artemyev, *Grain boundaries in metals and semiconductors*, (Les éditions de physique, Les Ulis, 1993).

[10.41] D. Wolf in, *Ceramic microstructures '86*, eds. J. A. Pask and A. G. Evans editors (Plenum, New York, 1987) p. 177.

[10.42] S. E. Babcock *et al.*, *Nature* **347**, 167 (1990).

[10.43] Y. Gao *et al.*, *Physica C* **173**, 487 (1991).

[10.44] C. L. Jia & K. Urban, *Interface Science* **1**, 291 (1993).

[10.45] Y. Gao *et al.*, *Physica C* **174**, 1 (1991).

[10.46] M. F. Chisholm & S. J. Pennycook, *Nature* **351**, 47 (1991).

[10.47] S. E. Babcock & D. C. Larbalestier, *Appl. Phys. Lett.* **55**, 393 (1989).

[10.48] S. E. Babcock *et al.*, *J. of Adv. Sci. (Japan)* **4**, 119 (1992).

[10.49] R. Gross *et al.*, *Phys. Rev. Lett.* **64**, 228 (1990).

[10.50] R. Gross *et al.*, *Appl. Phys. Lett.* **57**, 727 (1990).

[10.51] R. Gross & B. Mayer, *Phys. C* **180**, 235 (1991).

[10.52] E. A. Early *et al.*, *Phys. Rev. B* **50**, 9409 (1994).

[10.53] J. A. Alarco *et al.*, *Ultramicroscopy* **51**, 239 (1993).

[10.54] C. Traeholt *et al.*, *Physica C* **230**, 425 (1994).

[10.55] S. J. Rosner, K. Char & G. Zaharchuk, *Appl. Phys. Lett.* **60**, 1010 (1992).

[10.56] N. D. Browning, M. F. Chisholm & S. J. Pennycook, *Interface Science* **1**, 309 (1993).

[10.57] V. P. Dravid, H. Zhang & Y. Y. Wang, *Physica C* **213**, 353 (1993).

[10.58] B. V. Vuchic *et al.*, *IEEE Trans. Appl. Supercond.* **5**, 1225 (1995).

[10.59] B. V. Vuchic *et al.*, *J. Appl. Phys.* **77**, 2591 (1995).

[10.60] B. V. Vuchic *et al.*, *Appl. Phys. Lett.* **67**, 1013 (1995).

[10.61] B. V. Vuchic *et al.*, *Physica C* **270**, 75 (1996).

[10.62] L. Liu *et al.*, *Phys. Rev. B* **51**, 16164 (1995).

[10.63] D. C. Larbalestier *et al.*, *Physica C* **185–189**, 315 (1991).

[10.64] Y. Zhu, Z. L. Wang & M. Suenaga, *Phil. Mag. A* **67**, 11 (1993).

[10.65] S. E. Babcock *et al.*, *Physica C* **227**, 183 (1994).

[10.66] Y. Gao *et al.*, *Physica C* **174**, 1 (1991).

11

The atomic structure and carrier concentration at grain boundaries in YBa$_2$Cu$_3$O$_{7-\delta}$

N. D. BROWNING, M. F. CHISHOLM
and S. J. PENNYCOOK

11.1 Introduction

It is clear from the wealth of transport measurements involving both thin films [11.1–11.3] and bulk materials [11.4] that grain boundaries have a strong effect on the transport properties of high T_c materials. Perhaps the most likely source of this effect is the small superconducting coherence length (5–15 Å), which makes the high T_c superconductors extremely sensitive to defects that distort the perfect crystal structure. Electron microscopy is the only experimental technique capable of analyzing these defects on the scale of the coherence length. In particular, the scanning transmission electron microscope (STEM) allows both Z-contrast imaging and electron energy loss spectroscopy (EELS) to be performed with atomic resolution ($\sim$2.2 Å). By using these techniques it is possible to relate the changes in hole concentration, measured from the energy loss spectrum, with defined atomic locations at grain boundaries observed in the image [11.5,11.6]. Such a combination of techniques therefore provides insight into the structure–property relationships of grain boundaries at the fundamental atomic level. In this chapter, the application of these techniques to investigate the dramatic changes in carrier concentrations associated with grain boundaries in YBa$_2$Cu$_3$O$_{7-\delta}$ will be discussed.

11.2 Imaging and microanalysis of boundary structures

The specimens used in this study are laser ablated thin films of YBa$_2$Cu$_3$O$_{7-\delta}$ (YBCO) [11.7] grown on SrTiO$_3$ bicrystal substrates. This method of boundary preparation has been chosen as it is known to produce clean grain boundaries, i.e. free from impurity segregation or copper enrichment, in which the oxygen stoichiometry can be well controlled through annealing. Such specimens therefore contain the inherent limits to the superconducting properties of YBCO grain boundaries, which can only be degraded further by copper enrichment and oxygen deficiency.

11.2.1 Atomic structure determination

Atomic structure determinations are made using the high-angle annular dark-field or *Z*-contrast imaging technique. *Z*-contrast images [11.8–11.11] are formed by collecting the high-angle scattering on an annular detector (see Chapter 4) and synchronously displaying its integrated output on a TV screen as the probe is scanned across the specimen. Detecting the scattering at high angles and over a large angular range, allows each atom to be considered to scatter independently with a cross section approaching a Z^2 dependence on atomic number. This cross section effectively forms an object function that is strongly peaked at the atom sites. The detected intensity thus consists of a convolution of this object function with the probe *intensity* profile, bypassing the phase problem that makes image interpretation difficult in conventional TEM (Fig. 11.1).

The small width of this object function ($\sim$0.1 Å) means that the spatial resolution is limited only by the probe size of the microscope. For a crystalline material in a zone-axis orientation, where the atomic spacing is greater than the probe size, the atomic columns can be illuminated individually. Therefore, as the probe is scanned over the specimen, an atomic resolution compositional map is generated in which the intensity depends on the mean square atomic number of the atoms in the columns. This result holds even for thicker specimens where dynamical diffraction results only in columnar channeling [11.9, 11.10], and simply scales the scattering cross sections of the object function according to thickness. With this methodology, changes in focus and

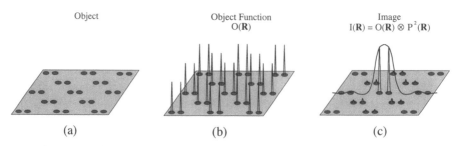

Fig. 11.1. The specimen consists of an array of atomic columns (a) for which the potential for high-angle scattering can be represented by an object function (b). The experimental image can be interpreted as a simple convolution of the experimental probe and the object function (c).

thickness do not cause contrast reversals in the atomic resolution image, so that atomic column sites can be identified unambiguously during the experiment. This enables atomic columns at grain boundaries to be located without the need for simulated images, even for column separations below the resolution limit which simply result in a single elongated image feature.

Figure 11.2 shows a typical *Z*-contrast image of a YBCO 30° [001] tilt boundary grown by laser ablation on a $SrTiO_3$ bicrystal substrate. In the image, the brighter columns consist of Y and Ba atoms, and the less bright columns consist of Cu(1), Cu(2) and O(4) atoms. Columns consisting solely of oxygen atoms, i.e. those containing O(2) or O(1) and O(3) atoms, are not imaged by this technique.

The high-*Z* atomic column positions at a grain boundaries can therefore be determined directly from the image during the experiment. However, on completion of the microscopy, a more accurate statistical method to obtain the column positions is to recover the *Z*-contrast object function through maximum entropy image analysis [11.12]. Maximum entropy image analysis is based on

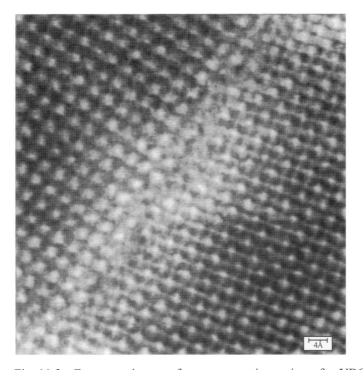

Fig. 11.2. *Z*-contrast image of an asymmetric section of a YBCO 30° [001] tilt boundary. For this boundary, grown on a similarly oriented $SrTiO_3$ substrate, an asymmetric boundary plane was the predominant feature.

Bayesian probability theory, and is an iterative technique designed to produce an object function which, when convoluted with the microscope probe, gives the best fit to the experimental image. The iterative process starts with an object function of uniform intensity, which makes no prior assumptions about the structure of the specimen, and moves towards the best fit to the experimental image by calculating the maximum entropy while minimizing the χ^2 fit with the experimental image (Fig. 11.3). This process allows the coordinates for the atom column positions to be determined to an accuracy of $\sim$0.1 Å [11.13].

In addition to unambiguous location of atomic columns, the Z-contrast technique has other distinct features that make it ideal to study high T_c grain boundaries. Z-contrast is extremely sensitive to the surface of the specimen, as the presence of an amorphous phase causes the probe to be broadened and the resolution to be degraded. Therefore, by simply obtaining a Z-contrast image, the specimen is known to be relatively free from surface damage effects that can be induced during the ion-milling process. Any changes in the resolution of the image towards the grain boundary are an indication of a varying amorphous layer. This is of crucial importance for microanalysis, as the signal

Fig. 11.3. The Z-contrast image of the YBCO asymmetric 30° [001] tilt boundary shown in Fig. 11.2 after maximum entropy processing.

detected is thickness integrated and significant errors can be induced from unknown surface compositions. However, perhaps the most important feature of the Z-contrast imaging technique, is that the image only uses the high-angle scattering. This means that it can be acquired simultaneously with an energy loss spectrum and, in fact, the Z-contrast image can be used to position the probe for spectroscopy with atomic precision. A knowledge of the exact location of the probe on the interface means that spectra can be correlated directly with the structure.

11.2.2 Hole concentration measurement

The structures of all of the known high T_c superconductors consist of CuO_2 sheets separated by insulating oxide layers. Superconductivity is induced by charge transfer between the insulating layers and the CuO_2 sheets and may be controlled by the presence of defects in these layers. For YBCO, the creation of charge carrying holes in the CuO_2 sheets is achieved primarily through oxygen doping, which results in the formation of unoccupied states near the Fermi level. As the bonding in YBCO occurs between the oxygen 2p states and the metal orbitals, the fine structure of the oxygen K-absorption edge spectrum accurately reflects the density of these unoccupied states [11.14–11.17]. In fact, the states responsible for superconductivity are immediately apparent in the spectrum in the form of a pre-edge feature to the main oxygen absorption edge (Fig. 11.4).

A detailed analysis of the electronic states in the pre-edge using dedicated EEL spectrometers has revealed the symmetry of these holes [11.14, 11.15]. These results can be used as a guide in interpreting the spectra obtained from the electron microscope, where the use of circular apertures to form the small probe averages the momentum transfers in all directions and precludes such an analysis. However, it must be stated that although it is not possible to address the symmetry of the hole states in the same detailed manner as by dedicated spectrometers, optimum collection conditions can be defined which reduce orientation effects on the accuracy of the hole concentration measurement [11.18]. Therefore, by fitting simple Gaussian functions to the features of the spectrum, i.e. the carrier states, the upper Hubbard band and the main absorption edge intensity, the local carrier concentration can be quantified to an accuracy of 5% [11.17].

Crucial in the use of EELS to study carrier concentrations at grain boundaries in high T_c materials is the ability to position the probe with atomic precision; a precision only afforded by the Z-contrast image. In order to achieve with EELS the same atomic resolution as the image, the range over

268   *N. D. Browning, M. F. Chisholm and S. J. Pennycook*

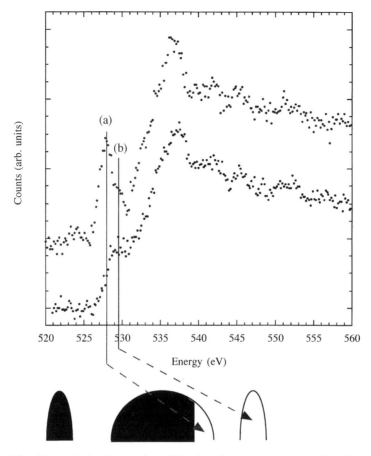

Fig. 11.4. A simple model of the band structure shows that for a nominally fully oxygenated sample of YBa$_2$Cu$_3$O$_{7-\delta}$ (i.e. $\delta \sim 0$); the holes responsible for superconductivity reside in the predominantly oxygen 2p valence band (a). As the oxygen deficiency, δ, is increased, the number of holes in the valence band is reduced until at $\delta \sim 0.6$ superconductivity is lost and a charge transfer insulator is created (b). Also shown are the corresponding oxygen K-edge spectra.

which a fast electron can cause an excitation event must be less than the interatomic spacing. Hydrogenic models [11.19–11.21] show that for the oxygen K-edge, the object function is localized within 1 Å of the atom core. Hence, like the Z-contrast image, we have an object function localized at the atom cores and an experimental probe of atomic dimensions. For zone-axis orientations, providing we maintain a large collection angle ($\sim$15 mrad), coherent effects will be averaged and the description of the spectrum in terms of a convolution of the probe with an object function is again valid (Fig. 11.1).

Probe channeling again preserves the spatial resolution, which allows atomic resolution analysis to be achieved even with large collection apertures.

The Z-contrast image can therefore be used to position the electron beam with atomic precision and the energy loss spectra can be obtained from defined crystallographic locations. Figure 11.5 shows quantified pre-edge profiles obtained from two different high-angle [001] tilt grain boundaries in a polycrystalline YBCO thin film. The spectra were recorded in single unit cell steps (~4 Å) across the boundary, and the pre-edge feature quantified as described earlier [11.16]. In any high resolution analysis of high T_c materials, electron beam induced damage is an important consideration. Previous studies of damage in the STEM have revealed that for probe sizes of atomic dimensions, acquisition times need to be limited to a few seconds [11.16]. In this experiment, two spectra, each with an acquisition time of 5 s, were obtained at each position of the probe. The two spectra from each position were quantified separately to ensure that no electron beam damage occurred during the spectroscopy [11.22].

These profiles show strikingly different hole concentrations for two high-

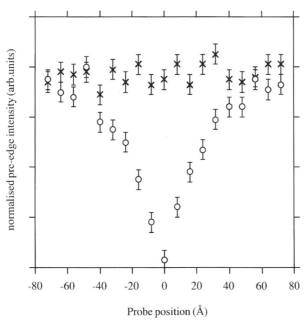

Fig. 11.5. Quantifying the change in pre-edge intensity by a normalized Gaussian fit, shows that within the errors of the experiment there is no hole depletion around the symmetric boundary (x), whereas the asymmetric boundary shows a hole depletion zone extending 40–50 Å either side of the boundary (o).

angle grain boundaries. Both boundaries are in the regime where the super-conducting transport properties are expected to be very poor [11.1–11.4], yet for at least one of them there is no measurable hole depletion. The most obvious difference between the two boundaries is that the broad hole depletion region corresponds to a 28° *asymmetric* [001] tilt boundary, and the non-depleted region corresponds to a 36° *symmetric* [001] tilt boundary. At this point it is important to remember that what is being measured by the energy loss spectrum is the *hole* concentration. For YBCO, it is often too easy to make the connection between hole concentration and oxygen stoichiometry, as in the bulk material oxygen doping creates the charge transfer between the insulating and CuO_2 planes that results in a superconductor. However, the holes reside in bands that are composed of hybridized Cu—O bonds. As such, a change in copper valence will also affect the hole concentration and therefore be meas-ured by the spectrum. What is clear from the profiles in Fig. 11.5 is that there is a fundamental difference between these asymmetric and symmetric boundary planes that results in one showing no hole depletion and another showing a broad depletion zone.

11.3 Structural models

The structure of grain boundaries has been the subject of extensive study for the last 50 years [11.23–11.25]. In this time, it has become convenient to classify grain boundaries into two broad regimes; low angle and high angle. Here, the application of grain boundary models in these two regimes for YBCO will be discussed specifically for [001] tilt boundaries.

11.3.1 Low-angle grain boundaries

The definition of low-angle grain boundaries generally covers the range of misorientations from 0° to 10°. In this regime, the grain boundary plane can be considered to be a linear array of separated dislocation cores. For [001] tilt boundaries in YBCO, the boundary plane will be composed of [100] or [010] dislocations, as shown in Fig. 11.6 (for YBCO the small distortion between the *a*- and *b*-axes needs to be incorporated for quantitative models, but structurally results in no observable differences in the dislocation cores).

The advantage of considering the boundary as an array of dislocation cores is that the strain field around the core can be calculated from linear elasticity theory [11.26]. Here, the lattice either side of the grain boundary is assumed to be unstrained and is used as a reference for the lattice positions in the vicinity of the dislocation. However, such models do not specifically take into account

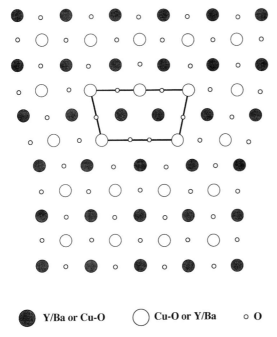

Y/Ba or Cu-O   Cu-O or Y/Ba   o O

Fig. 11.6. Schematic of a [100] or [010] dislocation core in YBCO. The plane of atoms that is removed to create the dislocation core may be either Cu—O or Y/Ba—O. Associated with the removal of a plane of atoms is a strain field that can be calculated from linear elasticity theory.

the atomic structure in the dislocation core and simply calculate the strain that would be present given the atom positions and a single plane being removed to create the dislocation core. In addition, for a multicomponent system like YBCO no distinction is drawn between the two possible sub-lattices, i.e. an [010] dislocation core centered on a copper column or one centered on an yttrium/barium column. Nevertheless, despite lacking a definition of the exact atomic structure in the dislocation core, linear elasticity quantitatively reproduces the critical current behavior of low-angle grain boundaries. In the model proposed by Chisholm & Pennycook [11.27], a strain of 1% was used as the cut-off between YBCO being superconducting and non-superconducting. This cut-off value was taken from the fact that a 1% strain causes YBCO to be tetragonal and no superconducting phase of YBCO is tetragonal. As grain boundary misorientation is increased, the strain field around the dislocation core increases and the separation of the dislocation cores decreases. Hence, progressively more of the grain boundary becomes non-superconducting until at the end of the low-angle regime the strain fields overlap. At this point it is impossible to use linear elasticity to calculate the grain boundary strain.

The dislocation core structure predicted in Fig. 11.6 has been observed experimentally in low-angle grain boundaries. In Fig. 11.7, two dislocation cores in YBCO are shown. In the first case the dislocation core is constructed from a missing Cu—O plane and in the second case by a missing Y/Ba—O plane. Dislocation cores in low-angle grain boundaries can therefore form on either sub-lattice. An interesting feature of both cores is that there exist atomic locations where the columns seem too close together. In such situations, like-ion repulsion would be expected to preclude such a structure.

However, if we remember that the Z-contrast image, like any transmission image, is simply a two-dimensional projection of the 3-dimensional crystal structure, a solution to this problem is for only one of the two sites in each

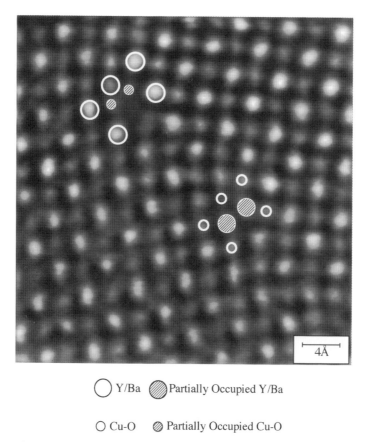

○ Y/Ba   ◉ Partially Occupied Y/Ba

○ Cu-O   ◉ Partially Occupied Cu-O

Fig. 11.7. Z-contrast image of two [100] dislocation cores in YBCO. The two dislocation cores have the same basic structure with the difference being that one is centered on the copper sub-lattice and the other is centered on the Y/Ba sub-lattice.

perovskite block to be occupied. If alternate sites are chosen, we would still see two columns in projection but avoid like-ion repulsion. An alternate view of these two 'half-columns' is that they represent a single atomic column that is distorted through the thickness of the crystal in a regular manner, a dislocation core reconstruction.

11.3.2 High-angle grain boundaries

At the point where the strain fields around dislocation cores begin to overlap, i.e. the end of the low-angle regime, models of grain boundaries based on linear elasticity become inapplicable. At this stage, an alternate methodology to describe the grain boundary structure is by structural units [11.28, 11.29]. These structural units are equivalent formally to a dislocation core model of the boundary and evidence suggests that in the case of the perovskites they are in fact the same core structures as seen in isolated dislocations [11.30]. The structural unit model has the advantage that once the structural units have been determined, it is possible to predict the structure of a grain boundary at any misorientation. However, similar to the dislocation core model for low-angle boundaries, structural units only define the type of structure that occurs at the grain boundary and on their own can not provide information on the structure–property relationships.

The structural unit model has been used successfully to predict the structures of grain boundaries in perovskite structured $SrTiO_3$ bicrystals [11.31–11.34]. The structural units observed for symmetric $SrTiO_3$ [001] tilt boundaries are shown in Fig. 11.8. In a similar manner to the isolated dislocation cores in YBCO, the structural units also appear to contain atomic positions where the cations are too close together. Again, depending on the structural unit, the close separation of the atomic columns can occur for either of the sub-lattice sites, i.e. the Ti—O columns or the Sr columns.

These bicrystals are the substrates that have been typically used for the preparation of individual grain boundaries in YBCO thin films [11.1–11.3]. The reason that $SrTiO_3$ is chosen as the substrate for YBCO is primarily due to the closeness of the lattice parameters (3.905 Å for $SrTiO_3$ compared with $a = 3.81$ Å and $b = 3.88$ Å in YBCO). In view of the fact that YBCO, and for that matter all of the high T_c superconductors, have structures that are essentially perovskite, it is reasonable to assume that the structure of the YBCO grain boundary would follow that of the $SrTiO_3$ bicrystal. Indeed, in terms of the dislocation core models, the only difference between $SrTiO_3$ and YBCO are that the Sr columns are replaced by Y/Ba columns and the Ti—O columns are replaced by Cu—O columns. The structure in the [001] projection is the

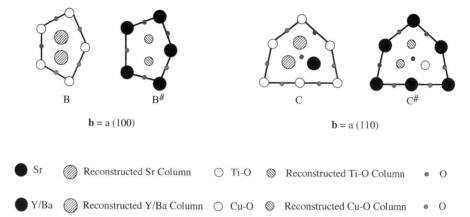

| ● Sr | ◍ Reconstructed Sr Column | ○ Ti-O | ◌ Reconstructed Ti-O Column | • O |
| ● Y/Ba | ◍ Reconstructed Y/Ba Column | ○ Cu-O | ◌ Reconstructed Cu-O Column | • O |

Fig. 11.8. The structural units for symmetric $SrTiO_3$ [001] tilt grain boundaries. Each of the units contains a reconstructed atomic column projecting as
two closely spaced sites, either a Ti—O column or a Sr column. Asymmetric
[001] tilt grain boundaries use only the units with the Ti—O reconstruction.

same and the lattice parameter is changed by less than 3%. Therefore, by being
able to predict the $SrTiO_3$ grain boundary structures it should be possible to
predict the structure of the thin-film YBCO boundary.

This simple argument of course neglects the kinetics of thin film growth. It
is well known that YBCO film growth occurs through the nucleation of three-
dimensional islands (see Chapter 4). The growth of the films takes place in
conditions far from equilibrium and, as such, it is hard to imagine a case where
the boundary in the film will follow the substrate boundary exactly. This is
consistent with TEM observations [11.35–11.37] which show the grain boundary plane to meander or facet around the boundary direction defined by the
substrate. Such boundaries are unlikely to contain the well defined symmetric
structure observed in the bicrystal substrate. In fact, even in the substrate,
which is prepared in bulk form and annealed under equilibrium conditions,
asymmetric facets are occasionally seen [11.31]. Facetting, when it occurs,
invariably creates segments of asymmetric grain boundary [11.38, 11.39].
Observations from $SrTiO_3$ show that these asymmetric facets are composed of
a subset of the structural units seen at symmetric grain boundaries, and again
occur in well defined and predictable sequences [11.33]. Only the units with
reconstructed Ti—O columns occur on asymmetric boundaries. There is therefore a critical difference between asymmetric and symmetric grain boundaries,
in that the structural units in asymmetric grain boundaries are centered on only
one sub-lattice. From the Z-contrast images of the dislocations in YBCO, it is
clear that YBCO and $SrTiO_3$ form the same type of isolated dislocation core

structures in low-angle grain boundaries (structural units B and B* in Fig. 11.8 are identical to the dislocation core structures in Fig. 11.7). For high-angle boundaries, the key question is whether there is any relationship between the structure of the facetted YBCO boundary and the structure of the bicrystal boundary. Figure 11.9 shows a magnified image of a section of the asymmetric boundary shown in Fig. 11.3, with the structural units superimposed upon it. The structural units observed for this YBCO grain boundary are the same as observed for SrTiO$_3$. For asymmetric YBCO grain boundaries this means the reconstructed atomic columns occur only on the copper sub-lattice. Therefore, while it is known that the YBCO grain boundary will facet around the orientation of the substrate, the structure of each individual facet can be predicted using the structural units in Fig. 11.9. The overall properties of the boundary will then simply be a sum of all the individual facets.

11.4 Predicting bulk structure–property relationships

By using the dislocation core model for low-angle grain boundaries and the structural unit model for high-angle grain boundaries, it is possible to predict the types of structures that will be present in a YBCO grain boundary at a given misorientation angle. However, these models alone cannot provide any information on the properties associated with each structure. To determine the effect that each structure has on, for example, the transport properties, it is necessary to examine the details of the atomic arrangements in the core of each structure.

One means of investigating the effect of atomic arrangement on the properties of the boundary is through bond-valence sum analysis. Bond-valence sums originate through a concept by Pauling [11.40], in which the formal valence of a given ion is distributed between its bonds to its nearest neighbors. The formal valence of an ion is therefore determined primarily by its bond length. Such analysis is routinely used in X-ray diffraction measurements to see if proposed crystal structures contain elements in reasonable valence states. This concept has been adapted by Altermatt & Brown [11.41, 11.42], to provide a simple expression by which the valence of an ion can be determined from its bond length. For the majority of known crystalline structures, the following expression predicts the valence to within 10% of the formal value

$$S = \exp[(r_0 - r_{ij})/B], \tag{11.1}$$

where r_0 is a constant characteristic of the elements in the bond, B is a constant which is assigned the value 0.37 by fitting to experimental data from a wide range of materials, and r_{ij} is the bond length.

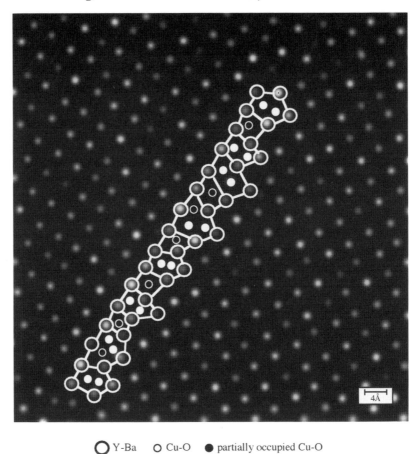

○ Y-Ba   ○ Cu-O   ● partially occupied Cu-O

Fig. 11.9. Maximum entropy image of a 30° asymmetric [001] tilt grain boundary in YBCO, obtained from Fig. 11.3 by convolution with a narrow Gaussian. The structural units of the boundary are the same as observed for $SrTiO_3$.

For application to YBCO, it has been pointed out that bond-valence calculations are empirical and cannot be used to determine the valences of the elements involved to better than around 10% [11.43]. However, in the case of the boundary structures observed here, the atomic positions can only be determined with an accuracy of 0.1 Å, making any errors induced by the bond-valence sum analysis second order. As such, the bond-valence sums can be used to indicate positions where the valence of the elements involved changes considerably, although relating the magnitude of the change to properties must be approached with care.

In a perfect unit cell of YBCO, the valences of most of the elements involved change very little between $YBCO_7$ and $YBCO_6$ (<10%). The exception to this is copper. In fully oxygenated YBCO the copper(I) valence is ~2.3, whereas in fully oxygen deficient YBCO the copper valence is ~1.2 [11.44] (note that the copper(II) valence changes by <10%). This implies that the copper(I) valence is a very sensitive measure of the number of charge carriers present in the structure. Of particular importance for the study of charge carriers at grain boundaries is the fact that the change in valence is large and should be easily measured by bond-valence sum analysis. Hence, for the structures observed in the Z-contrast image it is possible to perform bond-valence sum analysis to manipulate the atomic column positions (within the 0.1 Å error in their location) so that the valences of the yttrium, barium, oxygen and copper(II) atoms are within 10% of their expected values for fully oxygenated YBCO. The resulting copper(I) valence will thus give an estimate of the number of charge carriers present in the structure. Note that this analysis assumes the structure to be *fully* oxygenated. Oxygen deficiency will only degrade the carrier concentration still further.

Analysis of the copper valence around the two dislocation cores in Fig. 11.7 is shown in Fig. 11.10 (for all of the other elements, valences are within 10%

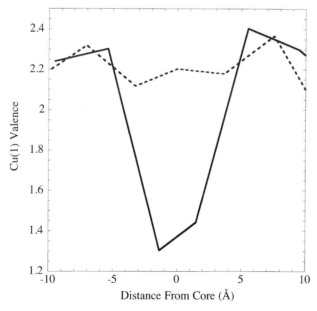

Fig. 11.10. A dramatic decrease in the copper(I) valence is observed (full line) at isolated [100] dislocation cores containing reconstructed copper columns. However, for the dislocation core containing reconstructed Y/Ba columns, the decrease in the copper valence is negligible (broken line).

of the values expected for perfect unit cells). It is immediately noticeable that the particular sub-lattice of the dislocation core has a strong effect on the copper valence. It may not be surprising that the reduction in copper valence occurs at the core centered on the copper sub-lattice, as reconstructing the copper columns perturbs the atoms most likely to change their valence, but it certainly has a profound effect on the superconducting properties of the dislocation cores. Dislocation cores that involve partial copper occupancy will have a far more deleterious effect on the number of charge carriers than those involving reconstructions on the Y/Ba site. In the case of high-angle grain boundaries, the decrease in the number of charge carriers around each structural unit may begin to overlap causing a non-superconducting barrier to the flow of current across the boundary. This will be particularly true for asymmetric grain boundaries where all of the structural units contain partially occupied copper sites.

The supposition that asymmetric grain boundaries will have a large carrier depletion zone associated with them is consistent with the EELS results shown in Fig. 11.5. To investigate this effect further, the structure of the asymmetric boundaries at various misorientation angles can be constructed from the structural units [11.33]. As a first approximation, the boundaries can be considered to be straight, i.e. consist of one particular grain boundary plane. For boundary misorientations of $11.4°$ (boundary plane $(100)/(510)$), $18.4°$ (boundary plane $(100)/(310)$), $26.6°$ (boundary plane $(100)/(210)$), $33.7°$ (boundary plane $(100)/(210)$), and $45°$ (boundary plane $(100)/(110)$), the copper(I) valence as a function of distance from the boundary core can be calculated from bond-valence sum analysis. In Fig. 11.11, the copper(I) valence is plotted as a function of distance from the boundary. Again, it is clear from this plot that the boundary perturbs the local electronic structure sufficiently to create a non-superconducting zone, and that the width of this zone increases with misorientation angle. These plots can now be used to define a grain boundary width. Obviously with the limits to the accuracy of the bond-valence calculations, the precise copper(I) valence defining the grain boundary width is debatable. Figure 11.12 shows the variation in grain boundary width as a function of misorientation angle using different copper(I) valences to define the width. It is clear from the plot that whatever criterion is used to define boundary width, the width increases with misorientation angle.

The majority of the copper sites in asymmetric grain boundaries are therefore non-superconducting. As was stated earlier, for thin films of YBCO grown on bicrystal substrates, the boundaries facet with predominantly asymmetric boundary planes. It is therefore reasonable to assume that the majority of

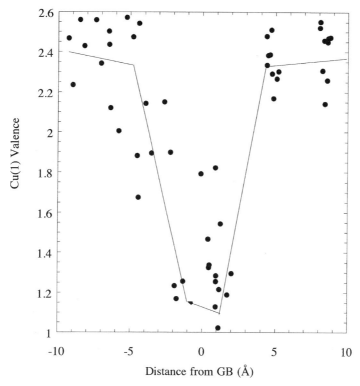

Fig. 11.11. The variation in copper(I) valence as a function of distance from the grain boundary core is plotted for a series of high-angle grain boundaries. The width of the copper(I) valence depletion region increases linearly with misorientation angle. Notice the similarity in the cusp-like shape of the depletion zone to that measured by EELS in Fig. 11.5.

transport measurements across high-angle grain boundaries were in fact measurements of transport across asymmetric grain boundaries. At these boundaries the current must flow by tunneling across the non-superconducting barrier. The magnitude of the tunneling current across a barrier can be calculated from [11.46]

$$J_{c} = J_{c0} \exp(-2\kappa\Delta) \tag{11.2}$$

where J_{c0} is the bulk critical current, Δ is the interface width and κ is the decay constant (7.7 per nm) [11.46, 11.47]. For each of the grain boundaries above, the width can be used to determine the tunneling current. In Fig. 11.13, the tunneling current as a function of misorientation angle for asymmetric boundaries is plotted and compared with a range of experimental critical current

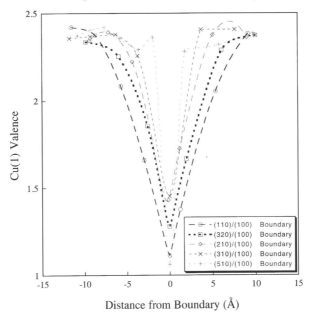

Fig. 11.12. The width of the copper(I) depletion zone as a function of the defining copper(I) valence for the experimentally determined 30° boundary and those simulated by bond-valence analysis. The width of this non-superconducting region increases with misorientation angle.

measurements [11.1–11.3]. For the boundary widths defined above, the structural unit model quantitatively reproduces the trend of exponentially decreasing critical current with increasing misorientation angle, and lies well within the range of experimental results. For comparison, the critical current behavior predicted by the d-wave symmetry of the superconducting order parameter is also shown in figure 11.13 [11.48].

The large variability in the J_c measurements shown in Fig. 11.13 is also naturally explained by the structural unit model. The TEM measurements have shown the universal presence of facets in thin film grain boundaries, which typically occur on length scales in the region of 10–100 nm. Each of these facets corresponds to a different grain boundary plane which will be characterized by a given set of structural units and will have an associated grain boundary width. The bulk scale J_c measurements therefore represent the sum of all facets whose individual behavior can be predicted from the structural unit model. As the growth of thin films is a non-equilibrium process, the range of boundary structures will be extremely sensitive to the growth parameters. This leads to the variability in measurements of J_c and the $J_c\rho_n$ product, where ρ_n

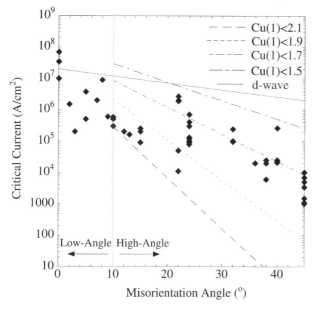

Fig. 11.13. Experimental observations of J_c ($T = 4.2$ K) as a function of misorientation angle from the results of several groups [11.1–11.3] show an exponential dependence. Where the results were reported at $T = 77$ K, the values at 4.2 K were extrapolated from the temperature dependence of J_c [11.45]. The grain boundary tunneling current calculated from eq. 11.2 using the grain boundary widths from Fig. 11.12 shows excellent quantitative agreement for a width defined by a copper(I) valence between 1.5 and 1.9. This copper valence corresponds to the copper(I) valence in bulk YBCO when it becomes non-superconducting. The predicted drop in J_c due to the symmetry of the superconducting order parameter is insufficient by two orders of magnitude to account for the observed behavior.

is the normal state resistivity, as well as a smoothing of the macroscopic current–voltage characteristics. Predictions of grain boundary properties must therefore take into account the degree of facetting that occurs in a given growth process.

11.5 Conclusions

The combination of *Z*-contrast imaging and EELS allows the hole concentration occurring at grain boundaries to be correlated with defined structural features. Using bond-valence sum analysis to interpret the results has highlighted the differences between structural units containing reconstructed atomic columns on different sub-lattice sites. In particular, it has been found that for reconstruction on the copper sub-lattice there is a broad region in which there is a suppression of the carrier concentration. Such a reconstruction on the copper sub-lattice occurs at all asymmetric grain boundaries; such boundaries are common in thin-film YBCO because of the tendency for one grain to facet on the {100} plane. In addition, as the boundaries used in the simulations are all fully oxygenated, these results strongly suggest that the dominant effect controlling the superconducting properties of grain boundaries is the perturbation in local electronic structure due to the strain surrounding the grain boundary cores. This assertion is further enhanced by a comparison with the expected results considering only the symmetry of the superconducting order parameter. In this case, the d-wave symmetry of the order parameter only results in a single order of magnitude decrease in J_c at 45° boundaries (Fig. 11.13) compared with the experimentally observed decrease of more than three orders of magnitude. Obviously effects such as the symmetry of the order parameter, oxygen deficiency and segregation of impurities or copper to the boundary will affect the experimentally observed J_c. However, these are second order effects that modulate the intrinsic superconducting properties of the boundary defined by its structure.

An interesting aspect of this result is that it predicts that symmetric grain boundaries should have better J_c behavior than asymmetric boundaries. The symmetric grain boundaries contain reconstructions predominantly on the Y/Ba sites (Fig. 11.32), and as such there should be unperturbed Cu sites at the boundary. This could explain the EELS observations from the symmetric 36° grain boundaries that show no hole depletion (Fig. 11.5), as well as the anomalously high J_c measurements from certain high-angle symmetric boundaries in thin-film YBCO [11.49]. The reconstructed columns represent ideal positions for the substitution of dopant atoms, which may be able to correct the copper valence and restore the carrier concentration. Obviously such mechanisms require extensive further study, but in principle they could lead to a fundamental understanding of the atomic-scale behavior of grain boundaries. This understanding may in turn aid the production of wires with greater current carrying capacity, and facilitate tailoring of boundary structures for applications such as superconducting quantum interference devices (SQUIDs).

Acknowledgments

We would like to thank D. P. Norton and J. T. Luck for the preparation of the samples used for this work, J. Halbritter and Z. G. Ivanov for valuable comments and J. Buban for performing the bond-valence sum calculations. This research was sponsored by NSF under grant No. DMR-9503877 and by the Division of Materials Sciences, US Department of Energy, under contract No. DE-AC05-96OR22464 with Lockheed Martin Energy Research Corp.

References

[11.1] D. Dimos, P. Chaudhari & J. Mannhart, *Phys. Rev. B* **41**, 4038 (1990).

[11.2] R. Gross & B. Mayer, *Physica C* **180**, 235 (1990).

[11.3] Z. G. Ivanov *et al.*, *Appl. Phys. Lett.* **59**, 3030 (1990).

[11.4] S. E. Babcock *et al.*, *Nature* **347**, 167 (1990).

[11.5] N. D. Browning, M. F. Chisholm, & S. J. Pennycook, *Nature* **366**, 143 (1993).

[11.6] P. E. Batson, *Nature* **366** 727 (1993).

[11.7] D. P. Norton *et al.*, *J. Appl. Phys.* **68**, 223 (1990).

[11.8] S. J. Pennycook & L. A. Boatner, *Nature* **336**, 565 (1988).

[11.9] S. J. Pennycook & D. E. Jesson, *Phys. Rev. Lett.* **64**, 938 (1990).

[11.10] D. E. Jesson & S. J. Pennycook, *Proc. R. Soc. Lond. A* **449**, 273 (1995).

[11.11] R. F. Loane, P. Xu, & J. Silcox, *Ultramicroscopy* **40**, 121 (1992).

[11.12] S. F. Gull & G. J. Daniell, *Nature* **272**, 686 (1978).

[11.13] A. J. McGibbon, D. E. Jesson & S. J. Pennycook, *J. Microscopy* in press.

[11.14] N. Nücker *et al.*, *Phys. Rev. B* **37**, 5158 (1988).

[11.15] N. Nücker *et al.*, *Phys. Rev. B* **39**, 6619, (1989).

[11.16] N. D. Browning, J. Yuan & L. M. Brown, *Supercond. Sci. Technol.* **4**, S346 (1991).

[11.17] N. D. Browning, J. Yuan & L. M. Brown, *Physica C* **202**, 12 (1992).

[11.18] N. D. Browning, J. Yuan & L. M. Brown, *Phil. Mag. A* **67**, 261 (1993).

[11.19] C. J. Rossouw & V. W Maslen, *Phil. Mag. A* **49**, 735 (1984).

[11.20] V. W. Maslen & C. J. Rossouw, *Phil. Mag. A* **47**, 119 (1983).

[11.21] R. H. Ritchie & A. Howie, *Phil. Mag. A* **58**, 753 (1988).

[11.22] N. D. Browning *et al.*, *Physica C* **212**, 185 (1993).

[11.23] W. T. Read & W. Shockley, *Phys. Rev.* **78**, 275 (1950).

[11.24] D. Hull & D. J. Bacon, *Introduction to dislocations* (Pergamon, 1984).

[11.25] A. P. Sutton & R. W. Balluffi, *Interfaces in crystalline materials* (Oxford, 1995).

[11.26] J. P. Hirth & J. Lothe, *Theory of dislocations* (John Wiley & Sons, New York, 1982).

[11.27] M. F. Chisholm & S. J. Pennycook, *Nature* **351**, 47 (1991).

[11.28] A. P. Sutton & V. Vitek, *Phil. Trans. R. Soc. Lond. A*, **309**, 1 (1983).

[11.29] A. P. Sutton, *Acta. Metall.* **36**, 1291 (1988).

[11.30] S. J. Pennycook *et al.*, in *Proceedings of microscopy and microanalysis* (San Francisco Press, 1996), p. 104.

[11.31] M. M. McGibbon *et al.*, *Science* **266**, 102 (1994).

[11.32] N. D. Browning *et al.*, *Interface Science* **2**, 397 (1995).

[11.33] M. M. McGibbon *et al.*, *Phil. Mag.* **A73**, 625 (1996).

[11.34] N. D. Browning & S. J. Pennycook, *J. Phys. D* **29**, 1779 (1996).

[11.35] C. L. Jia *et al.*, *Physica C* **196**, 211 (1992).

[11.36] A. F. Marshall & C. B. Eom, *Physica C* **207**, 239 (1993).

[11.37] C. Traeholt *et al.*, *Physica C* **230**, 425 (1994).

[11.38] B. Vuchic *et al.*, *Mat. Res. Soc. Symp. Proc.* **357**, 419 (1995).

[11.39] B. Kabius *et al.*, *Physica C* **231**, 123 (1994).

[11.40] L. Pauling, *J. Amer. Ceram. Soc* **51**, 1010 (1929).

[11.41] D. Altermatt & I. D. Brown, *Acta Cryst. B* **41**, 240 (1985).
[11.42] I. D. Brown & D. Altermatt, *Acta Cryst. B* **41**, 244 (1985).
[11.43] L. Jansen & R. Block, *Physica C* **181**, 149 (1991).
[11.44] I. D. Brown, *J. Sol. State Chem.* **82**, 122 (1989).
[11.45] J. Mannhart *et al.*, *Phys. Rev. Lett.* **61**, 2476 (1988).
[11.46] J. Halbritter, *Phys. Rev. B* **46**, 14861 (1992).
[11.47] H. L. Edwards *et al.*, *Phys. Rev. Lett.* **69**, 2967 (1992).
[11.48] H. Hilgenkamp, J. Mannhart & B. Mayer, *Phys. Rev. B* **53**, 14586 (1996).
[11.49] Z. G. Ivanov, private communication.

12

Microstructures in superconducting YBa$_2$Cu$_3$O$_7$ thin films

A. F. MARSHALL

12.1 Introduction

Thin films of the high-temperature superconductor, YBa$_2$Cu$_3$O$_7$, may be synthesized as highly oriented structures having a high degree of homogeneity as compared with bulk processed materials. These thin films exhibit some of the best transport properties of high-temperature superconductors (HTSC), particularly the highest superconducting critical currents as a function of temperature and magnetic field. The microstructure of films of different orientation is composed of arrays of specific crystallographic defects which may include, for example, low-angle grain boundaries, (110) twin boundaries, 90° domain boundaries, other misoriented regions or grains, stacking faults, antiphase boundaries, second phases, etc. The surface and interface structure may vary considerably, a smooth surface being essential for device fabrication. Thin-film synthesis also provides methods for fabricating or isolating localized crystallographic interfaces, such as grain boundaries, for transport studies. Therefore, in addition to their importance for technological applications, HTSC thin films may also serve as model systems for studying fundamental behaviors of these materials. In understanding properties it is essential to characterize the microstructure in as much detail as possible. In this chapter we will discuss the microstructure of YBa$_2$Cu$_3$O$_7$ (YBCO) thin films as characterized by transmission electron microscopy (TEM), with emphasis on the interface structure of films having a-axis and (103) orientation, and of individually fabricated boundaries synthesized using a-axis oriented films.

12.2 Grain boundaries

YBa$_2$Cu$_3$O$_7$ (YBCO) thin films are grown $in situ$ by a variety of vapor phase deposition techniques usually with either the c-axis or $a(b)$-axis normal to the plane of the film [12.1, 12.2]. We will refer to the films by their normal orientation. These oriented films are grown on the (100) face of cubic or pseudo-cubic oxide substrates; *a*-axis films typically form on substrates with a close lattice match and at deposition temperatures 50 to 100 °C below that of *c*-axis films. This is because formation of the low energy (001) surface is favored at higher temperatures where surface mobility is high, but is less favored for kinetic reasons at lower temperatures since its formation requires ordering along the growth direction. Buffer layers can also be used to modify the influence of the substrate orientation [12.3]. Films of other orientation can be grown using other cubic substrate symmetries, e.g. (110) and (103) films grow on (110) substrates and (133) films on (111) substrates [12.1]; this corresponds to symmetry matching of the YBCO pseudo-cubic subcell with the substrate and the three-fold index reflects the three-fold ordering of YBCO along the *c*-axis. We will briefly describe the general microstructural features of *c*-axis films for comparison with *a*-axis and (103) films which we will discuss in more detail.

The grain boundary microstructure of high quality *c*-axis films is composed mainly of low-angle boundaries. The presence of low-angle misorientations or mosaicity is reflected in X-ray rocking curve widths and widths of phi scan peaks, and is visible in both plan-view and cross-section TEM micrographs [12.4, 12.5]. Studies of individually fabricated boundaries on bicrystal substrates have shown that low-angle boundaries do not significantly degrade superconducting transport properties [12.6] and it has been suggested that these boundaries could in fact contribute to flux pinning [12.7]. *c*-Axis films grown on MgO and YSZ (Y-stabilized ZrO_2) where the film/substrate lattice match is poor, exhibit more than one in-plane orientation under some deposition conditions, resulting in high-angle grain boundaries [12.8–12.12]. These boundaries have specific misorientations across the boundary and are not generally coherent, although in some cases they may correlate with coincident or nearly coincident site lattice structures [12.13]. They are deleterious to the superconducting transport, consistent with the bicrystal studies.

c-Axis films also exhibit the twins that occur during the tetragonal-to-orthorhombic transition during post-deposition oxygenation and are crystallographically represented by a rotation of the *a*- and *b*-axes by approximately 90° across a shared (110) plane. Fig. 12.1 shows a TEM plan-view image of a *c*-axis film showing these twins which will be referred to as (110) twins. The direction of the *c*-axis is the same for both twin orientations and the superconducting CuO_2 layers remain aligned across the twin boundary. The atomic structure of these twins and their effect on transport have been the subject of many studies [12.14]. Systematic variations of the twin microstructure during

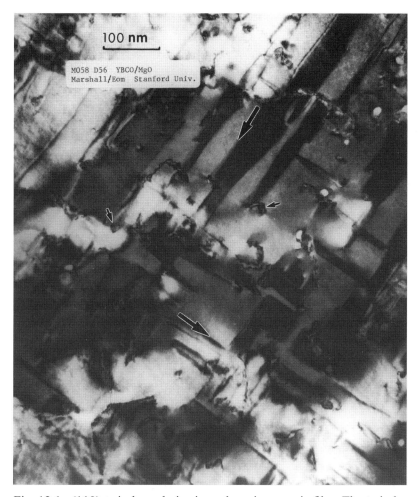

Fig. 12.1. (110) twin boundaries in a plan-view *c*-axis film. The twin boundaries are approximately along the two in-plane ⟨110⟩ directions (large arrows). Second phase particles are also present (small arrows).

processing have shown that the twin boundaries do not have a strong effect on flux pinning in thin films, indicating that other microstructural defects must be responsible for the large critical currents found in these films [12.15].

In contrast to *c*-axis films, YBCO films synthesized with an *a(b)*-axis or (103) normal orientation exhibit a domain or grain structure [12.16–12.19]. This is defined by two symmetrically equivalent orientations of the *c*-axis at 90° to each other. In the case of *a*-axis films, the *c*-axis lies completely in the film plane in two orthogonal directions; for (103) films the *c*-axis is at approximately ±45° out of the plane. This results in an array of 90° [100] or [010] grain boundaries in these films having tilt, twist, or mixed character. (The

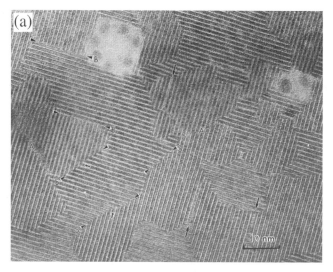

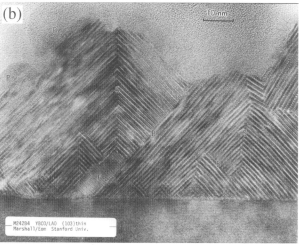

Fig. 12.2. The domain structure and 90° tilt boundaries in *a*-axis and (103) films. (a) Plan-view of an *a*-axis film. The *c*-axis has two in-plane directions resulting in symmetrical and BPF [100] 90° tilt boundaries (triangular arrows marked S and B respectively). Stacking faults (large straight arrows) and antiphase boundaries (A) are also observed. Facetting occurs in the symmetrical boundaries as marked by small straight arrows. These nanofacets effect a lateral shift of the (103) boundary plane which can be visually emphasized by sighting at a low angle along the boundary direction. (b) Cross-section of a (103) film viewed along the [010] direction. The *c*-axis has two orientations at ±45° to the substrate again resulting in 90° [010] tilt boundaries. The symmetrical tilt boundaries (S) propagate along the film growth direction. Second phases (P) are observed at BPF boundaries (B) and at the free (001) surfaces. Much of the surface is comprised of (001) and (100) rather than (103) planes, resulting in a high degree of surface roughness. Very small grains are observed at the interface.

boundary is designated by the axis common to both grains, and by the angular twist or tilt about this axis; mixed boundaries are a combination of tilt and twist misorientation.) TEM images of the domain structures are shown in Fig. 12.2 for a plan-view *a*-axis film and a cross-sectioned (103) film viewed in the [010] direction. The boundaries visible in Fig. 12.2 are 90° [100] or [010] tilt boundaries; they are easily imaged because the distinct *c*-axis lattice fringes are visible for both domains and the boundary is parallel to the viewing direction. They are further characterized as either symmetrical or basal-plane-faced (BPF) boundaries, depending on the position of the boundary plane. This is shown schematically in Fig. 12.3(b) and (c).

The [100] and [010] directions are not easily distinguished by TEM imaging and we will generally use the *a* and *b* indices interchangeably. We note, however, that X-ray diffraction confirms that some so-called '*a*-axis' films are, in fact, pure *a*-axis oriented [12.17] whereas others are a mixture of *a*- and *b*-

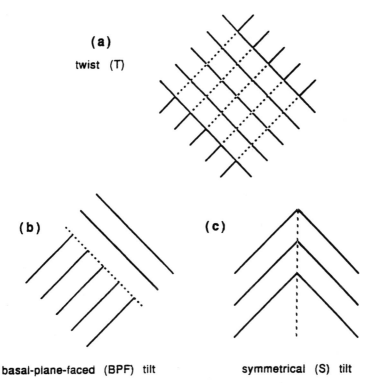

Fig. 12.3. Idealized representations of three boundary plane orientations observed for boundaries with a 90° misorientation of the *c*-axis. The lines represent the *c*-axis lattice planes with a spacing of 11.7 Å. Note that in (a) the boundary plane is normal to the viewing direction whereas in (b) and (c) it is viewed edge-on.

axis orientation [12.20]. The latter do not exhibit the typical (110) twin boundaries observed in *c*-axis films and the transition between regions of (100) and (010) has not been identified. On the other hand, (103) films do exhibit (110) twins, as shown in Fig. 12.4, and are in fact a mixture of (103) and (013) orientations. For simplicity we will continue to refer to these films as *a*-axis and (103), and to the in-plane direction as [010]. We note that the 90° grain boundaries in pure *a*-axis films are indeed [100] boundaries whereas for *a*-, *b*-axis films and (103) films there are both [100] and [010] boundaries. It is also possible that the *a*- and *b*-axes alternate across some of the boundaries in these films, resulting in an additional component of misorientation with the same character as the (110) twins.

In addition to the tilt boundaries shown in Fig. 12.2, 90° [100] or [010] twist boundaries, shown schematically in Fig. 12.3(a), also occur in *a*-axis and (103) films. A schematic of the (103) grain structure, showing the formation of twist boundaries along the [010] direction, as well as the tilt boundaries along the [$\bar{3}$01] direction is shown in Fig. 12.5. In a plan-view (103) film the twist boundaries are parallel to the viewing direction but are not readily visible because the projected orientation of both grains is the same.

The nature of the boundary is identified by microdiffraction of tilted speci-

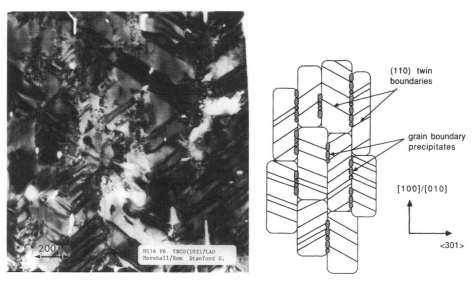

Fig. 12.4. Plan-view image of a (103) film. The domains or grains are elongated along [010]. (110) twins are observed within the grains on the crystallographically equivalent (110) and (1$\bar{1}$0) planes which are oblique to the film plane.

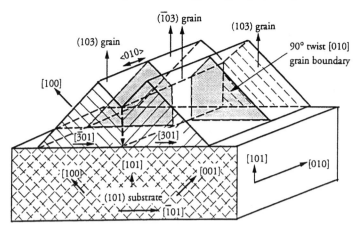

Fig. 12.5. Schematic of the (103) domain structure on the (101) cubic substrate. Tilt boundaries occur along the [$\bar{3}$01] direction; twist boundaries occur along the [010] direction.

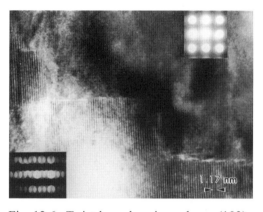

Fig. 12.6. Twist boundary in a planar (103) sample tilted 45° to show the *c*-axis fringes of one grain. Microdiffraction confirms that there is a 90° misorientation of the *c*-axis about the twist axis; the 45° specimen tilt is not compatible with lattice imaging of the *a*, *b*-plane.

mens as shown in Fig. 12.6. The specimen is tilted 45° about the [010] axis, so that the (001) planes of one grain are visible while those of the other are normal to the viewing direction, and the orientation is identified by lattice imaging and microdiffraction. In a transmission electron microscope, high-angle specimen tilt capability is generally compromised by decreased microscope resolution, so that these images are of lower resolution than those of the tilt boundaries.

Three types of boundaries have been identified between the 90° misoriented regions in *a*-axis and (103) films by TEM. High resolution TEM of these boundaries shows continuity of the lattice fringes across the boundary indicating coherency. If one considers that the cations in the YBCO structure are approximately arranged in a pseudo-cubic, body-centered sublattice, the possibility of a fourth type of boundary is realized, with (110) planes on one side meeting (103) planes on the other side (Fig. 12.7). This type of boundary has mixed tilt and twist character. Boundaries with this average orientation have not been observed to occur systematically in these films but may occur as facets within other boundaries and will be discussed further in the next section. All of these boundaries are semicoherent as opposed to coherent, since there are both small mismatch of the subcell and variations in the actual atomic positions within the planes of the three-fold unit cell which meet at, or comprise, the boundary. The misfit in semicoherent boundaries is typically accommodated by periodic misfit dislocations so that segments of the boundary between dislocations are perfectly coherent; in the case of YBCO thin films, stacking faults as well as dislocations appear to accommodate the misfit in these semicoherent boundaries [12.21–12.24].

The normal state and superconducting properties of *a*-axis films are gen-

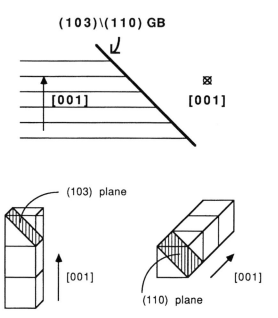

Fig. 12.7. Schematic of a 90° boundary with mixed tilt and twist character, designated (110)(103) for the planes that meet at the boundary plane. This type of boundary is not observed to occur systematically in these films, but may occur as facets.

erally inferior to those of c-axis films, whereas the properties of (103) films are anisotropic, being comparable to c-axis films when measured in the [010] direction across the twist boundaries and showing inferior transport when measured in the orthogonal [$\bar{3}$01] in-plane direction [12.16–12.19]. These properties are attributed mainly to the combined effect of the 90° grain boundaries and transport along the [001] direction (normal to the superconducting CuO$_2$ planes) in these films. The robust nature of the tilt boundaries along the [010] direction is of particular interest. Before discussing the properties in more detail we will discuss the microstructural variations within these average boundary orientations, and describe other structural interfaces in these films that must be considered in analyzing the transport behavior.

12.3 Boundary microstructures and facetting

Twist boundaries predominate along the [010] direction of (103) films, whereas tilt boundaries are observed in a-axis films and along the [$\bar{3}$01] direction of (103) films. However, none of these boundaries are oriented along a single crystallographic plane over a large area, but contain facets of other orientations which allow them to deviate locally from a nominal orientation [12.25]. The average orientation of the boundaries depends on such factors as the growth direction of the film, since boundaries tend to propagate along this direction, the growth anisotropy of YBCO, which is kinetically favored in the non-ordering a–b plane as compared with the c-axis direction, changes in grain size with film thickness and the energy of the boundary. c-Axis orientation is favored over a-axis at higher deposition temperatures where surface mobility is greater and is also preferred on polycrystalline and poorly lattice-matched substrates, indicating that the (001) surfaces are low-energy surfaces relative to other crystallographic planes. However [001] is also a slow growth direction for YBCO since it requires ordering.

The grain size of the (103) films is only tens of ångströms at the film interface, but increases significantly and stabilizes after the first several hundred ångströms (Fig. 12.2(b)). In this initial region there are both BPF and symmetrical tilt boundaries along [$\bar{3}$01]. After this the grain size is maintained and the symmetrical boundaries which propagate along the growth direction predominate. This can be seen in a very thick film at low magnification where all the boundaries are vertical (Fig. 12.8). The (001) planes and to some extent apparently the (100) planes, both of which are oblique to the growth direction, are preferred free surfaces leading to very rough film surfaces as seen in both scanning electron microscopy and cross-section TEM (Fig. 12.8 and 12.9). In a-axis films the BPF and symmetrical tilt boundaries form randomly in the film

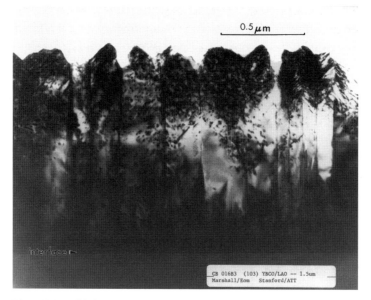

Fig. 12.8. Thick (103) film showing the stabilization of the grain size and propagation of the approximately symmetrical vertical tilt boundaries as the films become thicker. The surface roughness favoring (001) and (100) planes is also observed.

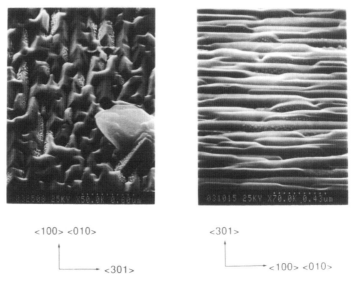

Fig. 12.9. SEM images of a (103) film showing the surface roughness and anisotropy. Fine structure on the growth faces is due to precipitates forming preferentially on the (001) planes.

plane as nucleating grains coalesce. When a [010] growth direction intersects a [001] direction a BPF boundary results; as this interface grows laterally it will become an approximately symmetrical boundary to accommodate the growth anisotropy of YBCO; also the symmetrical boundary will form when two [010] directions intersect directly. Both types of tilt boundaries are parallel to the film growth direction and therefore tend to propagate in this direction. They are randomly connected in series with respect to in-plane transport measurements.

The formation of many symmetrical boundaries in *a*-axis and (103) films, on the other hand, is favored by the crystal growth anisotropy and film orientation, and such boundaries are not necessarily low-energy boundaries. In fact they are observed to deviate considerably from their nominal orientation indicating facetting, although individual facets may be difficult to identify. BPF facets are observed, some extending only a single unit cell (Fig. 12.2(a)), and overlap of fringes through the TEM specimen thickness indicates twist or (110)(103) facets. Fig. 12.10 shows that these facets can also be quite small, on the scale of a few unit cells. In both the *a*-axis films and the (103) film measured along [$\bar{3}$01], twist facets are parallel to the transport measurements so that transport across them would involve indirect current paths. (110)(103) facets, on the other hand would intersect the macroscopic current path at an oblique angle for all the boundaries discussed here. In *a*-axis films, neither facet propagates parallel to the growth direction, but overlap of fringes is observed in the boundaries of thick planar *a*-axis TEM specimens, indicating some amount of these facets. In considering the structure of a (110)(103) boundary, it is observed that the intersection of planes at the boundary is very similar to the twist boundary. Although the two-dimensional lattice at the interface is square for the twist boundary and rectangular for the (110)(103) facet, the intersection of the CuO_2 planes and the projected change in ordering of the Y and Ba atoms is the same in traversing both boundaries (Fig. 12.11). All of the cations are coplanar for the (110)(103) facet, only one type of plane is possible and the accommodation of all the structural rotations within and across a single plane may make such a facet energetically unfavorable compared to the twist facet where the Cu atoms are not coplanar with the Y and Ba atoms at the boundary. However, given the structural similarity it seems reasonable to propose that the (110)(103) boundary, if it occurs over small regions, will have properties similar to the twist boundary.

In addition to facetting, the symmetrical tilt boundary also frequently exhibits a mismatched structure, with the bright fringes of the *c*-axis lattice images having a staggered arrangement where they meet at the boundary (Fig. 12.12). Such a structure requires a lateral shift in the boundary position by a single unit cell only, suggesting that the formation of this mismatched structure

Fig. 12.10. High magnification of an approximately symmetrical boundary of a (103) film shows that the twist component may extend over only one or two *c*-axis unit cells (regions circled). It also appears that overlap occurs in the thinner regions above those circled where a precise boundary plane is difficult to identify. The small extent of the facets suggests that the greater overlap observed in thicker regions may be due to a series of very small facets through the thickness rather than a few large facets.

may be energetically favored over a symmetrical matching of atomic positions at the boundary. The connectivity of the superconducting CuO_2 planes is decreased at the mismatched boundary which may have important consequences for the transport properties. Detailed TEM studies and image simulation by Jia and coworkers of BPF boundaries also show variations in the

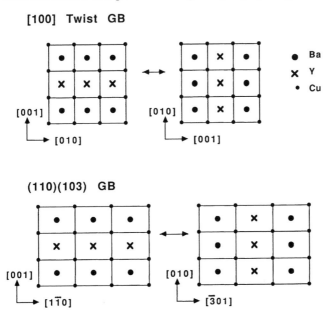

Fig. 12.11. Schematic of the planes which meet at a [100] twist boundary and at a (110)(103) facet. The similarities are in the connectivity of the CuO_2 planes and the change in ordering of the Y and Ba atoms across the boundary as illustrated. These similarities suggest similarities in properties. For the twist boundary the lattice is approximately square and the Y, Ba positions are not coplanar with the Cu atoms, whereas for the (110)(103) facet the lattice is rectangular and the atoms are coplanar.

composition of the atomic planes which meet at the boundary, suggesting that some of these may form insulating layers [12.23]. It is also interesting to note that the twist and mixed (110)(103) boundary structures do not have the potential to vary their cation structure to form insulating layers, or change the CuO_2 plane connectivity, as found for the two tilt boundaries.

Facetting may also occur in the macroscopic twist boundaries along the [010] direction in the (103) films. Figure 12.13 shows a low and high magnification of a planar (103) film tilted 45° about the [010] axis. The low-magnification image shows that the macroscopic orientation of these boundaries does not deviate significantly from the twist orientation. The higher-magnification image of Fig. 12.13, as well as Fig. 12.6, indicate the presence of BPF facets and possibly other facets. Tilt facets do not intersect the macroscopic transport measurement across the twist boundaries, whereas (110)(103) facets would intersect at an oblique angle.

Fig. 12.12. A symmetrical tilt boundary where the bright 11.7 Å *c*-axis lattice fringes appear staggered rather than matched at the boundary. This is a typical appearance for such boundaries (see also Fig. 12.2(a)) and may be a lower energy configuration than one with perfect matching.

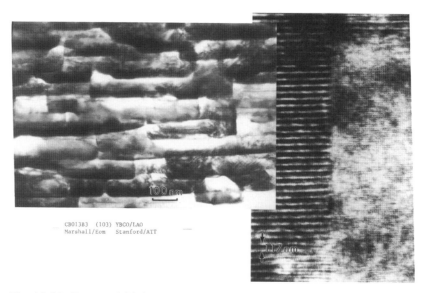

Fig. 12.13. Low and high magnification of a plan-view (103) film tilted 45° about the [010] axis. At low magnification it is observed that the grain boundaries are in the twist orientation (normal to [010]) whereas on a finer microscopic scale at high magnification it is observed that the boundaries may be faceted.

12.4 Stacking faults and antiphase boundaries

Other planar crystallographic defects occur in YBCO, particularly stacking faults (SFs) and antiphase boundaries (APBs). YBCO is a layered structure in the c-axis direction and many types of structural variation have been observed to occur in the stacking sequence, particularly in thin films. The most common of these is a double Cu—O chain layer replacing the single Cu—O chain layer of the 123 structure [12.26]. This is also the structural basis for two additional superconducting phases in the Y—Ba—Cu—O system, the $Y_2Ba_4Cu_8O_{16}$ or '248' phase [12.27–12.30] and $Y_2Ba_4Cu_7O_{15}$ or '247' phase [12.31]. A variety of other stacking variations have been reported, often designated by the local change in composition indicated by the fault structure, such as '224', having an extra Y—O layer between double Cu—O chain layers, and '125' having a triple Cu—O chain layer [12.32]. Structures with additional CuO_2—Y layers, analogous to the series of bismuth- and thallium-based cuprate superconductors, have not been identified, although stacking variations within the 123 unit cell (as opposed those occurring at the CuO chain layer) have been observed sometimes by TEM [12.33, 12.34]. These observations generally involve a single c-axis layer and do not indicate a tendency to form a distinct new phase.

Stacking faults are observed to some extent in all YBCO films, occurring as a universal feature of thin-film growth. Stacking faults may occur as equilibrium defects or to accommodate deviations of the average film composition from stoichiometry. C-Axis films deposited by laser ablation exhibit a higher density and variety of stacking defects than a-axis films using the same technique [12.35] or c-axis films made by sputtering and electron-beam evaporation. These may accommodate local composition variations which arise from the agglomerated nature of the laser-ablated species. Stacking faults occur in post-annealed thin films which are intermediate in composition and processing between the 123 and the 248 phase [12.34]. In these films, the stacking faults often terminate at a dislocation and the dislocation strain fields are observed to interact, producing a staggered array of stacking faults. Stacking faults occur in a-axis and (103) films due to a degeneracy in the choice of nucleation sites of the threefold ordered YBCO structure on the cubic substrate. Adjacent regions of the film, which are aligned in the plane, can nucleate and grow together with the in-plane component of the c-axis out of registry. This results in stacking faults and antiphase boundaries within the domains, as seen in for the a-axis film of Fig. 12.2(a).

Antiphase boundaries occur in ordered structures when two adjacent regions are related by a translation vector of the subcell. Here we use this term to describe boundaries intercepting the a- or b-axis characterized by a shift of the

(001) planes by $c/3$ or $c/6$. APBs have been observed in many thin films [12.4, 12.32]. In c-axis films $c/3$ APBs may nucleate at a step on the substrate or at a stacking fault; they tend to eventually to heal during c-axis film growth by interactions with stacking faults [12.32, 12.36]. Stacking faults, on the other hand, can terminate as dislocations and are not always associated with the presence of APBs [12.32, 12.34]. Since APBs do not extend through the thickness of c-axis films, they do not effectively intercept the macroscopic current path in these films. In a-axis and (103) films APBs occur due to the nucleation site degeneracy described above; only $c/3$ boundaries are observed in the a-axis films. Because images of the (103) boundaries are of lower resolution, as mentioned, $c/3$ and $c/6$ APBs are not clearly distinguished. However, since the substrate structure is the same for both a-axis and (103) films, and varies only in orientation, one might predict that APBs in (103) films are also of predominantly $c/3$ type. However, both $c/3$ and $c/6$ APBs occur in aligned a-axis films.

12.5 Aligned *a*-axis films

Recently, a-axis films with the c-axis almost completely aligned along one in-plane direction have been synthesized on non-cubic $LaSrGaO_4$ (100) substrates using a $PrBa_2Cu_3O_7$ (PBCO) buffer layer [12.37, 12.38]. PBCO has the same crystal structure as YBCO, with a close lattice match, and is similarly aligned on the substrate. The aligned a-axis films contain less than 4% volume fraction of 90° rotated grains [12.38] and the tilt and twist boundaries are thereby largely eliminated. This allows for study of the in-plane anisotropic transport behavior, particularly for comparison of c-axis transport with b-axis transport. However, even with a high degree of in-plane alignment, these films have a granular type of structure due to the nucleation misregistry described above. This results in a specific microstructure of APBs and SFs, as shown in Fig. 12.14. The SFs terminate on APBs so that relatively defect-free regions are enclosed by a combination of these planar defects. The connectivity of the boundaries indicates that most of the APBs and SFs are caused by nucleation misregistry rather than occurring as random defects in the film. These 'grains' are elongated in the [010] direction, the non-ordering, fast-growth direction, as indicated by both SEM and TEM, with a width of 100–500 Å and a length of several hundred to several thousand ångströms.

In the aligned a-axis films several SF structures and APBs with a shift of both $c/3$ and $c/6$ are observed as seen in Figs. 12.15 and 12.16. The central APB in Fig. 12.15 has $c/3$ displacement. It terminates on an unidentified SF with similar displacement which changes smoothly along its length to a 248

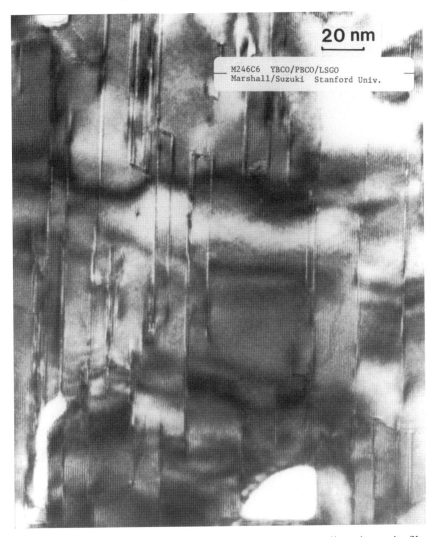

Fig. 12.14. Low-magnification plan-view image of an aligned *a*-axis film, showing a granular structure of interconnected SF and APBs.

SF, with $c/6$ displacement, which then connects with a $c/6$ APB. A continuous change in the nature of SFs along a c-axis layer has also been observed in laser ablated YBCO films [12.32]. In Fig. 12.16, rotated domains and APBs with $c/6$ displacement are imaged. The $c/6$ APBs are connected in two places to a SF with the stacking disruption occurring within the YBCO unit cell. The APBs also terminate in two places on the rotated domains. It is seen from these images that displacements due to the nucleation misregistry accommodate a

Fig. 12.15. A $c/3$ APB (A, across center of image) connects with an SF, also with $c/3$ displacement on the left (lower left arrow) which changes smoothly into a 248 SF with $c/6$ displacement which then connects to a $c/6$ APB in the upper left-hand corner (arrow). The $c/3$ APB also connects to an SF on the right; the image is smeared near the SF, suggesting that the APB is not parallel to the film plane in that region.

variety of interconnected structures which can transform smoothly from one to the other. The connectivity of $c/3$ and $c/6$ boundaries are quite different: for a $c/3$ APB, one out of every two CuO_2 planes connects with another such plane across the boundary. For a $c/6$ APB, none of the layers connect with a like layer. One might therefore expect different properties for these two types of APBs. Although the high-angle boundaries are eliminated from these films, the boundary structure is still complex and involves more than one type of interface along the transport directions.

Fig. 12.16. Some $c/6$ APBs (A), misaligned grains (R), and an SF, which disrupt the YBCO unit cell in the Y—CuO$_2$ layer region (two left arrows), as well as in the chain layer region (right arrow) are imaged in an aligned a-axis film.

12.6 Synthesis and properties

The types of grain boundaries and other interfaces occurring in these oriented thin films as described above are fairly limited and oriented with respect to macroscopic transport measurements so that the properties of specific boundaries can be analyzed. Random high-angle grain boundaries are known to have a significantly deleterious effect on superconducting transport properties of HTSC materials. This is true even when the boundaries are clean and the superconducting CuO$_2$ planes are aligned along the transport direction, as in bicrystal studies of [001] tilt boundaries [12.6] and in polycrystalline c-axis films on MgO and YSZ [12.11, 12.12]. In contrast to those studies, where the boundaries are incoherent, the boundaries measured here are semicoherent.

Also for these oriented films the transport measurements in some cases involve transport along the [001] direction (normal to the CuO_2 planes), which is not well understood and therefore complicates the situation. However, none of the interface structures in these oriented thin films, even the 90° boundaries, degrade properties to the extent of the random high-angle boundaries studied in bicrystal experiments and occurring in bulk material. The twist boundaries in (103) films, in particular, are found not to degrade J_c at all relative to good *c*-axis films [12.18]. Normalized magnetic field dependence of J_c does not show weak-link behavior in any of these boundary structures. In contrast, studies on nominal [100] 90° tilt boundaries formed across step edges in thin films do show degraded transport and junction-like behavior, so that the intrinsic boundary properties are still unresolved [12.22, 12.39, 12.40].

All of the films reported on here were made by off-axis magnetron sputter deposition from a stoichiometric target with 40/60 mTorr O_2/Ar pressure. The *c*-axis and (103) films were deposited at 720 °C and the *a*-axis film at 640 °C. In the case of aligned *a*-axis films, a PBCO buffer layer was deposited at 640 °C and the YBCO was deposited at 700 °C. Film thicknesses were 2000–4000 Å; the PBCO buffer layer was typically 500 Å. The normal state and superconducting transport properties of these films have been characterized by measuring resistivity vs. temperature, $\rho(T)$ and the critical current as a function of temperature and magnetic field, $J_c(T, H)$. In addition to the superconducting transition temperature, T_c, resistivity vs. temperature measurements provide the width of the transition (ΔT_c), the slope of the normal state resistivity $d\rho/dT$, and the zero temperature intercept, $\rho(0)$ when the slope is metallic and linear. A finite $\rho(0)$ indicates grain boundary resistance whereas an increase in $d\rho/dT$ suggests indirect paths and/or current cross-section reduction. The critical currents are measured by magnetization, which measures across all in-plane directions, or by direct transport measurement along only one macroscopic in-plane direction.

For high quality *in situ c*-axis films prepared by different synthesis methods, the T_c values may vary somewhat but other characteristics are comparable. High quality *c*-axis films made by laser ablation [12.41] and CVD [12.42] show optimized T_c values of >90 K with ΔT_c less than 1 K, as good as bulk material. Films made at lower O_2 pressure, e.g. by sputter-deposition, such as the films discussed here, tend to show somewhat depressed T_c values in the upper 80 K range, again with a sharp transition [12.16]. Other characteristics of *c*-axis films are comparable regardless of deposition technique: the slope is linear, $\rho(0)$ is zero, and $\rho(300 \text{ K})$ is comparable to single crystals, on the order of 150–200 µohm.cm. More importantly, *c*-axis films show the best critical currents for any YBCO material, with $J_c(4.2 \text{ K})$ greater than 10^7 A/cm^2 and

J_c(77 K) greater than 10^6 A/cm^2. $J_c(H)$ shows weak field dependence indicating an absence of weak-link behavior. For the films of different orientation discussed here the T_c values are of the order of 84–85 K for the two-domain a-axis films, 86–87 K for the c-axis film and aligned a-axis films, and 88 K for the (103) films.

Plots of $\rho(T)$ and $J_c(T)$ are shown in Fig. 12.17(a) and (b) for c-axis and a-axis films, and for (103) films measured along the two major in-plane directions. The same data for the aligned a-axis film are shown in Fig. 12.18(a) and (b). Along the [010] direction of the (103) film, i.e. measuring transport across the 90° [010] twist boundaries, there is no increase in resistivity or decrease in J_c as compared with the c-axis films. The [010] twist boundaries appear robust with respect to transport. This result is further confirmed by measurements on single twist boundaries, as will be discussed in the next section. Along the [010] direction of the aligned a-axis film, i.e. measuring across the APBs, there is a small but finite $\rho(0)$, indicating resistance due to these boundaries and some decrease in J_c; $d\rho/dT$ is the same as for the c-axis film in both cases indicating no current path lengthening or cross-section reduction. Note that for both the twist boundaries and the APBs the macroscopic current flow is along the superconducting CuO$_2$ planes on both sides of the boundaries, as it is for the c-axis film, and no change in $d\rho/dT$ is expected if the current flows directly across the boundaries. Measurements of the APBs in the aligned a-axis films do not distinguish between the properties of $c/6$ and $c/3$ boundaries, which may be quite different. The simplest conclusion from the TEM observations is that the twist boundaries and the $c/3$ APBs occur in series in the (103)/[010] measurements, where neither degrades transport, and that the $c/3$ and $c/6$ APBs occur in series in the aligned a-axis [010] measurement where it is the $c/6$ boundaries that degrade transport. Alternatively the $c/3$ APBs may be somewhat deleterious to transport, but their distribution in the (103) film is such that the twist boundaries are measured in parallel and short the effect of the APBs.

For a-axis films and for (103) films measured along [$\bar{3}$01], i.e. measuring across the tilt boundaries, the J_c values are lower and $\rho(T)$ and $d\rho/dT$ are higher as compared to the c-axis film. $\rho(0)$ is finite, indicating grain boundary resistance. There are several microstructural features to consider in interpreting this data. The (103) films have very rough surfaces and contain BPF and symmetrical tilt boundaries near the interface. The surface roughness indicates a reduced current cross-section relative to the average thickness, causing an increase in $d\rho/dT$ and a reduction in J_c. As discussed in detail by Eom and coworkers [12.16–12.19], $\rho(0)$ is greater for thinner (103) films where BPF boundaries and/or c-axis transport control the behavior. As the films become

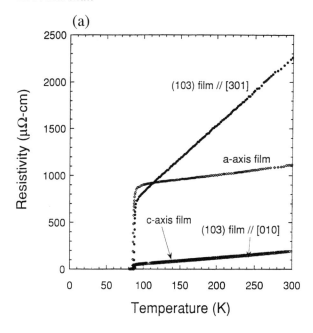

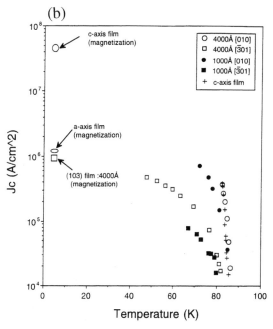

Fig. 12.17. Resistivity (a) and critical current (b) vs. temperature for *c*-axis, *a*-axis, (103) films.

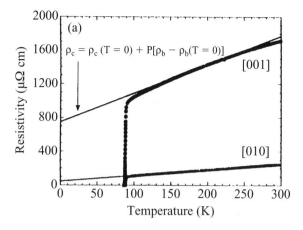

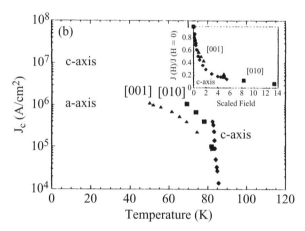

Fig. 12.18. Resistivity (a) and J_c (b) vs. temperature in the [010] and [001] directions of the aligned *a*-axis film. The J_c of a two-domain *a*-axis film and a *c*-axis film are shown for comparison. The inset of (b) shows the magnetic field dependence of the normalized J_c in the two directions at 77 K compared with a high-J_c *c*-axis film which is known to not be weak-link limited. The scaled field is in kG for the *c*-axis film and in kG/(mass ratio)$^{1/2}$ for the aligned *a*-axis film.

thicker symmetrical tilt boundaries containing many twist facets predominate. In thick films, a zigzag transport path is possible along [$\bar{3}$01] due to the connectivity of the CuO_2 planes across the boundaries. The $\rho(0)$ decreases significantly, but $d\rho/dT$ remains high. Again a simple interpretation is that in the thick (103) films *c*-axis transport is avoided and the nominally symmetrical tilt boundaries are not severely limiting transport. This latter situation may result from the presence of twist facets in the nominal tilt boundaries which transport the current through a relatively constant fraction of the total boundary area.

In *a*-axis films a zigzag transport path is, at first glance, also possible, but upon closer examination may be limited by the actual long-range connectivity of the planes across many grains. In *a*-axis films consecutive connectivity of individual planes does not continue for more than a few grains and some *c*-axis transport must occur. The current path through the *a*-axis films is potentially the most complicated of the films described here, because all of the 90° grain boundaries, stacking faults and APBs are present along with the probable *c*-axis transport, and these are mostly linked in series across the macroscopic transport path. Twist facets do not intercept the macroscopic transport path, but may become important if significant grain growth occurs with film thickness, as has been observed for some films. In the aligned *a*-axis films measured along [001], one measures *c*-axis transport in combination with SF separately from the effect of APBs but with some component due to the small fraction of misaligned grains. The resistivity of the two-domain *a*-axis films is better than that of the aligned *a*-axis film along [001] which reflects a larger component of [010] transport in series with *c*-axis transport. J_c in the two cases is less easy to

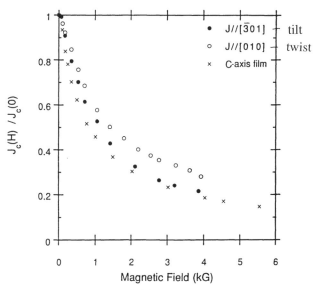

Fig. 12.19. Magnetic field dependence at 77 K of the normalized critical current density for both in-plane directions of a (103) film is comparable to a high-J_c *c*-axis film. The effective magnetic field in the (103) film is $(1/2)^{1/2}$ because the superconducting CuO_2 planes are at 45° to the substrate surface.

compare because the two-domain *a*-axis J_c is measured by magnetization, whereas the aligned *a*-axis film is measured by a direct transport measurement at higher temperatures. The *a*-axis film is therefore being measured in two dimensions in the film plane which complicates the connectivity of the different resistive components of the film.

The behavior of the superconducting critical current in a magnetic field is an indication of the weak-link behavior of boundaries in the films. Figure 12.19 compares the magnetic field dependence of the normalized critical current density for a *c*-axis film and both in-plane directions of the (103) film. Neither the twist nor the tilt boundary structures of the (103) film indicate weak-link behavior. The same is observed for the two in-plane directions of the aligned *a*-axis film, as seen in the inset of Fig. 12.18(b). The critical currents are normalized to zero field values and the field is scaled as noted. The absence of weak-link behavior for boundaries that show overall increased normal state resistivity or decreased superconducting critical currents may be interpreted as reflecting the behavior of a fraction of these macroscopically orientated boundaries where facetting of other structural variations makes the boundary robust for transport.

12.7 Single grain boundaries

Individual grain boundaries, having a nominal [100] 90° twist and BPF tilt orientation have been fabricated on LSGO using a bi-epitaxial technique by Lew *et al.* [12.43]. A PBCO buffer layer is required for the aligned *a*-axis growth already discussed. YBCO grows on a bare LSGO substrate in a *c*-axis orientation. By lithographically removing part of the PBCO buffer layer, regions of aligned *a*-axis and *c*-axis orientation are synthesized on the same substrate with a macroscopically aligned boundary direction, as shown in Fig. 12.20. Patterning the film as shown allows for transport measurements both across the boundary, and across the lead material only, thus distinguishing the properties of the boundary from those of the leads, which was not possible in previous measurements.

The effect of the two types of boundary on the measurements of resistivity vs. temperature are shown in Fig. 12.21. None of the leads, the *c*-axis material and the *a*-axis material measured in the [010] and [001] directions, shows ideal behavior. This is, at least in part because of the necessity of some compromise in the deposition temperature between ideal *a*-axis and *c*-axis conditions. However, the results for the twist grain boundary are consistent with that found previously for (103) films measured across many such grain boundaries: there

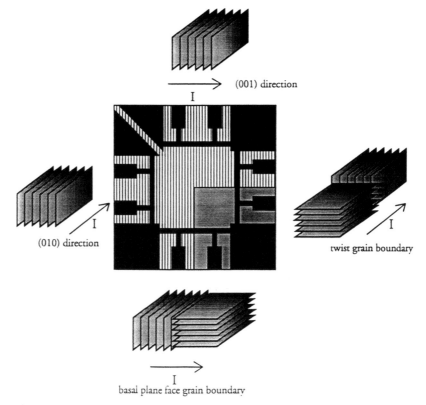

Fig. 12.20. Fabrication of single grain boundaries. The lines indicate the CuO_2 planes of the aligned a-axis region on a PBCO buffer layer and the shaded areas show the c-axis region on the bare substrate. The black pattern is the photolithographic mask used for defining the four measurement lines: the BPF and the twist grain boundaries, and the [010] and [001] leads in the a-axis film.

is no increase in resistivity due to the presence of the twist grain boundary and its resistivity is in fact consistently lower than that of the two leads combined.

The twist grain boundary is also found to be robust in terms of the critical current, showing no decrease in J_c across the boundary (Fig. 12.22) and no weak-link behavior. The BPF tilt boundary, on the other hand, increases the resistivity and limits the J_c as compared with the lead material (Figs. 12.21(b) and 12.22(b)). The ratio of the J_c of the a-axis lead material along [001] to the BPF boundary was consistently about 2–5. This effect is not large compared with measurements on random high-angle boundaries and $J_c(H)$ again shows the absence of weak-link character. These qualitative trends are consistent with what is observed in bi-domain a-axis films and (103) films where these

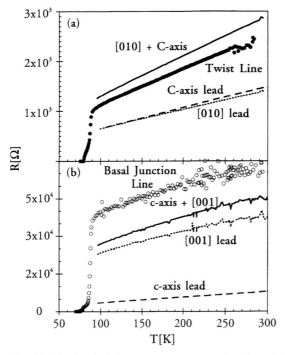

Fig. 12.21. Resistivity vs. temperature for the BPF and twist boundaries are compared with the lead material only. The sum of the leads is indicated by the solid line.

boundaries are measured in series with c-axis transport and other boundaries. In this case, however, the effect of the boundary is separated from that of c-axis transport (the lead material).

TEM of the BPF boundary, however, shows that the structure is far from ideal. When observed in SEM, or in low-magnification TEM (Fig. 12.23), both boundaries appear straight. On a microscopic scale there is some deviation in the boundary direction. Furthermore, as shown in Fig. 12.24, the BPF boundary contains a high density of precipitates. These are of the order of 100–200 Å in size and are identified as (110) Y₂O₃. These precipitates may occur due to small deviations of the deposition target from stoichiometry, or due to a non-equilibrium nucleation process. They nucleate preferentially at the BPF tilt boundary. As observed in Fig. 12.24, they extend into the c-axis side of the boundary, indicating that they nucleate on the (001) lattice planes of the a-axis aligned regions. Nucleation of second phases on (001) planes is also observed on the (001) surfaces and BPF boundaries of (103) films and along the (001)-faced sides of surface pits in a-axis films. Such compositional segregation has not been observed within a-axis oriented regions, along symmetrical bound-

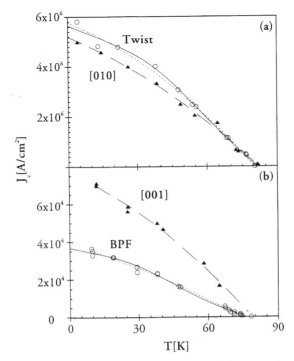

Fig. 12.22. Critical current density of the two single grain boundaries; J_c of the twist boundary is comparable to the [010] lead of the aligned *a*-axis material, whereas J_c of the BPF boundary is lower than that of the [001] lead.

aries or at the BPF boundaries of *a*-axis films. Clearly the precipitates are forming on (001) free surfaces during film growth. This explains the preferential formation of precipitates in the fabricated BPF boundary. Etching of the PBCO substrate prior to deposition produces a step at this boundary where two (001) free surfaces intersect and the nucleation of precipitates in this region is thus favored.

Because of the presence of the precipitates the intrinsic properties of the BPF boundary remain ambiguous. Furthermore, we must consider microscopic facetting. Segments of this boundary are very straight over a length of several hundred ångströms as seen in Fig. 12.24. However, some facetting is observed, indicating a small fraction of twist or (110)(103) facets. No evidence of facetting along the growth direction was observed although symmetrical tilt facets could form in this direction. The microscopic characterization indicates that synthesis of a clean boundary, free of precipitates and possibly with less facetting, would be optimized by devising a way of eliminating or reversing the step where the boundary forms, while maintaining the PBCO buffer layer.

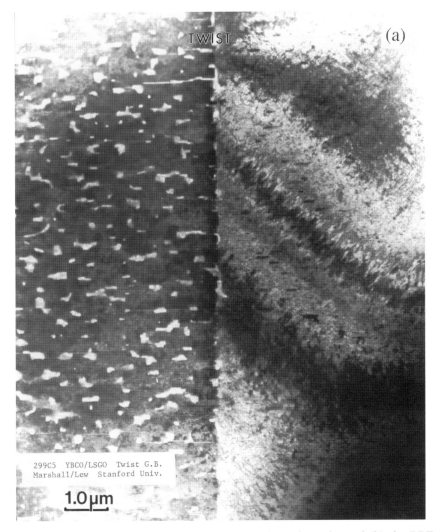

299C5 YBCO/LSGO Twist G.B.
Marshall/Lew Stanford Univ.

1.0 μm

Fig. 12.23. Low-magnification micrographs of (a) the twist and (b) the BPF fabricated boundaries. At this magnification both boundaries appear straight. Surface pitting is observed in the *a*-axis side of both boundaries; twins and bend contours are observed in the *c*-axis side (*continued overleaf*).

Facetting of the twist boundary must also be considered. In this case, BPF and symmetrical tilt facets do not intersect the macroscopic current path. Since we already have indication from the (103) films that the twist boundary does not degrade current, we do not expect that indirect paths crossing these boundaries will be preferred. However, when the nominal twist boundary deviates from its macroscopic direction it may form (110)(103) facets which do

Fig. 12.23 (*cont.*).

intersect the macroscopic current path. The transport measurements therefore may reflect the properties of both the twist and the (110)(103) facets. If this is the case it supports our previous suggestion that neither type of boundary orientation is detrimental to the transport behavior. Future work includes the fabrication of a boundary with nominal (110)(103) orientation using bi-epitaxy in order to investigate this type of boundary in more detail.

The twist boundary may exhibit a periodic array of dislocations and stacking faults as shown in Fig. 12.25. This appears to be an accommodation of the misfit that occurs at the deposition temperature where the structure is tetragonal and the c-axis is expanded relative to the $a(b)$-axis. Extra (010) fringes occur on the c-axis side of the boundary whereas stacking faults occur on the a-axis oriented side of the boundary, often in a stand-off position from the boundary. The periodicity is approximately 13 c-axis fringes, or about 150 Å, and is more regularly and closely spaced than the stacking faults occurring due to the

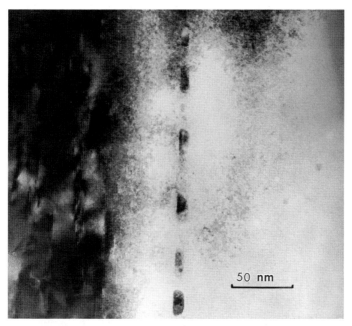

Fig. 12.24. Higher-magnification of the BPF boundary: the boundary is very straight and contains precipitates which appear to nucleate on the (001) planes of the *a*-axis region and grow into the *c*-axis side.

nucleation misregistry. The formation of a dislocation by such a periodic insertion of extra (010) planes will accommodate the misfit which exists at the deposition temperature of 700 °C (data obtained from Fig. 2 of Beyers & Shaw [12.44]). The role of the stacking faults is less clear. This periodic misfit structure was not observed in the twist boundaries of the (103) film, which may be related to the different growth orientation and to the limited extent of the boundaries in the latter case.

Finally we consider the large difference in transport behavior observed for the 90° boundaries in these films, which consistently show relatively small degradation of J_c and no weak-link behavior with that of step-edge structures containing nominally similar boundaries. The latter often show junction-like behavior with J_c degraded two orders of magnitude or more compared with the rest of the film. The reasons for this are not clear. However, we note that film thickness may be an important factor in obtaining optimal properties in these variously oriented films, with properties of *a*-axis films improving above about 1500–2000 Å. The films deposited on step-edge junctions are typically thinner than this, even when the average film thickness is comparable.

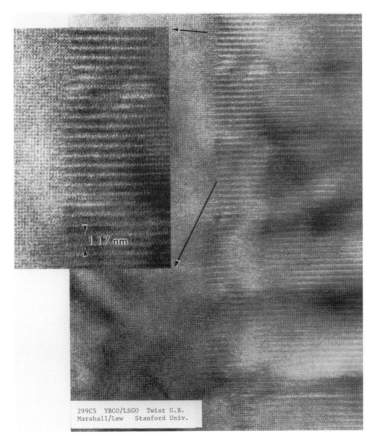

299C5 YBCO/LSGO Twist G.B.
Marshall/Lew   Stanford Univ.

Fig. 12.25. Higher-magnification of the twist boundary. Periodic stacking faults occur near or at the boundary; they are distinct from the stacking faults due to misregistry of nucleation. These faults occur every 13 *c*-axis lattice fringes along, with missing fringes on the *c*-axis oriented side of the boundary.

12.8 Summary

The oriented thin films and fabricated grain boundaries described here provide a way of quantitatively investigating the transport properties of individual grain boundaries and other planar defects having a well defined geometry, and of modeling the transport across a combined array of such defects. However detailed TEM characterization of the microstructures shows that even a highly aligned material with limited and specific boundary structures may still be much more complex than expected. The boundary structure must be analyzed both at the atomic scale level, involving such characteristics as nanofacetting and mismatch, and on a much larger scale to observe the relative distribution of different types of defects. Microstructural characterization gives insight into new ways of optimizing the synthesis and analysis of individual boundaries.

References

[12.1] C. B. Eom *et al.*, *Physica C* **171**, 354 (1990), and references therein.
[12.2] A. F. Marshall & R. Ramesh, in *Interfaces in high-T$_c$ superconducting systems*, eds. S. Shinde and D. Rudman (Springer-Verlag, New York, 1994) p. 71, and references therein.
[12.3] A. Inam *et al.*, *Appl. Phys. Lett.* **57**, 2484 (1990).
[12.4] S. K. Streiffer *et al.*, *Mat. Res. Soc. Symp. Proc.* **183**, 363 (1990).
[12.5] S. K. Streiffer *et al.*, *Phys. Rev. B.* **43**, 13007 (1991).
[12.6] D. Dimos, P. Chaudhari & J. Mannhart, *Phys. Rev. B* **41**, 4038 (1990); Z. G. Ivanov *et al.*, *Appl. Phys. Lett.* **59**, 3030 (1991).
[12.7] V. M. Pan *et al.*, *AIP Conference Proceedings* **251**, 603 (1992).
[12.8] R. Ramesh *et al.*, *Appl. Phys. Lett.* **56**, 2243 (1990).
[12.9] L. A. Tietz *et al.*, *J. Mater. Res.* **4**, 1072 (1989); M. G. Norton *et al.*, *Mat. Res. Soc. Symp. Proc.* **169**, 513 (1990).
[12.10] D. H. Shin *et al.*, *Appl. Phys. Lett.* **57**, 508 (1990).
[12.11] S. M. Garrison *et al.*, *Appl. Phys. Lett.* **58**, 2168 (1991).
[12.12] B. H. Moeckly *et al.*, *Appl. Phys. Lett.* **57**, 1687 (1990).
[12.13] T. S. Ravi, *Physical Rev. B* **42**, 10141 (1990).
[12.14] Y. Zhu & M. Suenaga, in *Interfaces in high-T$_c$ superconducting systems*, eds. S. Shinde & D. Rudman (Springer-Verlag, New York, 1994) p. 140, and references therein.
[12.15] B. M. Lairson S. K.Streiffer & J. C. Bravman, *Phys. Rev. B.* **42**, 10067 (1990).
[12.16] C. B. Eom *et al.*, *Phys. Rev. B* **46,** 11902 (1992), and references therein.
[12.17] C. B. Eom *et al.*, *Science* **249**, 1549 (1990).
[12.18] C. B. Eom *et al.*, *Nature* **353**, 544 (1991).
[12.19] C. B. Eom *et al.*, *Interface Science* **1**, 267 (1993).
[12.20] S. K. Streiffer *et al.*, *Phys. Rev. B* **47**, 11431 (1993).
[12.21] Y. Gao *et al.*, *Physica C* **173**, 487 (1991).
[12.22] C. L. Jia *et al.*, *Physica C* **196**, 211 (1992).
[12.23] C. L. Jia and K. Urban, *Interface Science* **1**, 291 (1993).
[12.24] A. H. Carim & T. E. Mitchell, *Ultramicroscopy* **51**, 228 (1993).
[12.25] A. F. Marshall & C. B. Eom, *Physica C* **207**, 239 (1993).
[12.26] H. W. Zandbergen *et al.*, *Phys. Status Sol. (a)* **105**, 207 (1988), H. W. Zandbergen *et al.*, *Nature* **331**, 596 (1988).
[12.27] A. F. Marshall *et al.*, *Phys. Rev. B* **37**, 9353 (1988).
[12.28] K. Char *et al.*, *Phys. Rev. B* **38**, 834 (1988).
[12.29] P. Marsh *et al.*, *Nature* **334**, 141 (1988).
[12.30] P. Bordet *et al.*, *Nature* **334**, 596 (1988).
[12.31] P. Bordet *et al.*, *Nature* **334**, 596 (1988).
[12.32] R. Ramesh *et al.*, *Science* **247**, 57 (1989), R. Ramesh *et al.*, *Nature* **346**, 420 (1990).
[12.33] Y. Matsui *et al.*, *Jpn. J. Appl. Phys.* **26**, L777 (1987).
[12.34] A. F. Marshall *et al.*, *Mat. Res. Soc. Symp. Proc* **169**, 785 (1990), A. F. Marshall *et al.*, *J. Mater. Res.* **5**, 2049 (1990).
[12.35] R. Ramesh, *et al.*, *Appl. Phys. Lett.* **57**, 1064 (1990).

[12.36] R. Ramesh, *et al.*, *J. Materials Res.* **5**, 704 (1990); R. Ramesh, *et al.*, *J. Materials Res.* **6**, 2264 (1991).

[12.37] S. Hontsu *et al.*, *Appl. Phys. Lett.* **59**, 1134 (1991).

[12.38] Y. Suzuki *et al.*, *Phys. Rev. B* **48**, 10642 (1993).

[12.39] C. L. Jia *et al.*, *Physica C* **175**, 545 (1991).

[12.40] J. Luine *et al.*, *Appl. Phys. Lett.* **61**, 1128 (1992).

[12.41] T. Venkatesan *et al.*, *Appl. Phys. Lett.* **54**, 581 (1989).

[12.42] J. Zhao *et al.*, *Appl. Phys. Lett.* **59**, 1254 (1991); H. C. Li *et al.*, *Appl. Phys. Lett.* **52**, 1098 (1988).

[12.43] D. J. Lew *et al.*, *Appl. Phys. Lett.* **65**, 1584 (1994).

[12.44] R. Beyers & T. M. Shaw. *Solid State Phys.* **42**, 135 (1989).

13

Investigations on the microstructure of $YBa_2Cu_3O_7$ thin-film edge Josephson junctions by high-resolution electron microscopy

C. L. JIA and K. URBAN

13.1 Introduction

The discovery of cuprates exhibiting superconductivity at relatively high temperatures has opened up new prospects for the application of superconductivity in many areas, in particular in sensor systems and in electronics [13.1, 13.2]. In this respect the superconducting quantum interference device (SQUID) is one of the most attractive developments. Many different designs have been fabricated and studied, and modern SQUIDs on the basis of $YBa_2Cu_3O_7$ have reached field sensitivity and performance levels not far different from those known for devices produced with classical low temperature superconductors [13.3, 13.4].

The physical properties of cuprate superconductors depend sensitively on the preparation conditions and the resulting microstructure. Owing to the essentially two-dimensional superconductivity and the very small coherence length in the cuprates, grain boundaries in general reduce the critical current density by orders of magnitude compared with the bulk value. Therefore much attention has been paid to the development of techniques for growing high quality epitaxial thin films of superconducting and appropriate non-superconducting materials required for active and passive electronic devices on suitable single-crystalline substrates. The remarkable progress achieved is closely related to both the development of proper materials preparation techniques and high quality materials characterization. In fact, device production requires atomic or close to atomic structural perfection. High-resolution transmission electron microscopy, which developed atomic resolution in many materials during the early 1980s, i.e. shortly before the new materials were discovered, has contributed substantially to the understanding of the structural properties of cuprate superconductors and to their use in electronic devices.

Besides instrumental resolution, two important technical factors are decisive for the enormous potential of high-resolution electron microscopy for super-

319

conductivity research. The first is the progress in numerical quantum mechanical and optical *image simulation*, representing the key point for the interpretation of the experimental images which in general cannot be understood intuitively. The second is *cross-sectional preparation* of layer systems, which permits the study of interfaces and related structural defects viewing parallel to the interface, i.e. perpendicular to the deposition direction.

In this chapter we present three examples demonstrating the potential of high-resolution transmission electron microscopy for the characterization of the microstructure of $YBa_2Cu_3O_7$ thin films and the related heterostructures that are the basis of superconducting electronic devices: (1) $YBa_2Cu_3O_7$ thin-film step-edge junctions [13.5–13.9], (2) $YBa_2Cu_3O_7/SrRuO_3$ triple-layer edge junctions [13.10, 13.11] and (3) $YBa_2Cu_3O_7/PrBa_2Cu_3O_7$ multilayer edge junctions [13.12, 13.13].

13.2 Experimental

Two types of sample were prepared for examination in the high-resolution electron microscope. *Cross-sectional specimens* were prepared by cutting the coated substrate wafers into slices. Two of these slices were glued together face-to-face joining the $YBa_2Cu_3O_7$ covered surfaces. The sandwich structure was cast in epoxy resin. After slicing perpendicular to the interface, 3 mm disc samples were cut. These samples contained, in the center, the joined wafer surfaces in edge-on orientation. After mechanical grinding and dimpling, these specimens were ion-milled to perforation on a stage cooled by liquid nitrogen. *Plan-view samples* were prepared from the original wafers by polishing and ion-milling from the substrate side.

All the experimental images presented in this chapter were taken on a JEOL 4000EX electron microscope with a Scherzer resolution of 0.17 nm at 400 kV. Image simulations for interpretation of the experimental images were carried out employing the EMS program package [13.14]. The parameters used for image simulation were 1 mm for the spherical aberration coefficient, 12 nm for the defocus spread, 1 mrad for the semi-convergence angle of the illumination and 11 nm^{-11} for the effective diameter of the objective lens aperture.

13.3 Microstructure of $YBa_2Cu_3O_7$

13.3.1 Introduction

It was first shown by Simon *et al.* [13.5] that narrow microbridges patterned in $YBa_2Cu_3O_7$ films deposited across a sufficiently steep step in a $SrTiO_3$

substrate function as Josephson junctions. Investigations by high-resolution electron microscopy by Jia *et al.* [13.8, 13.9] indicated that these junctions were created by two or more grain boundaries. Indeed it was found by Dimos *et al.* [13.15] that large-angle grain boundaries exhibit critical current densities orders of magnitude lower than in the bulk. On the other hand, Gross *et al.* [13.16] and Char *et al.* [13.17] demonstrated that isolated grain boundaries induced into epitaxial YBa$_2$Cu$_3$O$_7$ can be used as Josephson junction barriers. The technique of step-edge Josephson junctions on the basis of YBa$_2$Cu$_3$O$_7$ was developed by Cui *et al.* [13.6] and Herrman *et al.* [13.7] for use in rf-SQUIDs. These devices have reached excellent sensitivity values, which allow them to be used in investigations of brain activity by magnetoencephalography [13.18].

A typical step-edge junction is shown schematically in Fig. 13.1. In (a) the substrate is shown after ion-milling through a Nb mask producing a trench about 5 μm wide and 0.2 μm deep. The angle Θ between the flank and the bottom of the trench characterizes the step steepness. In (b) the junction is shown after deposition of the superconducting layer. The two pairs of 90°

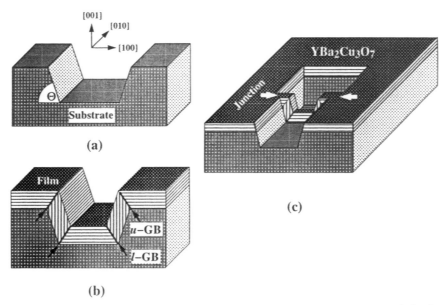

(a)

(b)

(c)

Fig. 13.1. Schematic of step-edge junctions. (a) Geometry of a trench in the perovskite substrate showing the step angle Θ and the indices of the crystallographic directions of the substrate. (b) The structure after deposition of the superconductor, the lines indicate the orientation of the YBa$_2$Cu$_3$O$_7$ lattice planes between the four 90° rotation grain boundaries (arrows; *u*-BG and *l*-GB stand for upper and lower grain boundary). (c) Illustration of the geometry and electrical connections of a step-edge junction used in an rf-SQUID.

rotation grain boundaries which form the electrically active junction of the device are denoted by GB. The prefix *u* refers to the upper and the prefix *l* to the lower boundary of a pair. In (c) the use of these step-edge junctions for an rf-SQUID is indicated schematically.

13.3.2 Preparation of step-edge samples

Test structures were fabricated following the same procedures employed in device production [13.19]. Multiple trenches were produced by ion-milling in (001) $SrTiO_3$ and $LaAlO_3$ substrates. The ion beam direction was either parallel to the substrate plane normal or made an angle with it. The edge of the step in the substrate extended along the [010] direction. After ion-milling, the substrates were annealed in flowing oxygen at 1000 °C for 1 h to reduce the surface damage caused by this process. The $YBa_2Cu_3O_7$ films were subsequently grown on the step-edge substrates by pulsed laser deposition. The deposition temperature of 750 °C was chosen in order to obtain high quality films whose *c*-axis was parallel to the substrate normal.

13.3.3 Dependence of the microstructure upon step angle

Across the steps, the microstructure of the film was found to vary with the steepness of the steps. Figure 13.2 shows lattice images of the films deposited on $SrTiO_3$ substrates across steps of different angle Θ. The small arrows in the figures mark the *c*-axis of the $YBa_2Cu_3O_7$ film. The two images show that the crystallographic orientation of the film on the flank of the step depends upon

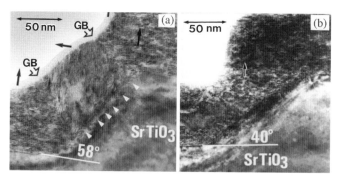

Fig. 13.2. Lattice images of $YBa_2Cu_3O_7$ films grown on steps showing the dependence of the film microstructure upon the step angle Θ. (a) The film over a 58° step. Open arrows mark two grain boundaries and the triangles mark the 90° domains. (b) The continuous film over a 40° step indicates no grain boundary.

Θ. On the flank of the 58° step, shown in Fig. 13.2(a), the film rotates its *c*-axis by a 90° angle (about an axis parallel to the viewing direction) with respect to the film grown on the flat substrate surface. A structure consisting of multiple 90° domains exists in the film near the interface between the film and the substrate. The domains are marked by triangles. When the thickness of the film exceeds about 30 nm, these domains are shunted by a larger domain with the *c*-axis perpendicular to that of the flat film. In this case, two 90° grain boundaries (marked by open arrows) are formed near the step edges. This holds for all the observed steps with $\Theta > 50°$. The small 90° domains directly at the substrate edge could not be clearly identified in all cases.

Across the low-angle step, $\Theta = 40°$, of Fig. 13.2(b) the film grows with a uniform *c*-axis orientation (parallel to the substrate normal) and no grain boundary is formed. This also holds for the steps with an angle smaller than this value. When the angle Θ is close to 45°, the film grows in the form of 90° multidomains throughout the film thickness. Hence, multiple 90° grain boundaries are present. This type of morphology is also found for films deposited on a (110)-oriented SrTiO$_3$ substrate.

Figure 13.3 shows a lattice image of a film on a step in a LaAlO$_3$ substrate whose flank surface has a variable slope. This step was produced using an ion beam which made an angle of 30° with the normal of the substrate surface. It can be seen that the slope decreases smoothly from 90° to a value below 45°. In this case, the flank of the step is formed by a curved surface and the step has no evident lower edge as in Fig. 13.2. Nevertheless, two grain boundaries (denoted by open arrows) are visible in Fig. 13.3. The upper grain boundary starts at the

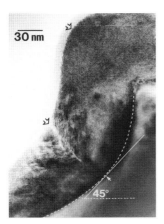

Fig. 13.3. YBa$_2$Cu$_3$O$_7$ film over a step with a flank of variable slope. Two grain boundaries are denoted by open arrows and a white arrow marks the starting point of the lower boundary.

upper step edge and its morphology is similar to that shown in Fig. 13.2(a). Although there is no evident lower edge of the step, the lower boundary is found to start at the position close to the point of the curved surface where the slope of the tangent is equal to 1, corresponding to $\Theta = 45°$, as indicated by the small white arrow. In the surface area with the large slope of the tangent, corresponding to $\Theta > 45°$, the film grows with the c-axis perpendicular to that of the flat surface film, while the film in the region with $\Theta < 45°$ keeps its c-axis unchanged. This indicates a critical value of the step slope for the formation of the grain boundaries in the film across the step.

It is known that high deposition temperatures favor the growth of the $YBa_2Cu_3O_7$ film with its c-axis perpendicular to the {100} substrate surface. The presence and abundance of local domains nucleated on the flank surface are related to the ratio of two kinds of {100} facet areas which compose the step flank. The two {100} facets are (100) and (001) when the step edge runs along [010] of the substrate, as shown in Fig. 13.4. The ratio of the two facet areas (A_{100} and A_{001}) is a function of the step angle Θ:

$$R = \frac{\sum A_{100}}{\sum A_{001}} = \tan \Theta. \tag{13.1}$$

The area of the vertical facets, A_{100}, is dominant when the slope of the flank is large ($R > 1$), corresponding to $\Theta > 45°$. The probability of a film nucleating on these facets is then correspondingly larger than on the others. During further growth of the film, the domains on the dominant facets will eventually shunt the domains with different orientations. Above a certain thickness, only the domains with a uniform orientation remain and therefore only two grain boundaries run through the whole thickness of the film. This phenomenon may

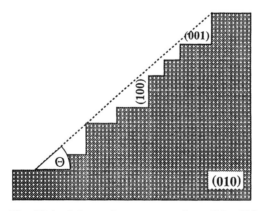

Fig. 13.4. Schematic representation of the {100} facets on a step flank.

also be kinetically enhanced since the atomic flux during deposition is not entirely isotropic, i.e. higher for normal incidence than for off-normal incidence.

For a low-angle step ($\Theta < 45°$, i.e. $R < 1$), the situation is just the opposite. The area of the facets parallel to the flat surface dominates. Consequently, the orientation of the film does not change across the step. In the critical case of a 45° step, the ratio $R = 1$. Both growth directions are equally favorable and this results in a zigzag multidomain structure with 90° boundaries.

It was possible to correlate the measured electrical properties of the films deposited on different types of steps with the value of the step angle [13.20, 13.21]. Optimum properties for the fabrication of Josephson devices require high-angle steps in the substrate. In contrast, low-angle steps allow the fabrication of electrical interconnects and crossovers where a reduction of the critical current density has to be avoided.

13.3.4 Grain boundaries in step-edge junctions

1. Morphology of grain boundaries in films over large-angle steps

Figure 13.5 shows a lattice fringe picture of a $YBa_2Cu_3O_7$ film across a steep step in LaAlO₃. The upper grain boundary exhibits a relatively regular habit

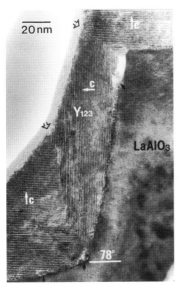

Fig. 13.5. Lattice fringe image of the $YBa_2Cu_3O_7$ film across a high-angle step on a LaAlO₃ substrate. The two grain boundaries are denoted by large arrows. The small arrows indicate the *c*-axis direction.

and, on the average, traces out a (103)-type plane. The lower grain boundary is very irregular and exhibits a tendency to tilt the whole habit plane toward the *ab*-plane of the flank film. Both grain boundaries consist of a chain of (103)(103)-type and (100)(001)-type segments. The (103)(103)-type boundaries have a habit plane parallel to the (103) or (013) plane of the two adjoining grains, while for the (100)(001)-type boundaries a (100) or a (010) plane of one grain faces a (001) plane of the other.

Figure 13.6 shows a film area including an upper boundary (a) and a lower boundary (b) in plan view, i.e. we view through the layer arrangement. The white and black arrows indicate the projected upper and lower grain boundary regions, respectively. At this low magnification, only lattice fringes corresponding to the *c*-lattice parameter that indicate the flank film area ('F') can be recognized. The zigzag pattern of bright and dark contrast in the upper film ('U') is a Moiré pattern originating from the overlap of the $YBa_2Cu_3O_7$ and the $LaAlO_3$ lattices along the viewing direction. This allows the twin structure of the $YBa_2Cu_3O_7$ film to be recognized. The twin boundaries across which the Moiré fringes change their direction run into the area where the flank film overlaps with both the top $YBa_2Cu_3O_7$ film and the substrate. By viewing the picture at a glancing angle along the *c*-fringes one can also recognize a shift of these fringes at the projected location of the twin boundaries. In the lower horizontal film ('L') the twin boundaries terminate again at the lower (100)(001)-type boundary. This image shows no overlap of the lower film with the flank film across the boundary area. This indicates that only the (100)(001)-type boundary segment is left over after the ion-milling procedure necessary for electron microscopic specimen preparation. Along the step edge the

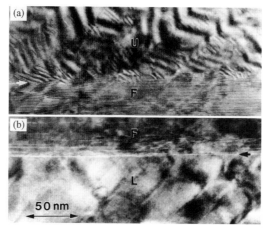

Fig. 13.6. Plan view of the film areas of an upper boundary (a) and a lower boundary (b).

(100)(001)-type boundaries are sharp and straight, only occasionally disturbed by a facet perpendicular to the edge. By combining the plan-view images with the cross-sectional one of Fig. 13.5 it is possible to reconstruct the three-dimensional morphology of the film over such a step.

2. Atomic structure of the grain boundaries

Although the general shape of the two boundaries is, as demonstrated by Fig. 13.5, quite different, both the upper and the lower grain boundaries consist of (103)(103)-type and (100)(001)-type segments. Figure 13.7 shows a lattice image of an upper grain boundary. The boundary is indicated by a sequence of small white circles. It consists of a (103)(103)-type segment on the upper left and a (100)(001)-type segment on the right hand side. The unit cell of $YBa_2Cu_3O_7$ is indicated by frames in both grains. The imaging parameters were determined by quantitative comparison of the experimental image with sets of images of $YBa_2Cu_3O_7$ calculated for an extended range of objective-lens focus and thickness values. This yielded a sample thickness of 2.3 nm and a defocus value of -30 nm. In the image the cation columns appear dark on a bright background. Moreover, the geometric positions of the cations are correctly represented by the dark dots. This means that the position of the Cu-atoms, the central Y-atom and the two Ba-atoms can be recognized as 'points' of dark contrast. The exact position of the unit cell can be safely determined on the basis of the inequality of the c-lattice parameter of the three perovskitic sub-units of $YBa_2Cu_3O_7$.

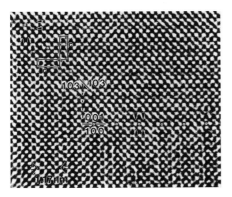

Fig. 13.7. Lattice image of an upper grain boundary. The $YBa_2Cu_3O_7$ unit cells are indicated by frames.

2.1 (100)(001)-type grain boundary In (100)(001)-type grain boundaries there is a mismatch between c in one grain and $3a$ in the other grain (or, if the small orthorhombic distortion is taken into account, between c and $3b$). Here $a = 0.382$ nm, $b = 0.388$ nm and $c = 1.168$ nm denote the lattice parameters of the orthorhombic lattice. Furthermore, a perovskite sub-lattice mismatch at the boundaries exists in the direction parallel to [100] (or [010]) of one grain and the [001] direction in the other grain. As shown in Fig. 13.8, the spacing between Cu atoms in the first grain is equal to the a (or b) lattice parameter everywhere along the boundary. In the second grain two different spacings occur along the grain boundary: along the c-axis, the spacing between the CuO-chain plane and the CuO_2 plane is 0.415 nm while the spacing between the two CuO_2 planes is 0.338 nm.

This sub-lattice mismatch is detectable for proper imaging conditions (such as those used for Fig. 13.7) which yield an image with a good correspondence to the geometric positions of the cations. In this case, the position of the grain boundary can be directly localized by observing the sub-lattice mismatch. Figure 13.9 shows the (100)(001) boundary segment in the center part of Fig. 13.7 at a larger magnification. The sub-lattice mismatch is found by following the double vertical CuO_2 planes which exhibit the small spacing from the bottom of this image. The pairs of arrowheads indicate the position of the Cu atom columns. The distance of the upper pair corresponds to the a-lattice

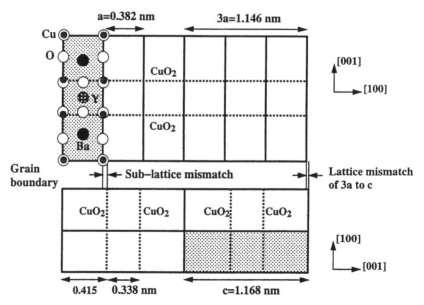

Fig. 13.8. Schematic representation of the lattice and the sub-lattice mismatch at a (100)(001) grain boundary. Shaded frames outline the unit cell.

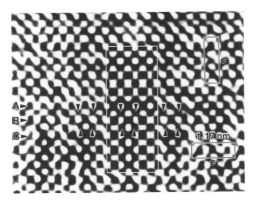

Fig. 13.9. Lattice image of a (100)(001) boundary. The calculated image (inset) of the boundary based on the model shown in Fig. 13.10(a) agrees well with the experimental one. The cations appear dark on a bright background. The letters A, B and C refer to the line scans in Fig. 13.11.

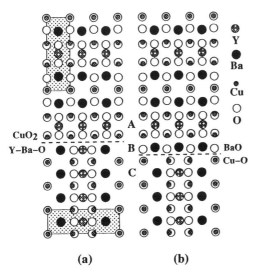

(a)   (b)

Fig. 13.10. Two structure models of the (100)(001) grain boundaries. (a) A (001) CuO_2 plane of the upper grain faces a (100) Y—Ba—O plane of the lower grain. (b) A (001) BaO plane of the upper grain faces a (100) Cu—O plane in the lower grain. Broken lines mark the boundaries.

parameter of the upper grain, while the distance of the lower pair corresponds to the spacing of the double CuO_2 planes of the lower grain. The grain boundary should be situated between these two types of spacings at B in Fig. 13.9.

Accordingly, two basic structure models can be constructed for this boundary (Fig. 13.10). In model (a) the (001) CuO_2 plane of the upper grain faces the

(100) Ba—O—Y—O—Ba plane of the lower grain. In model (b) the (001) BaO plane of the upper grain faces the (100) CuO plane of the lower grain. Image simulations were carried out based on these two models. The inset in the center of Fig. 13.9 shows the simulated image which yields the best match. It is based on model (a). The agreement between the two types of images is very good; also the spacing difference which arises from the perovskite sub-lattice mismatch is reproduced in the calculated image. Nevertheless, for these imaging conditions, no significant contrast difference was found between the simulated images of model (a) and model (b). However, the two models differ with respect to the lateral sequence of cations in the boundary layer B. In model (a), the horizontal B layer exhibits a Ba—Y—Ba—Ba—Y—Ba sequence while the respective sequence is Ba—Ba—Ba for model (b). This means that on the basis of model (a) we expect to measure, along A in the upper grain, a uniform value for the lateral cation separation of $a = 0.382$ nm. Along B, in the lower grain, we expect a large value (0.430 nm, in the following denoted by L) for the Ba—Ba pairs and a smaller spacing (0.368 nm, denoted by S) for the Ba—Y—Ba sequence in the lower grain. In model (b) we expect, in layer B, a uniform separation only.

Actual measurements were carried out by densitometer line scans in the experimental image (Fig. 13.9). Part (a) of Fig. 13.11 shows the results. Part (b) depicts the position of the intensity minima which indicate the position of

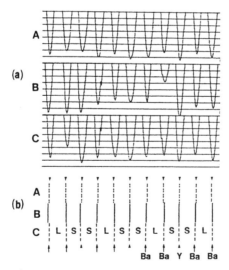

Fig. 13.11. (a) Intensity in densitometer line scans along A, B and C in the direction denoted by an arrow in Fig. 13.9. (b) The minimum positions of the intensity profiles in (a). Arrows and arrowheads indicate the ideal positions of Ba and Y atoms.

the cations. The expected uniform spacing in layer A of the upper grain and the LSSLSS sequence in layer C of the lower grain are obtained. Within experimental error, the layer B clearly shows the same sequence of large and small distances as layer C. From this we conclude that the boundary layer B is a Ba—Y—Ba layer, in agreement with model (a).

For the (100)(001)-type boundaries in *a*-axis-oriented thin films, other structure models have also been reported in the literature [13.12]. The type of terminating plane of the grains in the (100)(001)-type boundary affects the local strain level. From our experimental images of the boundaries with the structure of model (a), shown in Fig. 13.9, no relaxation of atomic positions is inferred in the first (100) Cu—O layer of the lower grain which is adjacent to the boundary. The normal spacing of the neighboring CuO_2 planes is preserved. Hence the effect of the boundary layer on this spacing must be small. Indeed, according to model (a) a Y—Ba—O boundary layer is expected to produce a small lattice distortion since it belongs to the lower grain. The effect of the Y atoms on the upper grain can be neglected since the spacing between the Cu atoms in the first CuO_2 plane above the boundary is mainly determined by the oxygen atoms located in between (see Fig. 13.10(a)). In contrast, because of their large atomic radius, the Ba atoms in a Ba—O boundary layer are expected to expand the spacing of the Cu atoms in the two adjacent CuO_2 planes below the boundary (see Fig. 13.10(b)).

2.2 (103)(103) grain boundary Figure 13.12 shows a part of the (103)(103) boundary segment of Fig. 13.7 (upper left) at high magnification. In order to achieve this representation the original image was low-pass Fourier-filtered in order to eliminate photographic noise. Frames mark the unit cell of $YBa_2Cu_3O_7$ in both grains. White circles denote the Y positions while the Ba and Cu positions are marked by large and small white dots, respectively.

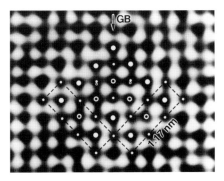

Fig. 13.12. Lattice image of a (103)(103) boundary.

All along the boundary in Fig. 13.7 we find that, along the [100] direction in both grains, the Cu atom positions (in the CuO_2 planes) next to the boundary plane deviate from those expected from a simple boundary model obtained by a mirror operation in the (103) plane. This is indicated in Fig. 13.12 by a pair of open (unrelaxed model position with respect to Fig. 13.13) and full circles (real position as deduced from the center of the dark contrast region representing the Cu atom column in the image). This shift of the Cu position can most easily be detected by a comparison of the spacing between the Cu and Ba positions (vertical spacing between a small and a large solid dot) with that between the Cu and Y positions (vertical spacing between a small dot and an open circle). In the first (103) atomic plane adjacent to the boundary the Cu—Ba spacing is smaller than the Cu—Y spacing while the opposite holds further away from the boundary, i.e. in the ideal $YBa_2Cu_3O_7$ lattice.

If the boundary is constructed schematically by a simple mirror symmetry operation located in the (103) plane of one grain, the positions of the Cu atoms in the CuO-chain plane and the positions of the Y atoms in the two adjacent grains coincide, as shown in Fig. 13.13(a). However, the other atoms are obviously misplaced with respect to the boundary plane. The formal operation creates 'double' atoms (denoted by pairs of arrowheads) and 'imperfect' atoms (single arrowheads). Replacing these artifacts, in a realistic model, by a single atom located in the boundary plane induces a periodic sequence of tensile

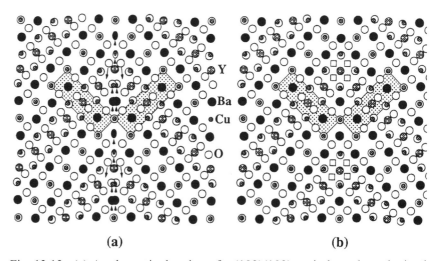

(a)   (b)

Fig. 13.13. (a) A schematic drawing of a (103)(103) grain boundary obtained by a simple mirror operation at the (103) plane. (b) A structural model of a (103)(103) grain boundary in which relaxation of the atoms near the boundary is introduced; the positions marked by squares are assumed to be partially occupied by oxygen.

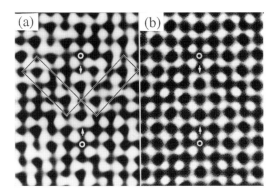

Fig. 13.14. The calculated image (b) based on the model of Fig. 13.13(b) in comparison with the experimental one (a).

stress at the position of the 'double' atoms and compressive stress for the 'imperfect' atoms along the common [$\bar{3}$01] direction.

It is to be expected that a redistribution of the most mobile atomic species, oxygen, induces a reduction of the local stress level. As indicated in this model, the compressive stress indicated by the 'imperfect' atoms (single arrowhead) should induce a migration of the oxygen atoms to the unoccupied position in the dilatation region indicated by the 'double' atoms (pair of arrowheads). This should locally adjust the position of the nearby atoms, i.e. it should be possible to induce, as observed, a shift of the Cu atoms towards the Ba atoms. The final model is shown in Fig. 13.13(b). Small black dots mark the positions around the Y atom positions in the grain boundary which are assumed to be partially occupied by oxygen atoms in order to keep the valency of the cations unchanged.

A calculated image based on this model is shown in Fig. 13.14(b) together with the experimental image (Fig. 13.14(a)). It indeed reproduces the prominent features of the experimental image, e.g. the square-shaped arrangement of bright dots around the two circular marks and the bright 'bridges' denoted by arrows. This figure shows two repetition units along the [$\bar{3}$01] direction. Clearly the characteristic contrast feature around the upper circular mark also appears around the lower one.

13.3.5 Lattice defects and distortions near grain boundaries in step-edge junctions

Lattice distortions and defects were frequently found at or near the grain boundaries induced by the step geometry of the substrates. Figure 13.15 shows

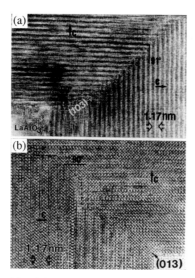

Fig. 13.15. Two upper boundaries exhibiting different morphologies and different levels of distortion.

two upper boundaries. In both cases we find that, at the LaAlO$_3$ edge, the grain boundary angle is 90°. In Fig. 13.15(a), however, the angle increases with the distance from the edge to about 91°. By viewing the pictures at a glancing angle one can verify that the boundary in Fig. 13.15(a) shows stronger contrast irregularities, due to larger strain, than the boundary shown in Fig. 13.15(b). This sometimes makes it quite difficult to localize the grain boundary habit plane. In Fig. 13.15(a) the plane, on the average, deviates from an exact (103) plane orientation by about 5° clockwise. Steps due to facetting are rare in this boundary. In contrast, the boundary in Fig. 13.15(b) includes a larger number of facets. As indicated by a dashed line, the boundary segments are either of the (103)(103)-type or the (100)(001)-type. In Fig. 13.15(b), the overall habit plane is tilted with respect to the (013) plane by about 5° counterclockwise.

Because of the different ratios of c to $3a$ and $3b$, the misorientation angle of a (103)(103) boundary differs from that of a (013)(013) boundary. In the (013)(013) boundary, the two b-axes make an angle of 90.2° because of $c/b = 3.01$. For the (103)(103) boundary a 91.2° angle between the a axes in the two grains is obtained from $c/a = 3.06$. In Fig. 13.15, evidence for both the (103)(103) and the (013)(013) boundaries is found since the angles of 91° and 90° occur. The large strain observed around the (103)(103) boundary may originate from the constraints exerted by the cubic substrate. For epitaxy the upper film and the flank film have to grow in two orthogonal directions, i.e. the substrate–lattice structure forces the a-axes of the two grains to make a 90°

angle. Lattice distortion or defects are required in order to accommodate the 1° angular deviation for the (103)(103) boundary. The strong facetting of the [100]-tilt boundary can be explained by the fact that the misorientation angle (90.2°) of the (013)(013) boundary segment is very close to the angle of the (010)(001) boundary segment (90°). In this case the lattice distortion is relatively small.

Fig. 13.16(a) shows a part of the lower boundary of Fig. 13.5, at a large magnification. We see that the boundary consists to a large extent of sharp (100)(001)-type segments. The steps (marked by an open arrow in the upper part of the picture) between subsequent segments have a height equal to the full (001) interplanar spacing of the flank film. Misfit dislocations are indicated by arrowheads with a projected Burgers vector of $a/2$ [100]. In other parts of the boundary area strong lattice distortions and '124' stacking faults can be seen on the side of the flank film (right-hand side).

Figure 13.16(b) shows an area of a lower grain boundary which includes a high density of defects. The lower horizontal film extends to the right while the film on the step flank extends to the left. The lower film is again much more

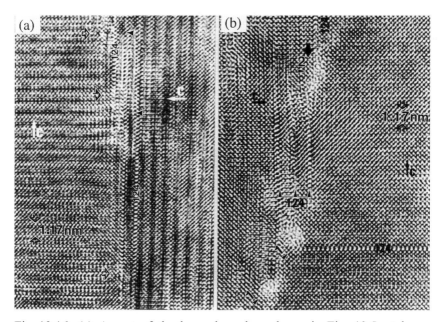

Fig. 13.16. (a) A part of the lower boundary shown in Fig. 13.5, at larger magnification. The open arrows mark the segment steps and the arrowheads indicate the misfit dislocations. (b) Another lower grain boundary which contains '124' stacking faults. Arrowheads mark boundary dislocations. A bold black arrow denotes the termination of a set of a–b planes at a boundary step.

perfect than the flank film. Along the boundary, the flank film is heavily faulted. To the left of the boundary, extended '124' faults were formed in connection with a heavy distortion of the $YBa_2Cu_3O_7$ lattice in the environment. In the upper part of the picture, the bold arrow marks the termination of a set of $a–b$ planes. This causes heavy local distortion which is in turn partly accommodated by the adjoining '124' faults. By viewing the picture at a glancing angle along the direction indicated by the inclined arrowheads denoted '1' in the lower left part of the picture, partial dislocations of the $YBa_2Cu_3O_7$ lattice can be recognized. By tracing Burgers circuits around these dislocations a projected Burgers vector of $[a/2\ 0\ a/6]$ was determined. It can be recognized from the fringe contrast in the center of the picture that, in spite of the introduction of the defects, large areas exist which exhibit high strain levels. A set of fringes in the direction from the lower left corner of the figure to the upper right corner shows an enhanced contrast in this area. This originates from lattice bending around an axis perpendicular to the direction just mentioned. Arrowhead '2' marks a misfit dislocation which has the same properties as those shown in Fig. 13.16(a). The irregular contrast close to dislocations denoted by '1' and '2' indicates local strain at the dislocation cores.

In our work on step-edge junctions we found that misfit dislocations accommodate the mismatch between c and $3a$ (or $3b$) in the case of (100)(001)-type boundaries. This is in contrast to the results of Ref. [13.22] where it was found that the formation of stacking faults was the way to accommodate this mismatch in 90° grain boundaries in a-axis films. We found that, for the boundary surrounding an a-axis-oriented grain in a c-axis-oriented film, the combination of the dislocations at the boundary with the stacking faults can accommodate this mismatch [13.24]. In general the misfit dislocations are located in the boundary, or close to it, and occur with quite regular spacing. In the boundaries of step-edge films, however, these dislocations are irregularly distributed. This may be a consequence of the boundary irregularities introduced by the step geometry.

On average, the density of defects such as stacking faults and dislocations is higher at the lower grain boundary than at the upper one. Nevertheless, there are also segments in the lower grain boundary which appear free of defects. Overall, the density of boundary defects and distorted areas is considerably higher than that observed for the 90° boundary structure in the a-axis-oriented films [13.22, 13.23] and also for the boundaries surrounding a-axis grains in otherwise c-axis-oriented films. This can be attributed to the complicated non-equilibrium film growth conditions resulting from the step geometry. The irregular contrast observed in the boundary plane, or around the dislocation cores, can be explained by strain and atomic disorder on a sub-unit cell level.

13.4 Interfaces in YBa$_2$Cu$_3$O$_7$ multilayer edge junctions

13.4.1 Introduction

In the case of multilayer systems the Josephson junction is produced in the more classical way by sandwiching a thin layer of non-superconducting material or a material with lower T_c or lower critical current density between two superconducting layers. In this case, constraints arise for the materials employed and for the fabrication processes used, owing to the need to maintain perfect epitaxy throughout the multilayer system. In order to obtain optimum characteristics of the junctions, a possible degradation of the electrical properties of the superconducting layers must be limited to a very narrow zone at the interfaces. This demands high structural perfection of the interfaces and a very low degree of interdiffusion between the superconducting and non-superconducting materials. For multilayer junctions on the basis of the superconductor YBa$_2$Cu$_3$O$_7$, barrier materials with perovskite-related structures are generally employed [13.25–13.28] since their structure is very similar to that of the superconductor and the lattice mismatch is small. In these systems, a detailed investigation of the interfaces on an atomic scale is particularly important in order to monitor the results of the film deposition process and to understand and control the structure-dependent electrical behavior of different types of devices.

Multilayer junctions are in general fabricated as edge junctions [13.29]. This technique offers the advantage that the superconducting current flows along the *a–b* planes in the two superconducting films on both sides of the non-superconducting barrier. Figure 13.17 shows a schematic of the edge junction developed by Faley *et al.* [13.12, 13.13], which as far as field sensitivity and noise properties are concerned is one of the most advanced modern designs. The following sections will be concerned with the problems encountered in the

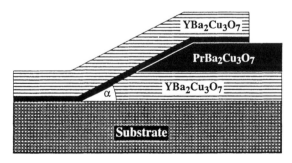

Fig. 13.17. Edge-type multilayer Josephson junction as developed by Faley *et al.* [13.12, 13.13].

barrier production for this type of junction. The first example deals with SrRuO₃ as a barrier material. Electrically this perovskite is not superconducting but metallic. Its structure was investigated by Dömel *et al.* using X-ray techniques [13.10]. The second example concerns PrBa₂Cu₃O₇ barriers which were first introduced by Poppe *et al.* [13.25] and by Rogers *et al.* [13.30] and are now the most frequently employed barriers in superconducting multilayer Josephson junctions.

13.4.2 Interfaces in triple-layer films of YBa₂Cu₃O₇/SrRuO₃

Figure 13.18 shows a low-magnification overview of a YBa₂Cu₃O₇/SrRuO₃/YBa₂Cu₃O₇ triple-layer film on a SrTiO₃ substrate. The junction was prepared using an off-axis sputtering technique. The three different layers can easily be identified by their different image contrast. The lower interface, between the first YBa₂Cu₃O₇ layer and the SrRuO₃ layer, is sharp but wavy. In contrast, the upper interface, between the second YBa₂Cu₃O₇ layer and the SrRuO₃ layer, is relatively flat but exhibits pronounced dark strain contrast. An additional band-like contrast can be seen in the YBa₂Cu₃O₇ layers (e.g. in the area denoted by an arrow). It is caused by a Moiré fringe pattern, which will not be discussed in detail here. We only note that the pattern results from the overlap, along the viewing direction, of two parts of YBa₂Cu₃O₇ that are separated by a spiral growth center, a screw dislocation, and therefore have a slight deviation in orientation.

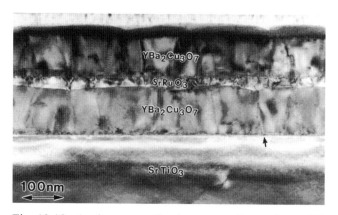

Fig. 13.18. A low-magnification overview of a YBa₂Cu₃O₇/SrRuO₃/YBa₂Cu₃O₇ triple-layer film on a SrTiO₃ substrate. A small arrow indicates one of the defects which produces a band-like contrast in the image.

1. Interface between the SrRuO$_3$ layer and the first YBa$_2$Cu$_3$O$_7$ layer

Figure 13.19 shows a lattice image of a part of the lower interface between the first YBa$_2$Cu$_3$O$_7$ layer and the SrRuO$_3$ layer. As mentioned, this interface looks wavy at lower magnification. The high-magnification picture indicates that this is caused by a high density of steps. Nevertheless, the interface is atomically sharp. Across a step the interface is displaced by 1.17 nm, i.e. by a full *c*-axis lattice parameter of YBa$_2$Cu$_3$O$_7$. As a consequence the type of terminating atomic plane is the same for all the interface segments separated by these steps. Misfit dislocations were occasionally observed at the interface. They compensate for the lattice mismatch between the two compounds; one of them is indicated by the vertical arrow.

In order to clarify the atomic structure of the interface, a series of lattice images with different defocus values was recorded from the same region of a specimen where the interface plane is parallel to the electron beam and there is no interface step along the beam direction within the thickness of the specimen. Two images from such a series taken with the electron beam parallel to the [010] direction of YBa$_2$Cu$_3$O$_7$ are shown in Fig. 13.20. According to the image simulations for both the YBa$_2$Cu$_3$O$_7$ and the SrRuO$_3$ structure, the two images were taken at a sample thickness of 3.1 nm and defocus values of -30 nm (a) and -60 nm (b).

For the defocus value of -30 nm (Fig. 13.20(a)), the cation columns are imaged as dark dots on a bright background. In the image with the defocus value of -60 nm (Fig. 13.20(b)) a reversal of the image contrast takes place, i.e. the cation columns give rise to bright dots on a dark background. For these imaging conditions all the cation positions are correctly imaged, in so far as they appear at the geometrical position of the atoms. This allows the three perovskite sub-unit cells of YBa$_2$Cu$_3$O$_7$ to be identified on the basis of the

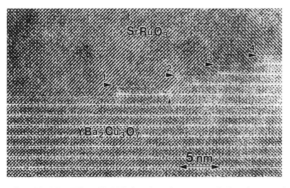

Fig. 13.19. The [010] lattice image of the lower interface between the first YBa$_2$Cu$_3$O$_7$ layer and the SrRuO$_3$ layer.

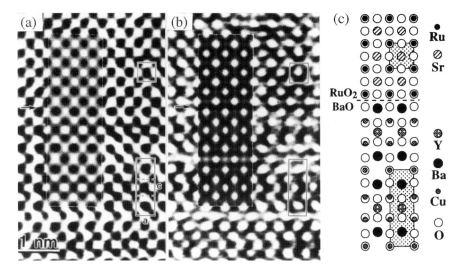

Fig. 13.20. Two images of the (001) interface of $YBa_2Cu_3O_7/SrRuO_3$ recorded using different defocus values. The insets show simulated images based on the structural model in (c) calculated for a sample thickness of 3.1 nm and the defocus values (a) −30 nm and (b) −60 nm. (c) A model of the interface with a BaO plane of $YBa_2Cu_3O_7$ facing a RuO_2 plane of $SrRuO_3$, viewed along the [010] direction of $YBa_2Cu_3O_7$. The dashed line marks the interface.

inequality of their dimensions along the [001] direction as already discussed in chapter 3. As shown in Fig. 13.20, the square and rectangular frames mark the unit cells of $SrRuO_3$ (upper part of the pictures) and $YBa_2Cu_3O_7$ (lower part), respectively.

By inspecting the stacking sequence of the sub-units of $YBa_2Cu_3O_7$ along the c-axis direction from the $YBa_2Cu_3O_7$ layer into the $SrRuO_3$ layer, the terminating atomic planes of the $YBa_2Cu_3O_7$ layer at the interface can be identified as either a (001) BaO plane or a CuO-chain plane. Accordingly four possible structure models were designed with four different types of connection between the two compounds: (i) BaO/RuO_2, (ii) BaO/SrO, (iii) CuO/RuO_2 and (iv) CuO/SrO. Image simulations were carried out based on these models. The best fit of the calculated images with the series of experimental images was obtained for model (i), as shown in Fig. 13.20(c). The corresponding simulated images are shown as insets in the experimental images of Figs. 13.20(a) and (b). The interface plane is marked by arrows. All the calculated images are in very good agreement with the corresponding experimental ones. In particular, for the images of Fig. 13.20(b), the interface gives rise to a pronounced special feature: a dark gap between the BaO and the RuO_2 planes is obvious in both the calculated and the experimental images. A double row of large bright dots representing the Ba—O and Ru—O columns appears for these imaging conditions. The important element of model (i) is that the $YBa_2Cu_3O_7$ layer terminates with a BaO rather than a CuO-chain plane.

The atomic planes that terminate at the flank of the interface steps can also be determined. This is possible by observing the different dimension of the perovskite sub-cells of $YBa_2Cu_3O_7$ in comparison with the cubic $SrRuO_3$ in the [001] direction of $YBa_2Cu_3O_7$ across the step. Figure 13.21(a) shows two of the interfacial steps which are traced out by a dashed line. According to the image simulation this image was recorded at a defocus value of −28 nm in a 3.1 nm thick sample. For these conditions, a faint bright contrast at the position of the (001) CuO-chain planes distinguishes the $YBa_2Cu_3O_7$ layer from the $SrRuO_3$ layer. This contrast is, as confirmed by image simulation, an indication of oxygen ordering in the basic CuO-chain planes. When the CuO chains are perpendicular to the electron beam, they show this faint extra image contrast for these special imaging conditions.

The pronounced change in image contrast (compare the contrast at the

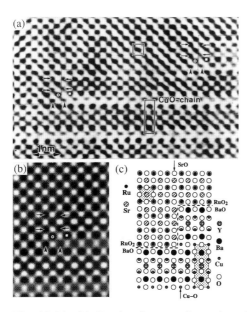

Fig. 13.21. (a) Lattice image of two interfacial steps. The first $YBa_2Cu_3O_7$ layer and the $SrRuO_3$ layer are separated by a dotted line. The other details are described in the text. (b) Calculated image based on the structural model in (c) for a sample thickness of 3.1 nm and a defocus value of −28 nm; (c) A structure model for the interface step viewed down the [100] direction of $YBa_2Cu_3O_7$.

positions marked by the left and right arrowheads across each step) indicates a change from a Ru—O column to a Cu column across the interface step. Above these points of contrast change two geometric features are detectable which help to clarify the atomic structure of the step flank. The first is the change in spacing from the value between the two CuO_2 planes (0.338 nm, marked by the pairs of left-pointing arrows) to the value between two RuO_2 planes (0.393 nm, marked by the pairs of right-pointing arrows). This change in atomic plane spacing coincides with the contrast difference indicated by the arrowheads and therefore identifies the vertical (010) Cu—O plane and RuO_2 plane. The second geometric feature concerns a vertical shift from the Ba position (marked by squares) to the Sr position (marked by circles) across the (010) Cu—O plane. This shift originates from the smaller ionic radius of Sr and Ru in comparison with Ba and Cu, respectively. This indicates that the vertical atomic plane between the two pairs of arrows is clearly a SrO plane.

A structure model for the interface step can be designed on the basis of the above analysis, as shown in Fig. 13.21(c). The calculated image (Fig. 13.21(b)) based on this model agrees well with the experimental one: both the geometric features and the details of the image contrast of the interfacial step are reproduced. We therefore conclude that the structure of the interface step is characterized by a (010) Cu—O plane of $YBa_2Cu_3O_7$ facing a Sr—O plane of $SrRuO_3$.

We found that in all cases of the lower interfaces studied here in which the $SrRuO_3$ layer is on top of the $YBa_2Cu_3O_7$ layer the latter is terminated by a BaO plane. The structure of the interface is controlled by various factors; the chemistry of the joined compounds, the growth kinetics and the presence of defects. A spiral growth of c-axis-oriented film has been derived from the results of scanning tunneling microscopy, the minimum growth unit being the full $YBa_2Cu_3O_7$ unit cell [13.31]. However, to our knowledge, the atomic structure of the surface plane of the c-axis-oriented $YBa_2Cu_3O_7$ films has not been determined as yet. For this purpose growth steps were investigated in cross-sectional samples of $YBa_2Cu_3O_7/PrBa_2Cu_3O_7$ superlattices [13.32]. However, the terminating plane at the interfaces could not be unambiguously identified. The reason for this is that the two compounds, $YBa_2Cu_3O_7$ and $PrBa_2Cu_3O_7$, have the same structure and also a very similar composition. If, in the present investigation, we make the plausible assumption that the interfaces were once the growth surfaces, the last atomic layer formed just before $SrRuO_3$ deposition should be identical with the interface layer. For the c-axis-oriented films of $YBa_2Cu_3O_7$, the surface is composed of horizontal (001) BaO planes and vertical (100) and (010) Cu—O planes (at steps) for the deposition conditions used in this work. The reported horizontal Cu—O plane termination of c-axis films may be explained by differences in the experimental conditions under which the interfaces or surfaces were produced [13.34, 13.35].

The atomic plane of the $SrRuO_3$ layer which starts on the BaO surface plane of the $YBa_2Cu_3O_7$ layer was always found to be a RuO_2 plane. Indeed, if a SrO plane were the starting plane, an oxygen co-ordination for both the Ba- and the Sr-cations would be formed, which is not likely to be energetically favorable. In addition, both Ba and Sr have a larger ionic radius than the others and direct joining of them would introduce large mechanical stresses in and close to the interface. In contrast, the combination of the BaO plane with the RuO_2 plane in the interface results in an oxygen environment for the Ba and the Ru cations quite similar to that in the respective undisturbed matrix structure. In addition, a stacking sequence in which a cation with a large radius follows a cation with a small radius minimizes the stress level and thus leads to a low interface energy.

2. Interface between the second $YBa_2Cu_3O_7$ layer and the $SrRuO_3$ barrier

Figure 13.22 shows a lattice image of the interface between the second $YBa_2Cu_3O_7$ layer (top) and the $SrRuO_3$ layer. This interface is relatively flat. Nevertheless, lattice imperfections such as stacking faults and antiphase boundaries were frequently observed at or near the interface. Such a fault is shown in Fig. 13.22 (marked by 'F'). A double CuO-chain plane on the right-hand part of the interface produces an irregularity in the image contrast in comparison with the left-hand part where a perfect and regular connection of the two compounds is obvious. The higher density of lattice defects is the

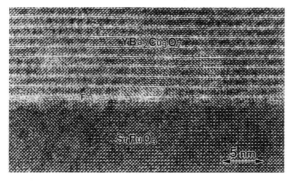

Fig. 13.22. A lattice image of a part of the interface where $YBa_2Cu_3O_7$ layer is on top of the $SrRuO_3$ layer. A stacking fault with a double CuO-chain plane on the left part of the interface is marked by the letter 'F'.

origin of the higher level of strain in the upper grain boundary in comparison with the lower one.

The structure of the interface in a defect-free area is investigated first, for which purpose this area in Fig. 13.22 is further magnified in Fig. 13.23. In the image simulation for the ideal structure of the two compounds, the best fit of the calculation with the experiment can be obtained for a sample thickness of 3.1 nm and a defocus value of −75 nm. Under these conditions, the cations gave rise to bright image contrast on a dark background. For the $SrRuO_3$ compound the Ru—O columns correspond to the larger dots than the Sr columns. In the $YBa_2Cu_3O_7$ layer, the large dots are coincident with the position of Ba. The CuO-chain planes are characterized by a faint contrast between two bright dots. Under these conditions the position of the cations in the two structures is correctly imaged.

An examination of the sequence of image contrast features of the different type of atoms and the inequality of the contrast of the different perovskite sub-units of $YBa_2Cu_3O_7$ suggests two possible models for the interface structure. In the first model, a RuO_2 plane faces a BaO plane, as in the lower interface (Fig. 13.20(c)). In the second a SrO plane faces a CuO_2 plane. Image simulations were carried out on the basis of these two models. The results confirm the model with a RuO_2 plane facing a BaO plane. The corresponding calculated image is shown as an inset in Fig. 13.23. A crucial feature of the image for the structure identification of the interface, which appears in both the calculated and the experimental images, is that the two interfacial planes,

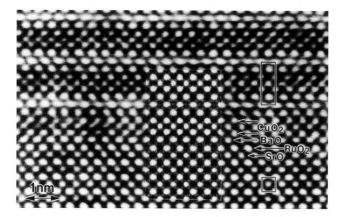

Fig. 13.23. A perfect segment of the interface. The upper and the lower frames indicate the unit cell of $YBa_2Cu_3O_7$ and $SrRuO_3$. The interface structure is indicated by the arrangement of the atomic planes. The inset is a calculated image based on the same model as shown in Fig. 13.20(c) for a sample thickness of 3.1 nm and a defocus value of −75 nm.

the RuO$_2$ plane and the BaO plane, give rise to a double row of large bright dots. The second model did not yield such an image. This also holds for the interface part with the stacking fault.

Figure 13.24 shows a lattice image of an interface with a stacking fault. This image was recorded under conditions such that black dots correspond to the cation locations. There is an interface step which is denoted by an arrowhead. This step is marked by a change in lattice spacing. Viewing at a glancing angle guided by the two pairs of arrows from the left and the right side of the image, it can be seen that the two types of spacing, one corresponding to the double CuO$_2$ plane (small) and the other corresponding to the *a*-lattice parameter of SrRuO$_3$ (large), terminate at the arrowhead. However, the interface plane cannot be unambiguously identified using this image only. The reason is that, for the imaging conditions used, the large dots appear at the Sr, Ba and Y positions and there is no apparent difference in the image contrast between them. Nevertheless, the joining of a BaO plane to a SrO plane or a Cu—O plane to a RuO$_2$ plane at the interface, which would have produced a direct joining of two large or two small dot rows, can be ruled out by the image contrast. In the picture a small-dot row is followed by a large-dot row across the interface area. In this case the possible structure for the interface is a BaO plane facing a RuO$_2$ plane or a CuO-chain plane facing a SrO plane for the left part and a CuO$_2$ plane facing a SrO for the right part of the interface. On the basis of the above two possibilities for the interface structure one finds that the maximum value for the step height is still smaller than the *c* lattice parameter of YBa$_2$Cu$_3$O$_7$. Since this step does not induce an additional defect into the YBa$_2$Cu$_3$O$_7$ lattice, it must be accommodated by a change of the terminating atom plane of YBa$_2$Cu$_3$O$_7$. This means that as a result of the presence of

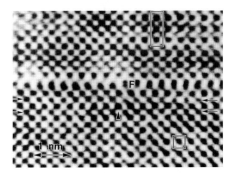

Fig. 13.24. A lattice image of an interface where a stacking fault (F) exists. An interface step is denoted by an arrowhead. Two pairs of small arrows mark a double CuO$_2$ plane (right) and two RuO$_2$ planes (left).

surface steps in the SrRuO$_3$ layer, different types of interface structure occur in the second YBa$_2$Cu$_3$O$_7$/SrRuO$_3$ interface.

3. General aspect of this investigation

We emphasize a special aspect of this investigation which is of general importance for the fabrication of any type of epitactic system in which a simple single unit cell perovskite is combined with a multi-sub-unit cell compound. We find in the present case that the upper interface contains a higher density of defects and a higher level of strain than the lower one. This originates from the different surface geometry, as shown schematically in Fig. 13.25.

The surface steps on the top of the first YBa$_2$Cu$_3$O$_7$ layer always have a height of a full unit cell in the *c*-axis direction. Since this is approximately three times larger than the cubic unit cell of SrRuO$_3$ the strain remains small and no lattice defects are introduced into the deposited SrRuO$_3$ layer. In contrast, the upper interface, where the YBa$_2$Cu$_3$O$_7$ layer is on top of the SrRuO$_3$ layer, contains a high level of elastic strain and has a non-uniform lateral structure. Indeed the surface steps on the SrRuO$_3$ layer can be *n*

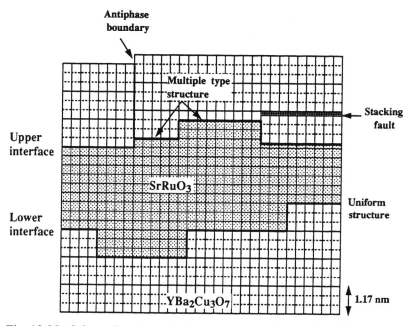

Fig. 13.25. Schematic view of the two types of interface. The variety of structures including vertical antiphase boundaries and horizontal faults can be explained by the growth of the YBa$_2$Cu$_3$O$_7$ on a SrRuO$_3$ surface exhibiting steps of different heights which are not always multiples of the *c*-axis dimension of the superconductor.

perovskite unit cells high ($n = 1, 2, 3, \ldots$). On the other hand, the $YBa_2Cu_3O_7$ unit comprises three such units. As a consequence, problems are unavoidable if the step height on $SrRuO_3$ cannot be expressed in full multiples of 3. This leads to vertical antiphase boundaries [13.33] or horizontal stacking faults in the $YBa_2Cu_3O_7$ layer when, as illustrated in Fig. 13.25, a grain nucleated on a $SrRuO_3$ terrace grows laterally into the area of another terrace a few perovskite sub-units higher. The misfit can, as we have seen in this study, also be accommodated by changing the starting layer of the next $YBa_2Cu_3O_7$ layer on top. This variety of possibilities not only leads to an interface structure which varies along the interface but also to elastic strain. Nevertheless in this case, we never found a direct connection of BaO to SrO planes and of Cu—O to RuO_2 planes. The reasons for this are the same as those discussed for the lower interface.

This leads us to an important conclusion concerning the structural constraints for epitaxial junctions produced on the basis of $YBa_2Cu_3O_7$ and single unit cell non-superconducting materials. The interface between the first $YBa_2Cu_3O_7$ layer and $SrRuO_3$ contains steps with heights that are multiples of 3 perovskite units. The latter can be accommodated by the single-perovskite $SrRuO_3$. Apart from a low density of defects such as misfit dislocations, the interface is structurally and electrically homogeneous. The upper interface, however, is critical since the growth of the $YBa_2Cu_3O_7$ comprising three perovskite unit cells unavoidably leads to severe defect formation in the superconductor as long as the interface is not atomically flat. This may have serious consequences for all junctions for which the particular function, as is generally the case, requires a high degree of structural perfection of both interfaces. In the design of superconducting electric-field effect devices based on $YBa_2Cu_3O_7/SrTiO_3$, it appears advantageous to grow the superconductor first and then deposit on it the dielectric $SrTiO_3$. This should benefit the reproducibility of the function of the devices. A reverse sequence would produce an interface resembling our upper $SrRuO_3/YBa_2Cu_3O_7$ interface whose electrical properties should be much less favorable.

13.4.3 Interfaces of $YBa_2Cu_3O_7/PrBa_2Cu_3O_7$ edge junctions

1. Introduction

Figure 13.26 shows a picture of a triple-layer edge junction produced by Faley *et al.* [13.12, 13.13]. It is obvious that although the overall shape of the $PrBa_2Cu_3O_7$ barrier is as indicated in Fig. 13.17, the very low angle of 3° only induces an essentially flat interface on the atomic scale which is occasionally

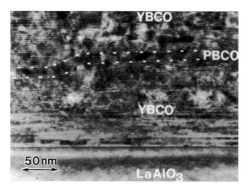

Fig. 13.26. High-resolution picture of a triple-layer edge junction produced by Faley *et al.* [13.12]. The white broken lines indicate the position of the $PrBa_2Cu_3O_7$ layer in between the two $YBa_2Cu_3O_7$ layers.

interrupted by more or less regular steps. During the fabrication of electronic devices and circuits, *ex situ* processing steps, in particular patterning, of the multilayer system are unavoidable. The structure and properties of the interfaces thus produced can differ considerably. In particular, the structural perfection, damage level and contamination of the surfaces largely depend on the processing technology. Patterning can be performed involving either ion-milling [13.29] or chemical etching [13.12]. The electrical properties reported for $YBa_2Cu_3O_7/PrBa_2Cu_3O_7/YBa_2Cu_3O_7$ junctions and for $YBa_2Cu_3O_7/YBa_2Cu_3O_7$ connections (without $PrBa_2Cu_3O_7$ barrier layer) fabricated using different patterning techniques [13.13, 13.37] differ considerably. Since the performance of the chemically etched junctions is in general superior to those fabricated using ion-milling it is interesting to investigate and compare the structure of the interfaces produced by the two techniques. This is described in the following sections.

Samples with $YBa_2Cu_3O_7/PrBa_2Cu_3O_7/YBa_2Cu_3O_7$ triple layers for investigation in the high-resolution electron microscope were prepared on $SrTiO_3$ substrates by the high-oxygen pressure d.c.-sputtering technique. Both types of interfaces were produced by *ex situ* processing. The first, lower interface was formed by deposition of the $PrBa_2Cu_3O_7$ layer on a surface prepared by employing a non-aqueous Br-ethanol etch (Faley *et al.* [13.12]) to the first $YBa_2Cu_3O_7$ layer. The second interface was based on an Ar-ion-etched surface of the $PrBa_2Cu_3O_7$ layer. The energy of the ion beam used was 600 eV and the ion-current density $50 \, \mu A/cm^2$. This resulted in an etching rate of about

2 nm/min. The samples were annealed in an oxygen atmosphere at 800 °C for 30 min prior to the deposition of the subsequent layer. For comparison, triple-layer systems produced by an *in situ* technique in a dual-target sputtering system under otherwise identical conditions were also investigated.

Figure 13.27 shows a low-magnification [100] lattice fringe picture of a cross section of the triple-layer system whose c-axis is perpendicular to the (001) surface of the SrTiO$_3$ substrate. The lower interface, between the first YBa$_2$Cu$_3$O$_7$ layer and the PrBa$_2$Cu$_3$O$_7$ layer, is that obtained after chemical etching. The upper interface, between the PrBa$_2$Cu$_3$O$_7$ layer and the second YBa$_2$Cu$_3$O$_7$ layer, is the one obtained after ion etching. A general difference in the image contrast of the two interfaces is evident. The lower interface looks much sharper than the upper one, which includes an interface layer with a thickness of a few nanometers (pair of white arrows). The c-lattice period (1.17 nm) of the two compounds is uninterrupted across the lower interface from the first YBa$_2$Cu$_3$O$_7$ layer to the PrBa$_2$Cu$_3$O$_7$ layer, although faint strain contrast is visible. However, this period is interrupted in the second interface by the above mentioned interface layer. In fact, the interface layer separates the second YBa$_2$Cu$_3$O$_7$ layer from the PrBa$_2$Cu$_3$O$_7$ layer. Nevertheless, the epitaxial growth mode is maintained throughout all three layers. In some regions of the upper interface, a-axis-oriented grains of YBa$_2$Cu$_3$O$_7$ were found to grow upward from the thin interface layer to a thickness of about 10 nm. Then they were over-grown by the c-axis-oriented film.

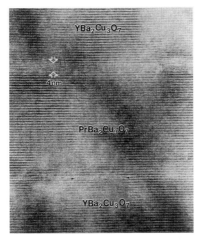

Fig. 13.27. A general view of the experimental YBa$_2$Cu$_3$O$_7$/PrBa$_2$Cu$_3$O$_7$/YBa$_2$Cu$_3$O$_7$ triple-layer system. The upper interface is based on the ion-etched surface and the lower on the Br-ethanol-etched surface.

2. The chemically etched interface

The pattern of lattice fringes in Fig. 13.27 indicates that the $YBa_2Cu_3O_7$ and $PrBa_2Cu_3O_7$ layers connect well. The detailed investigation showed that the structure is essentially perfect over more than 80% of the interface area. The observed fault structures are as follows. Occasionally contrast irregularities occur over areas extending about 30 nm horizontally (arrows). The irregularity is however restricted to a single c-axis lattice parameter vertically. Stacking faults with a double CuO-chain plane are also occasionally observed. This type of defect is also well known and frequently observed in homogeneous $YBa_2Cu_3O_7$ films. In addition, such stacking faults are also frequently found at or near the interfaces in $YBa_2Cu_3O_7/PrBa_2Cu_3O_7$ multilayer or superlattice films prepared *in situ* [13.33]. A possible origin of this problem is a slight deviation of the initial composition of the deposited material from that of the exact $YBa_2Cu_3O_7$ (or $PrBa_2Cu_3O_7$) stoichiometry after changing the target for the deposition of the subsequent heterolayer. Misfit dislocations at the interfaces are found to accommodate the lattice mismatch in the a–b plane. In general, no difference between the best *in situ* interfaces and the Br-ethanol-etched interfaces was found.

3. The ion-etched interface

Jia *et al.* [13.40] investigated both experimentally and theoretically the possibilities for chemically sensitive imaging of $YBa_2Cu_3O_7/PrBa_2Cu_3O_7$ heterostructures. They found that under suitable illumination and focusing conditions Y- and Pr-atom rows could be distinguished by their individual contrast behavior. Figure 13.28 depicts an image taken along the [010] zone axis of $YBa_2Cu_3O_7$ with the imaging parameters chosen for optimum chemical

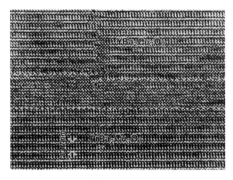

Fig. 13.28. Chemically sensitive image of the interface layer related to the ion-etched surface. The interface layer has a structure which results in a square image pattern.

sensitivity (corresponding to a sample thickness of 7 nm and a defocus value of −50 nm as deduced by the image calculation). The Y-atom positions appear bright on a dark background while the Pr-atom positions exhibit background intensity only. This can be verified by an inspection of the contrast around the areas indicated by the two small arrows in the YBa$_2$Cu$_3$O$_7$ and the PrBa$_2$Cu$_3$O$_7$ regions. There is no evidence for amorphous phases. On the other hand, it is obvious that an additional thin layer with different structure is located between the YBa$_2$Cu$_3$O$_7$ and the PrBa$_2$Cu$_3$O$_7$ layers. The lattice-fringe image of this layer exhibits a square pattern with a periodicity close to that of the *a* or *b* lattice parameter of the YBa$_2$Cu$_3$O$_7$ compound.

The image of the additional layer is similar to that of YBa$_2$Cu$_3$O$_7$ viewed down the *c*-axis direction, i.e. it is in principle compatible with an *a*-axis-oriented grain in the *c*-axis-oriented film matrix. Therefore one cannot *a priori* rule out this possibility. By employing image simulation we can, however, clarify the structure of the layer. Figure 13.29 shows simulated images of YBa$_2$Cu$_3$O$_7$ and PrBa$_2$Cu$_3$O$_7$ along [010] (a) and [001] (b) zone axes calculated for the sample thickness of 7 nm and the defocus of −50 nm. Along the [010] zone axis (a), the calculated images match the experimental ones very well for both YBa$_2$Cu$_3$O$_7$ and PrBa$_2$Cu$_3$O$_7$. Both the bright dots at the Y-atom positions and the background intensity at the Pr-atom positions are evident. As shown in (b), the calculated images of the two compounds along the [001] zone axis are clearly different from the experimental image of the thin layer. The

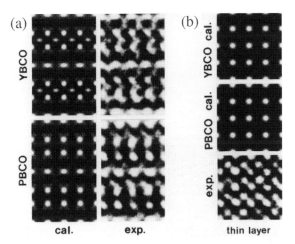

Fig. 13.29. (a) Calculated [010] images of YBa$_2$Cu$_3$O$_7$ and PrBa$_2$Cu$_3$O$_7$ for sample thickness of 7 nm and defocus value of −50 nm in comparison with corresponding experimental images. (b) Calculated [001] images of YBa$_2$Cu$_3$O$_7$ and PrBa$_2$Cu$_3$O$_7$ for sample thickness of 7 nm and defocus value of −50 nm in comparison with the experimental image of the interface layer.

calculated images of $YBa_2Cu_3O_7$ and $PrBa_2Cu_3O_7$ show a bright dot pattern with a and b lattice parameter periodicity while the experimental image of the additional layer exhibits a pattern with a smallest spacing of about $(\sqrt{2}/2)a$, indicating a clearly different structure.

The structural difference of the thin layer can be attributed to a loss of cation ordering in the $PrBa_2Cu_3O_7$ structure as a result of radiation damage during the ion-etching process. Indeed it could be shown that irradiation of $YBa_2Cu_3O_7$ by ions can damage the crystal structure and suppress the superconductivity, depending on the energy and the fluence of the ions used [13.41–13.43]. The first effect, the structural damage, is also expected for the isomorphous $PrBa_2Cu_3O_7$ compound. The ion beam used in the present study had a very low energy. Therefore only a very thin $PrBa_2Cu_3O_7$ layer below the surface is expected to contain damage at the end of ion etching.

Image simulation of the layer structure was carried out on the basis of a cation-disorder model for the imaging conditions of Fig. 13.29. In this model the cation positions of the $YBa_2Cu_3O_7$ structure are partially occupied by different atoms corresponding to the measured stoichiometry. The result is displayed in Fig. 13.30. We see a pattern similar to the experimental one. The irregularities in the experimental image can be attributed to local ordering of the cations.

Our results on the structural properties can be used to explain the difference in electrical properties of the $YBa_2Cu_3O_7/PrBa_2Cu_3O_7$ junctions and $YBa_2Cu_3O_7/YBa_2Cu_3O_7$ connections prepared by chemical etching [13.13] and ion etching [13.37], if the ion-etching conditions are assumed to be similar to those used in the present study. The quality of the interface layer controls the quality of the junction and of the connections. We find that this interface layer has almost ideal epitaxial structure in the case where chemical etching was used. On the other hand, the interface structure is not only less perfect in the

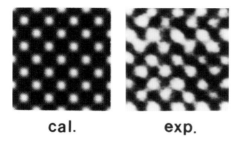

cal.   exp.

Fig. 13.30. A simulated image (left) of cation-disordered $Y_{0.4}Pr_{0.6}Ba_2Cu_3O_7$ for sample thickness of 7 nm and defocus value of -50 nm in comparison with the experimental image (right) of the interface layer.

ion-etched case but we even find an intermediate layer where the structure of the patterned compound is destroyed.

13.5 Summary

The microstructure of YBa$_2$Cu$_3$O$_7$ thin films and related heterostructures, which are the basis for several types of edge-based Josephson junction were investigated by high-resolution transmission electron microscopy. The results show that the microstructure, interfaces, grain boundaries and defect configurations in these films are strongly influenced by the substrate and the often complex structure of the film surfaces on which the subsequent layers are deposited.

The examples chosen indicate that devices based on epitaxial systems of cuprate superconductor films have to be fabricated at atomic or close to atomic precision. The orthorhombic structure of YBa$_2$Cu$_3$O$_7$ and the fact that the three perovskitic sub-units making up the unit cell have different dimensions induces problems in grain boundary formation which affect the weak links in step-edge junctions and lead to fault structures if YBa$_2$Cu$_3$O$_7$ is deposited on simple perovskitic substrates such as SrTiO$_3$.

The examples also indicate that device production on the basis of cuprate superconductors demands an essentially atomically resolving technique for structure investigation and for proper inspection and control of the results of the individual fabrication processes. This can be provided by modern transmission electron microscopy combined with cross-sectional specimen preparation.

Acknowledgments

The authors would like to thank A. I. Braginski, M. Faley, U. Poppe, R. Dömel, K. Herrmann and B. Kabius for providing samples and for stimulating discussions.

References

[13.1] J. Clark, *Nature* **333**, 29 (1988).
[13.2] A. I. Braginski, *Physica C* **153-155**, 1598 (1988).
[13.3] A. I. Braginski, in *The new superconducting electronics*, NATO ASI series E, **251** (Kluwer Academic Publishers, Dordrecht, 1993) p. 89.
[13.4] C. Heiden, in *Multicomponent and multilayered thin films for advanced microtechnologies: techniques, fundamentals and devices*, NATO ASI series E, **234** (Kluwer Academic Publishers, Dordrecht, 1993) p. 567.
[13.5] R. W. Simon *et al.*, *IEEE Trans. Magnetics* **27**, 3209 (1991).

[13.6] G. J. Cui *et al.*, in *Non linear superconductive electronics and Josephson devices* (Plenum, New York, 1991), p. 109.

[13.7] K. Herrmann *et al.*, *Superconductor Sci. & Technol.* **4**, 583 (1991).

[13.8] C. L. Jia *et al.*, *Physica C* **175**, 545 (1991).

[13.9] C. L. Jia *et al.*, *Physica C* **196**, 211 (1991).

[13.10] R. Dömel *et al.*, *Superconductor Sci. & Technol.* **7**, 277 (1994).

[13.11] C. L. Jia, R. Dömel & K. Urban, *Phil. Mag. Lett.* **69**, 253 (1994).

[13.12] M. I. Faley *et al.*, *Appl. Phys. Lett.* **63**, 2138 (1993).

[13.13] M. I. Faley *et al.*, *IEEE Trans. Appl. Supercond.* **7**, 3702 (1997).

[13.14] P. Stadelmann, *Ultramicroscopy* **21**, 131 (1987).

[13.15] D. Dimos, P. Chaudhari & J. Mannhart, *Phys. Rev. B* **41**, 4038 (1990).

[13.16] R. Gross *et al.*, *Appl. Phys. Lett.* **57**, 727 (1990).

[13.17] K. Char *et al.*, *Appl. Phys. Lett.* **59**, 723 (1991).

[13.18] Y. Zhang *et al.*, *Brain Tomography* **5**, 397 (1993).

[13.19] K. Hermann, Doctoral Thesis, U. Giessen (1993).

[13.20] M. Siegel *et al.*, *IEEE Trans. Appl. Supercond.* **3**, 2369 (1993).

[13.21] K. Hermann *et al.*, *J. Appl. Phys.* **78**, 1131 (1995).

[13.22] Y. Gao *et al.*, *Physica C* **173**, 487 (1991).

[13.23] H. W. Zandbergen *et al.*, *Physica C* **166**, 255 (1990).

[13.24] C. L. Jia & K. Urban, *Interface Science* **1**, 291 (1993).

[13.25] U. Poppe *et al.*, *Solid State Comm.* **66**, 661 (1988).

[13.26] R. Dömel *et al.*, *Superconductor Sci. & Technol.* **7**, 277 (1994).

[13.27] R. Soltner *et al.*, *Physica C* **191**, 1 (1991).

[13.28] X. D. Wu *et al.*, *Appl. Phys. Lett.* **62**, 2434 (1993).

[13.29] J. Gao *et al.*, *Appl. Phys. Lett.* **59**, 2754 (1991).

[13.30] C. T. Rogers *et al.*, *Appl. Phys. Lett.* **55**, 2032 (1989).

[13.31] P. Lang *et al.*, *Physica C* **194**, 81 (1992).

[13.32] S. J. Pennycook *et al.*, *Phys. Rev. Lett.* **67**, 765 (1991).

[13.33] C. L. Jia *et al.*, *Physica C* **210**, 1 (1993).

[13.34] S. J. Pennycook *et al.*, *Physica C* **202**, 1 (1992).

[13.35] W. Zhou, D. A. Jefferson & W. Y. Liang, *Material Science Forum* **129**, 65 (1993).

[13.36] C. L. Jia *et al.*, *Physica C* **182**, 163 (1993).

[13.37] Y. Buguslavskij *et al.*, *Physica C* **194**, 268 (1992).

[13.38] U. Poppe *et al.*, *J. Appl. Phys.* **71**, 5572 (1992).

[13.39] M. I. Faley *et al.*, *IEEE Trans. Supercond.* **5**, 2608 (1995).

[13.40] C. L. Jia *et al.*, *Ultramicroscopy* **49**, 330 (1993).

[13.41] G. J. Clark *et al.*, *Appl. Phys. Lett.* **51**, 139 (1987).

[13.42] S. J. Pennycook *et al.*, *Nucl. Instr. and Meth. B* **79**, 641 (1993).

[13.43] J.-P. Krumme *et al.*, *J. Mater. Res.* **9**, 2747 (1994).

14

Controlling the structure and properties of high T_c thin-film devices

E. OLSSON

14.1 Introduction

The highly anisotropic crystal structures of the layered high T_c superconducting cuprates induce the necessity of using epitaxial films in most devices based on high T_c thin films. Each individual application demands a specific crystallographic orientation of the film as well as a certain combination of substrate, high T_c superconducting film and non-superconducting layer materials. The intention of this chapter is to provide examples of what aspects need to be considered when designing the device and predicting its behavior. The behavior depends on the detailed microstructure of the thin films. The examples are therefore discussed in terms of microstructure and how it can be controlled and manipulated.

As in all epitaxial structures, the interfacial interactions are crucial for the resulting microstructures. The direct interaction between the substrate and a single-layer high T_c thin film is illustrated in the following section. The description is followed by a discussion of different aspects of the use of buffer layers. The text is restricted to high T_c YBa$_2$Cu$_3$O$_{7-x}$ (YBCO), mainly owing to the fact that the vast majority of published data concern this superconductor. However, there are common characteristics of epitaxial film growth between different high T_c superconductors. Results from the YBCO films can thus be used when considering the other high T_c superconducting thin films which is pointed out in Section 14.10.

The mechanical interaction between the different epitaxial layers may result in the formation of misfit dislocations. Nucleation and propagation of cracks can ensue if the mismatch in thermal expansion coefficient is relatively large. The defects significantly influence the physical properties of the thin films. Examples from different material combinations and models of how to predict the numbers for critical thicknesses are provided in Section 14.4.

The introduction of grain boundaries and fabrication of other types of junctions are key issues in the development of high T_c devices such as sensors. Different methods of introducing grain boundaries and growing multilayer structures are addressed. The microstructural studies show that there are common characteristics of the resulting grain boundary structures which are introduced by manipulation of the film growth. Integrated circuits require multilayer structures where the superconducting thin films are separated by non-superconducting layers. There are generic characteristics to be considered when growing the multilayer structures, which are addressed in a separate section.

It should be noted that most of the microstructural data to be presented rely on different characterization techniques using electron microscopy in combination with other scanning probe microscopy techniques, e.g. scanning tunneling microscopy (STM) and atomic force microscopy (AFM). The details of specimen preparation and the analysis techniques of electron microscopy are described in other chapters of this book and are therefore not further covered in this chapter. The references provided in the chapter are merely a selection, since the field of research is vivid and the number of publications is large. The goal is to show the salient features which can be further explored by use of the referenced literature.

14.2 Single-layer films

The first step towards the fabrication of thin-film devices is the growth of a single-layer film. The anisotropic layered structure and short superconducting coherence lengths of most high T_c superconductors introduce aspects of film growth that need not to be considered for conventional low-temperature superconductors.

The anisotropic crystal structure, see Fig. 14.1, gives rise to an anisotropy of the superconducting properties. Examples of how the anisotropy manifests itself is that the critical current density, J_c, is considerably lower along the *c*-axis than in perpendicular directions. It is therefore advantageous to align the grains of the superconductor crystallographically in order to maximize the J_c. In addition, due to the short coherence lengths grain boundaries act as weak links [14.1–14.7].

The detailed behavior of an individual grain boundary depends on its geometry, see Fig. 14.2, and the presence of doping elements. Epitaxial growth where the crystal structure unit cell of the deposited material has a fixed orientation relationship with respect to the unit cell of the substrate is most often desired when depositing thin films where a high J_c is required.

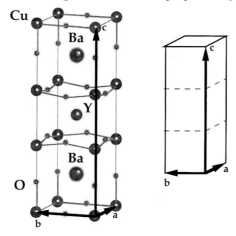

Fig. 14.1. The crystal structure of $YBa_2Cu_3O_{7-x}$. The lattice parameters are $a = 0.383$ nm, $b = 0.388$ nm and $c = 1.17$ nm. The lattice parameters depend on the oxygen content [e.g. 14.11]. In the following figures, the structure is represented by a rectangular box shown to the right.

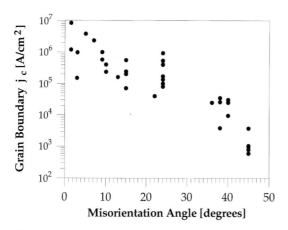

Fig. 14.2. Critical current density as a function of misorientation angle of [001]-tilt YBCO boundaries on $SrTiO_3$ bi-crystal substrates. Data are extracted from [14.1–14.4].

14.2.1 Epitaxial growth of c-axis-oriented films on planar surfaces

One of the most investigated high T_c superconductors is $YBa_2Cu_3O_{7-x}$ (YBCO). The discovery of this superconductor initiated a worldwide effort to optimize the quality of thin films on different types of planar substrate using various deposition techniques, for example pulsed laser deposition, magnetron sputter deposition, co-evaporation and molecular beam epitaxy. It should be noted that many basic characteristics of the YBCO thin films are shared by other layered high T_c superconductors.

1. c-Axis-oriented YBa$_2$Cu$_3$O$_{7-x}$ (YBCO)

The vast majority of the reported work on YBCO thin films concerns c-axis films since the growth kinetics and surface energies usually promote this orientation [14.8–14.10]. This is illustrated in Fig. 14.3 where the surface is assumed not to interact with the YBCO film. The YBCO spontaneously forms platelets with the c-axis along the short axis of the plates. The a- and b-axes are the rapid growth directions and the low-energy surfaces are (001), (100) and (010). The preferred orientation with respect to the substrate can be manipulated by the interaction between the film and the surface on which it is growing.

2. Well lattice-matched film/substrate configurations

The orientation of the film is determined by the surface/film interaction in the initial stage of film growth. The lattice match between the two crystal structures, the interaction between the interface atoms and the surface morphology determine which one of the orientation configurations has the lowest energy and thus dominates. YBCO is a perovskite-like structure with three building blocks along the c-axis, see Fig. 14.1. Epitaxial growth is conse-

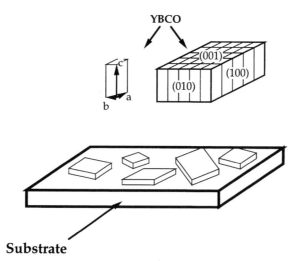

Fig. 14.3. Schematic illustration of YBCO grains which have nucleated on a surface where the interfacial interactions can be neglected. The (100), (010) and (001) low-energy crystallographic planes of YBCO constitute the boundary surfaces of the crystals.

quently promoted on perovskite-type substrates like $SrTiO_3$, cubic $LaAlO_3$, $NdGaO_3$ and $PrGaO_3$ [14.11]. The lattice match of the former two substrates favors a cube-on-cube orientation relationship where $[001]_{YBCO}//[001]_{substrate}$ and $[110]_{YBCO}//[110]_{substrate}$, see Fig. 14.4. In contrast, the YBCO unit cell is rotated $45°$ about the c-axis with respect to the unit cells of the two latter substrates and the orientation relationship is $[001]_{YBCO}//[001]_{substrate}$ and $[100]_{YBCO}//[110]_{substrate}$, $[010]_{YBCO}//[1\bar{1}0]_{substrate}$. A closer look at the crystal structures shows that the oxygen sublattices are of crucial importance for the orientation. All orientation relationships above include an alignment of these sublattices.

The in-plane lattice mismatches along the a- and b-axes of the YBCO between the above mentioned crystal structures do not exceed 1% at the deposition temperature [14.11]. There are other commonly used substrates such as MgO and Y—ZrO_2 where the in-plane lattice mismatch is considerably larger, 9% and 4% respectively [14.11]. The increase in lattice mismatch increases the interfacial energy and the difference between the configuration of lowest interfacial energy and other alternative orientation relationships decreases significantly [14.12]. The probability of forming YBCO nuclei of other orientations in the initial stage of film growth simultaneously increases. It is therefore more difficult to grow single crystal films on substrates with larger lattice mismatch.

3. Graphoepitaxy

Epitaxial growth, with one orientation relationship, in substrate/film systems with relatively large lattice mismatch can be promoted by use of graphoepitaxy.

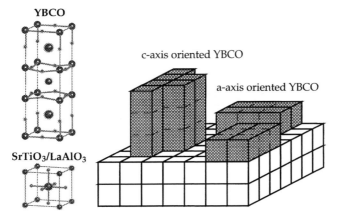

Fig. 14.4. Schematic representation of the epitaxial orientation relationships for YBCO on (001) $SrTiO_3$ and (001) cubic $LaAlO_3$ substrates.

360   *E. Olsson*

The surface morphology of the substrate is modified in order to align the nuclei of the YBCO, see Fig. 14.5. The growth kinetics and surface energies of the YBCO favor the formation of platelets with (100), (010) and (001) surfaces, see Fig. 14.3. Surface steps provide nucleation sites and the [100] and [010] axes of a YBCO nuclei align with a surface ledge if present. A single orientation relationship is thereby induced by the presence of ledges in specific orientations. Such ledges are spontaneously formed on vicinal surfaces after heat treatment of both MgO and YSZ. The ledges run along the $\langle 100 \rangle$ direction of MgO [14.13] and the $\langle 110 \rangle$ direction of Y—ZrO_2 [14.14]. Their orientation is consistent with the orientation of the YBCO obtained on such surfaces.

4. Effect of deposition temperature

In situ thin films need to be deposited at temperatures in the range from 700 to 800 °C in order to obtain high quality superconducting properties [14.11]. The *in situ* requirement is introduced by future applications which are based on integrated circuits. The relatively high temperatures increase the probability for detrimental chemical reactions at the film/substrate interface. YBCO films grown on Si substrates are an example of severe chemical interactions in a system which would be highly interesting for semiconductor/high T_c hybrid structures [14.15]. The chemical interaction is prevented by the use of buffer layers prior to deposition of YBCO [14.16].

YBCO

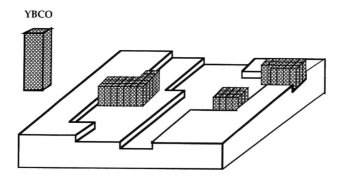

Fig. 14.5. Schematic illustration of the graphoepitaxial effect of YBCO film growth on MgO and Y—ZrO_2 substrates. The $[100]_{YBCO}$ and the $[010]_{YBCO}$ align with the crystallographic ledges which form on annealed substrate surfaces. The ledges are parallel to the $\langle 100 \rangle$ direction on MgO and to $\langle 110 \rangle$ on Y—ZrO_2.

14.3 Buffer layers

Similar to the case of semiconductor thin films and quantum well structures, there is a need to deposit buffer layers prior to deposition of the superconducting thin-film, see Fig. 14.6. The role of a buffer layer is to prevent detrimental interactions between the film and substrate and to diminish the effect of surface defects on the film growth. In addition, there are examples of film/substrate combinations where the mismatch in thermal expansion coefficients is severe enough to cause cracking in the thin films [14.16, 14.17]. A suitably chosen intermediate buffer layer can reduce the stress caused by such a mismatch.

14.3.1 Chemical interaction

Si is mentioned above as an example of a chemically reactive substrate for YBCO. Despite the fact that Si not only reacts with YBCO even below 550 °C but also exhibits a relatively large mismatch in thermal expansion coefficient, there is a considerable interest in developing techniques for growing YBCO films on Si substrates [14.18–14.23]. The reason is that a large fraction of the current semiconducting technology is based on Si substrates. Most often Y—ZrO_2 buffer layers are used to prevent the chemical interaction between YBCO and Si. Another technologically important substrate is Si-on-sapphire (SOS) where Y—ZrO_2 also is used as a buffer layer to prevent the chemical reaction between YBCO and Si [14.24–14.26]. The buffer layer should act as a diffusion barrier. It is therefore essential to avoid the presence of grain boundaries in the buffer layer since they can act as rapid diffusion paths for the reacting species [14.21].

It should be noted that there is a minor chemical reaction between YBCO and Y—ZrO_2. The reaction occurs at the initial stage of film growth and results in the formation of $BaZrO_3$ [14.9, 14.19–14.23, 14.25–14.28]. The reaction is slow and the thickness of the $BaZrO_3$ layer is of the order of 5–10 nm for film deposition temperatures in the range between 700–800 °C [14.27]. The thickness increases with increasing temperature. Provided the YBCO films are relatively thick the effect on the superconducting properties can be neglected. However, a second buffer layer could be used to prevent the limited interaction between the YBCO and Y—ZrO_2 [14.22, 14.26]; see Fig. 14.6.

GaAs is another common substrate material in the semiconductor technology. Surfaces of GaAs need to be encapsulated by a buffer layer which must be deposited at temperatures below 500 °C. Otherwise, Ga agglomerates into droplets which cause thermal etching of the substrate and give rise to severe surface roughness [14.29]. The requirement of a low deposition temperature

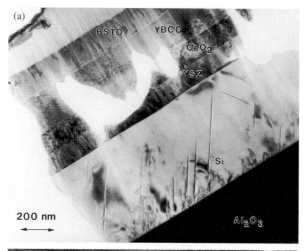

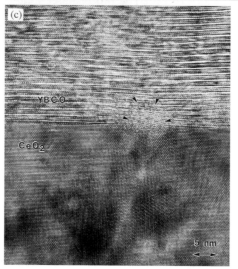

Fig. 14.6. TEM micrographs of an epitaxial YBCO film deposited on a silicon-on-sapphire substrate $((001)\mathrm{Si}/(1\bar{1}0\bar{2})\mathrm{Al}_2\mathrm{O}_3)$ with a $\mathrm{CeO}_2/\mathrm{Y}{-}\mathrm{ZrO}_2$ (YSZ) buffer layer. The role of the YSZ layer is to prevent the chemical reaction and interdiffusion between the YBCO and Si. The CeO_2 prevents the limit reaction between YBCO and YSZ and also has a beneficial influence on the YBCO film evolution. (a) A low-magnification micrograph providing an overview. (b) High-resolution TEM (HRTEM) of the Si/YSZ interface showing an amorphous layer between the two crystalline layers. (c) HRTEM of the $\mathrm{CeO}_2/\mathrm{YBCO}$ interface.

for the buffer layer limits the number of possible buffer materials. The use of MgO renders acceptable quality of the YBCO [14.30-14.31].

14.4 Mechanical interactions

An epitaxial film is strained in the initial stages of film growth. The strain energy increases with film thickness and may eventually be relaxed by the introduction of misfit dislocations [14.32–14.35], see Fig. 14.7, or by forma- tion of (110) twins in the YBCO [14.36]. The critical thickness at which the misfit dislocations form depends on the lattice mismatch and the elastic properties of the film. The misfit in epitaxial c-axis-oriented YBCO films is accommodated by the formation of twins and edge dislocations with Burgers vectors $[100]_{YBCO}$ and $[010]_{YBCO}$ [14.37].

The expected critical film thickness for introduction of misfit dislocations as a function of lattice mismatch can be calculated using the expression of Freund [14.38] and elastic parameters for the YBCO [14.39]; see Fig. 14.8. It should be noted that c-axis-oriented films on $SrTiO_3$ substrates are expected to relax at a thickness of 6 nm. The introduction of misfit dislocations is one mechanism of relieving the strain. Prior to the introduction of misfit dislocations the lattice mismatch can also partially be accommodated by coherent Stranski-Krastanow growth [14.40, 14.41]. Elastic deformation of the crystal lattice surrounding the islands is the relieving part of the epitaxial strain.

Fig. 14.7. A TEM micrograph of a YBCO thin film on a (001) MgO substrate. Dislocations (d) and twin boundary (t) planes are indicated by arrows.

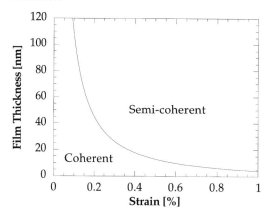

Fig. 14.8. The critical thickness for introduction of misfit dislocations in a *c*-axis-oriented film according to [14.38]. The Burgers vector [100] is 0.389 nm at 700 °C, the biaxial modulus is 248 GPa, the Poisson's ratio is 0.281 and the shear modulus is 70 GPa [14.39].

The YBCO films are deposited at relatively high temperatures, i.e. about 700 °C, and are subsequently cooled to room temperature. The T_c of YBCO is about 90 K depending on the quality of the thin-film, and film devices are therefore cycled between room temperature and temperatures below 90 K during use. It is thus essential to have a good match in thermal expansion coefficient between the film and substrate. The magnitude of the strain due to thermal expansion mismatch increases with deviation from the deposition temperature. Other mechanisms relieving the strain may be activated as the temperature decreases, since the formation probability and mobility of a misfit dislocation decrease with temperature.

An alternative stress relief mechanism is crack formation and propagation [14.42]. The phenomenon is, for example, observed in (110) oriented YBCO films on (110)SrTiO$_3$ substrates [14.17]; see Fig. 14.9. In this system, misfit dislocations are formed at the deposition temperature while the cracks are introduced during cooling to room temperature. The crack spacing, l, depends on the film thickness, h, according to the expression

$$l \approx 5.63(h\,K_f)^{1/2}/(M\varepsilon),\tag{14.1}$$

where K_f is the mode-I fracture toughness along the cleavage plane of the film, M is the biaxial elastic modulus of the film and ε is the elastic strain. Thermodynamically, no cracks are allowed to propagate at film thicknesses below a critical film thickness h_c. The critical film thickness for crack

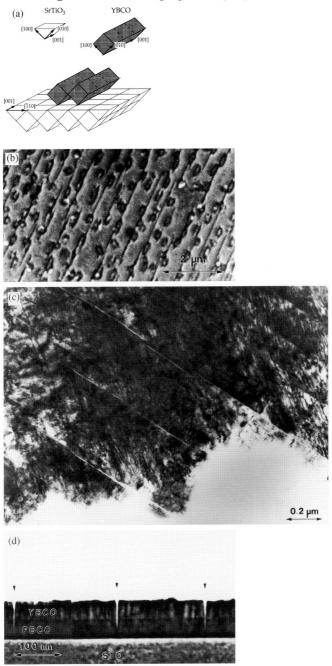

Fig. 14.9. (a) Schematic showing the geometry of (110)YBCO on (110)SrTiO$_3$. (b) SEM micrograph of a PrBa$_2$Cu$_3$O$_{7-x}$ on (110)SrTiO$_3$ substrate. The cracks run along the (001) planes of PBCO. (c) Plan-view TEM micrograph of the film in (b). (d) Cross-section TEM micrograph of a (110)PBCO/YBCO film on (110)SrTiO$_3$ substrate.

propagation as a function of strain is

$$h_c = 0.50[K_f/(M\varepsilon)]^2. \tag{14.2}$$

Cracks are frequently observed in YBCO films deposited on Si substrates owing to the relatively large mismatch in thermal expansion coefficient [14.19]; see Fig. 14.10. The cracks propagate along the (100)- and (010)-planes of the YBCO which are the most probable propagation planes after the (001)-planes [14.43]. The critical film thickness for crack propagation is here increased by the presence of a Y—ZrO$_2$ buffer layer. This is a second reason why the use of Y—ZrO$_2$ buffer layers are beneficial for films on Si substrates [14.16].

There is often an amorphous layer containing oxygen and Si between the Si substrate and the Y—ZrO$_2$ [14.19–14.23]. The layer is either caused by a remaining native oxide layer on the Si substrate or by diffusion through the buffer layer. The latter is usually the reason in films intended for devices since a high quality buffer layer needs to be epitaxial with a single orientation relationship with respect to the substrate. A native oxide layer would prevent the interfacial interaction and disrupt the epitaxial growth. The presence of the amorphous layer between the Si substrate and the Y—ZrO$_2$ layer in combination with the BaZrO$_3$ layer at the upper interface towards the YBCO introduces additional interfaces at which mechanisms other than crack formation and

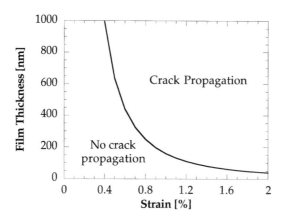

Fig. 14.10. Critical film thickness for crack propagation in a (110)YBCO film on (110)SrTiO$_3$ according to eq. (14.2). The fracture toughness K_f is 0.8 MPa$^{1/2}$ and M is 102 GPa [14.42].

propagation can act to relieve the strain, thus increasing the critical thickness for crack propagation.

14.5 Grain boundaries

The next step in the development of high T_c thin-film devices after the optimization of single layer films is the introduction of defects in precisely defined positions in order locally to alter the properties of the superconductor. The use of grain boundaries as Josephson junctions in SQUIDs (Superconducting QUantum Interference Devices) is an example of how defects can be utilized.

The properties of the grain boundaries depend on their fine scale microstructure and it is well known that the relative misorientation angle between adjacent grains in, for example, YBCO affects the critical current density, J_c. In [001]-tilt grain boundaries the J_c increases with increasing angle, see Fig. 14.2. Analyses of the effects of misorientation angle indicate that the grain boundary energy is related to the J_c where it decreases with increasing energy [14.44, 14.45]. The grain boundary energy generally increases with misorientation angle. There are, however, special low energy grain boundary configurations that give rise to local minima of energy [14.45–14.47]. This is evidenced by special grain boundaries that exhibit anomalous behavior [14.48, 14.49].

Grain boundaries with special geometries can be introduced by control of the epitaxial orientation relationships between the film and the surface on which it is growing. Four types of methods to artificially introduce grain boundaries in YBCO thin films can be distinguished, see Fig. 14.11. One method is to use bi-crystal substrates where the substrate boundary is inherited by the thin film [14.1–14.4, 14.6, 14.50–14.56]. A second method is to use template layers that locally change the orientation of the film [14.49, 14.57–14.59]. The latter method is more flexible since the position of the boundary is determined in a patterning step and not by a substrate boundary with a macroscopic extension. The third method is an alternative patterning procedure to make step-edges on the substrate surface prior to the YBCO film deposition [14.60–14.64]. A fourth method is a combination of the second and third techniques. The surface is patterned and a surface modification is achieved using ion-milling. The surface perturbation caused by the ion-milling gives rise to a change in the epitaxial orientation relationships [14.65–14.67].

14.5.1 Bi-crystal substrates

Up to now, most bi-crystal substrates are used to make YBCO grain boundaries

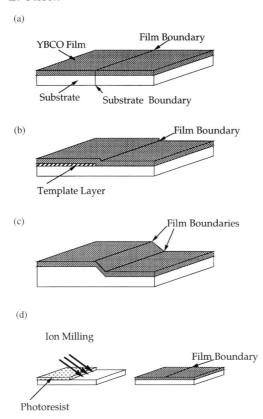

Fig. 14.11. Schematic illustration of four techniques to introduce grain boundaries in YBCO thin films. (a) Bi-crystal substrate technique. (b) Bi-epitaxial technique where a template layer is used to change the epitaxial orientation of the YBCO film with respect to the substrate. (c) Step-edge on the substrate. (d) Surface modification.

with the [001]-tilt geometry. Common bi-crystal substrates are Y—ZrO_2 [14.6, 14.50], and $SrTiO_3$ [14.1–14.4, 14.52–14.54, 14.56], but there are also reports of MgO [14.55] and $NdGaO_3$ [14.51] bi-crystal substrates. The bi-crystal substrates offer the possibility of a free variation of misorientation angle between the two adjacent YBCO grains. Both symmetrical and asymmetrical boundaries are investigated. It should be noted that the above mentioned bi-crystal substrates (except $NdGaO_3$) are of cubic crystal structure. This allows the large variety of misorientation angles, whereas less symmetric crystal structures would limit the number of possible orientations. This is because differences in thermal expansion coefficients would give rise to cracks.

As discussed in Section 14.2.1.2, $SrTiO_3$ is a material which promotes epitaxial growth of YBCO. The $[001]_{YBCO}$ always aligns with the $[001]_{SrTiO_3}$.

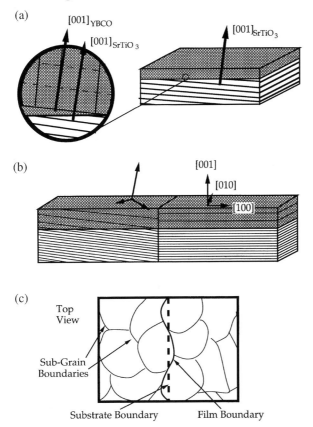

Fig. 14.12. (a) The [001]YBCO aligns with the [001] SrTiO$_3$. An out-of-plane tilt of the substrate is inherited by the YBCO film which is schematically illustrated in this figure. (b) The effect shown in (a) affects the YBCO grain boundary geometry obtained on SrTiO$_3$ bi-crystal substrates. (c) The island growth results in a meandering YBCO grain boundary.

This manifests itself also on SrTiO$_3$ bi-crystals where not only the [001]-tilt is inherited by the YBCO grain boundary but also the out of plane tilt, see Fig. 14.12. Most bi-crystal substrate boundaries are straight, without pronounced deviations from the predetermined boundary plane. However, the YBCO film boundary exhibits wavy morphologies and deviates from the location of the substrate boundary [14.50, 14.52–14.54, 14.56]. This behavior is explained by an island growth mechanism [14.50, 14.54, 14.56]. It is also suggested that a step flow growth mode is induced by vicinal SrTiO$_3$ substrates which would be faster than the growth on substrate surfaces where the deviation from the (001) plane is negligible [14.54]. The location of the boundary would then be shifted from a vicinal substrate surface side towards a more true crystallographic one because of the difference in ledge movement velocity of the YBCO.

The YBCO boundary morphology is invariably wavy on bi-crystal $SrTiO_3$ substrates. A detailed structural investigation of the boundary plane geometry reveals that the boundary consists of (100) and (110) facets [14.54]. The explanation for their appearance is that the YBCO films grow as islands on either side of the substrate boundary. The YBCO islands are facetted with ledges along the $[100]_{YBCO}$, $[010]_{YBCO}$ and the $[110]_{YBCO}$ directions [14.28] which correspond to low-energy surfaces of the YBCO. In the vicinity of the substrate boundary the island ledges surpass the substrate boundary until they encounter a YBCO island from the opposite side of the substrate boundary. This is the reason for the wavy morphology of the boundary. There are exceptions to the (100) and (110) facetting of the YBCO boundary. In these cases, the relative misorientation angle allows the boundary to consist of low-index habit planes of the YBCO [14.53, 14.54].

There are significant differences between the structure of YBCO grain boundaries obtained on $Y—ZrO_2$ bi-crystals compared to those on $SrTiO_3$ bi-crystals. These are caused by the differences in interfacial interactions between the two substrates. One difference is the chemical reaction that takes place at the $Y—ZrO_2$ interface but not at the $SrTiO_3$ one. The reaction results in the formation of $BaZrO_3$ and thus provides an excess of Y and Cu [14.19–14.23, 14.27, 14.28]. These species are free to diffuse, and a preferential diffusion path is the boundary. This may influence the physical properties of the YBCO grain boundary since they are sensitive to the exact composition of the boundary.

A characteristic property of YBCO films on $SrTiO_3$ substrates is that the [001] axes always align for c-axis-oriented films (see Fig. 14.4), while the $[001]_{YBCO}$ is perpendicular to the substrate surface in c-axis-oriented films on $Y—ZrO_2$ substrates [14.68]. The situation illustrated in Fig. 14.12(b) will thus not arise on $Y—ZrO_2$ substrates. Instead, the effect of substrate grooving at the boundary strongly influences the structure of the YBCO in the vicinity of the substrate boundary, see Fig. 14.13. The bi-crystal substrates are generally heat treated at temperatures above 1000 °C prior to film deposition in order to promote the epitaxial growth of YBCO and improve the film quality. Thermal etching may give rise to grooving at the substrate boundary during the anneal. The depth of the groove increases with increasing boundary energy. As the YBCO film nucleates during subsequent film deposition it will encounter different orientations of the substrate surface normal with respect to the $[001]_{Y—ZrO_2}$. Provided that YBCO islands nucleate in the groove there will be other grain boundaries in addition to the one introduced by the bi-crystal boundary. The relative misorientation angle between the YBCO grains will be determined by the local angle of the groove at the nucleation site. The density

(a)

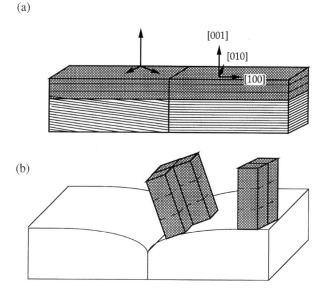

(b)

Fig. 14.13. (a) The [001] YBCO aligns with the substrate surface normal on YSZ substrates. (b) The bi-crystal substrate is thermally etched during heat treatment and a groove develops at the substrate boundary. Its presence can locally change the orientation of the YBCO.

and size of YBCO nuclei will determine the extent of additional YBCO grains in the grain boundary region. The probability of nucleation in the groove can be smaller because of a higher interfacial energy in the groove, and it is thus possible to control and even avoid undesired smaller misaligned grains of YBCO. In the absence of misaligned grains, YBCO grains nucleated on the flat surfaces away from the grain boundary cover the groove as well and bridge the bi-crystal grain boundary.

There are also similarities between the YBCO boundaries obtained on Y—ZrO_2 and $SrTiO_3$ bi-crystal substrates. They both exhibit wavy YBCO boundary morphologies with (100), (010) and (110) facets. Their formation mechanism is the same in both systems. It is thus likely that the interaction between the impinging species and the nucleated YBCO is stronger than that with the substrate. Otherwise, the bottom part of the YBCO grain boundary plane would coincide with the position of the substrate boundary. MgO substrates exhibit similar characteristics to Y—ZrO_2 in terms of the $[001]_{YBCO}$ orientation with respect to the substrate normal. There is, however, no chemical reaction observed at the MgO/YBCO interface. It is thus reasonable to assume that YBCO films grown on MgO bi-crystal substrates will have the same principal structural characteristics as films grown on Y—ZrO$_2$ bi-crystals except for the chemical reaction.

It should be noted that it is the relative misorientation between the adjacent grains that is inherited by the films grown on bi-crystal substrates, whereas the exact boundary geometry is determined by nucleation and growth mechanisms. Special misorientation angles which enable the substrate boundary to coincide with low-index habit planes of the YBCO or with coincidence site lattice boundaries with high symmetries would provide a straight geometry of the film boundary. However, these boundaries are expected to have low energies and would probably not serve as efficient Josephson junctions for devices.

14.5.2 Bi-epitaxial grain boundary junctions

The bi-crystal YBCO grain boundaries with suitable misorientation angles, i.e. in the range of 24° to 36°, provide electrical characteristics that are suitable for Josephson junctions in sensors [14.6]. However, the bi-crystal geometry restricts the device to one single grain boundary that extends throughout the bi-crystal substrate. A more flexible geometry would be obtained if the YBCO boundary could be introduced using patterning methods. The bi-epitaxial grain boundaries are one example of such a method [14.49, 14.57–14.59].

The principle of these junctions is to use a template layer which locally changes the orientation of the YBCO layer with respect to the bottom substrate, see Fig. 14.11. The only reported type of grain boundaries using this method are 45° [001]-tilt boundaries. In addition, microstructural characterization using transmission electron microscopy showed that there were also out-of-plane tilt boundaries in the vicinity of the template layer edge [14.59]. It appears as though the small step at the edge of the template layer, about 2 nm in height, provides nucleation sites for YBCO grains having other orientations than those formed on the surrounding flat interfaces.

The bi-epitaxial method is interesting even though there are remaining aspects that need to be addressed. It is, for example, necessary to avoid the formation of misaligned YBCO grains at the template layer edge. In addition, the 45° type of YBCO grain boundary is not always the optimum one and other misorientation angles are therefore desired. The development of a bi-epitaxial method for other misorientation angles is not a trivial task. Further information about the control of epitaxial orientation relationships will contribute to the improvement of the flexibility of the bi-epitaxial method.

14.5.3 Step-edge YBCO junctions

Another method to locally change the orientation of the YBCO is to pattern step-edges on the substrate surface [14.60–14.64]. The most common steps are about 100–200 nm high. Again the substrates can be divided into two groups with respect to their influence on the type of YBCO boundary that is formed. One group consists of the substrates on which the $[001]_{YBCO}$ aligns with the $[001]_{substrate}$, for example $SrTiO_3$ and $LaAlO_3$. The other group consists of substrates like MgO and Y—ZrO_2 where the $[001]_{YBCO}$ aligns with the substrate normal.

Steps which have a surface normal that forms an angle, θ, less than 45° with the [001] axis of $SrTiO_3$ or $LaAlO_3$, see Fig. 14.14(a), do not give rise to a change in orientation of the YBCO growing on the step [14.61, 14.62]. When the angle θ exceeds 45° the region covering the step is *a*-axis oriented, see Figs. 14.14(b) and (c). The presence of the *a*-axis-oriented YBCO region at the step-edge gives rise to two YBCO grain boundaries which run along the top and the bottom part of the step respectively, see Fig. 14.14(b). Another important aspect is that the presence of two grain boundaries depends on the nucleation of an *a*-axis grain. The substrate surface structure and the area of the step-edge must promote the nucleation, otherwise the *c*-axis-oriented grains on either side of the step-edge will merge without any grain boundary formation.

MgO surface step-edges have another influence on the orientation of the YBCO film [14.63]. The $[001]_{YBCO}$ generally aligns with the substrate normal of MgO and consequently the $[001]_{YBCO}$ is tilted on a step-edge. The tilt angle corresponds to the angle θ, see Fig. 14.15. The step-edge shown in the transmission electron micrograph of Fig. 14.15 exhibits a number of different θ. The nucleation site determines the tilt angle of the YBCO on the step-edge and if several YBCO grains nucleate on the step-edge there will be more than two grain boundaries on the path between the upper and lower part of the YBCO film. A well defined geometry and reproducibility of the YBCO grain boundaries require a step-edge with one single θ. The use of carbon masks during ion-beam patterning of the surfaces improves the quality of the step-edge morphology for $LaAlO_3$, see Fig. 14.14(b), and is likely to provide the same result for other types of substrates as well [14.64].

14.5.4 Surface modification

Grain boundaries can also be introduced by locally changing the orientation relationships similar to the bi-epitaxial junction method but without template layers. MgO is a suitable substrate for this method since the large lattice

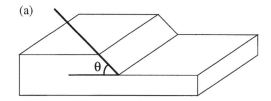

(a)

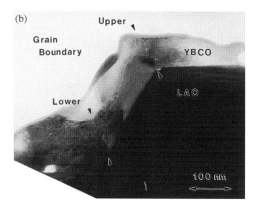

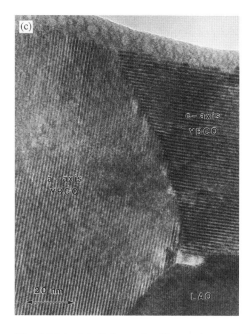

Fig. 14.14. (a) Schematic illustration of the geometry of a step-edge on a substrate and the definition of θ. (b) Low-magnification TEM micrograph of a YBCO film on a step-edge on a (001) LaAlO$_3$ substrate. (c) HREM micrograph of the upper grain boundary arrowed in (b).

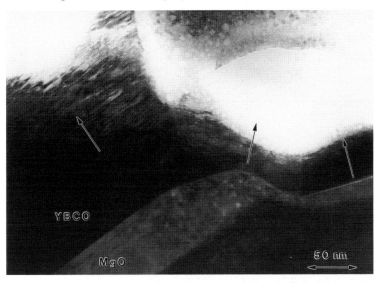

Fig. 14.15. TEM micrograph of a YBCO film on a (001)MgO substrate with a step-edge. The [001]YBCO aligns with the substrate surface normal. The surface normal at the nucleation site thus determines the orientation of the YBCO. Several slopes on the step-edge can thus give rise to several grain boundaries along the step.

mismatch to YBCO provides alternative orientation relationships with interfacial energies that do not deviate extensively from that of the cube-on-cube orientation relationship.

The preferred orientation of the YBCO with respect to the MgO lattice can, for example, be altered by modifying the surface using ion-beam milling [14.65, 14.66]. One type of modification is achieved by protecting one part of the substrate with photoresist and bombarding the uncovered surface with ions. The YBCO film grown on the resulting surface has a cube-on-cube orientation relationship on the surface which has been covered by the photoresist, while on the ion-beam milled surface the $[001]_{YBCO}$ is parallel to the $[001]_{MgO}$ and $[110]_{YBCO}$ is parallel to $[100]_{MgO}$. A 45° [001]-tilt boundary is thus formed at the line separating the ion-beam milled MgO surface from the as-received one.

There is also an alternative method to introduce similar YBCO grain boundaries on (001) MgO substrates using ion-beam milling [14.67]. Again a part of the substrate is covered by photoresist and a small step is produced by the milling procedure, see Fig. 14.16. There are conditions where both the ion-milled and the protected surface render cube-on-cube orientation relationships for the YBCO. The morphology of the YBCO on the ion-milled surface is affected by the ion-milling conditions (see Fig. 14.16). There is a small portion

(a)

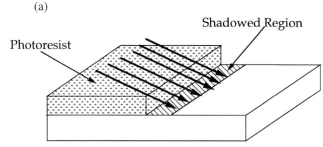

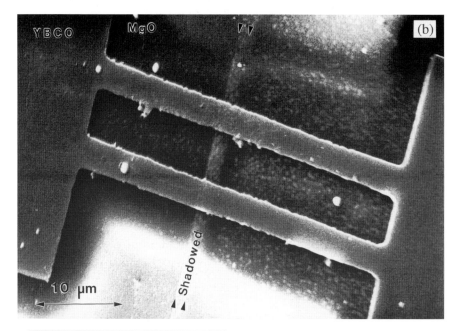

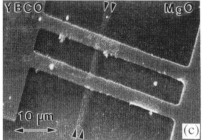

Fig. 14.16. (a) Shallow steps on MgO are produced by covering part of the substrate with photoresist and subsequently ion-milling the uncovered part of the substrate. (b) SEM micrograph of a YBCO film surface. The film is grown on a MgO substrate prepared according to (a). The location of the step is arrowed. (c) SEM micrograph showing the surface of a YBCO film where the substrate is milled using a higher acceleration voltage. The voltage affects the surface morphology of the substrate and the YBCO film growth.

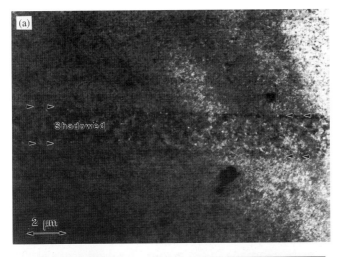

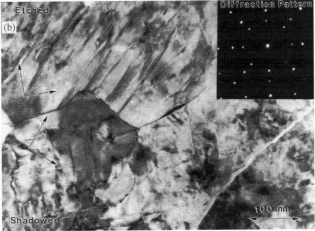

Fig. 14.17. Plan-view TEM micrographs of YBCO grain boundaries in the YBCO film shown in Fig. 14.16(b). (a) The shadowed region indicated in Fig. 14.16 is reflected in the YBCO film morphology. The width of the arrowed region in the TEM micrograph corresponds to the width of the shadowed region. (b) The dark region in (a) consists of YBCO where the unit cell is rotated 45° along the [001] axis. The rest of the film has a cube-on-cube orientation relationship with respect to the MgO substrate. (c) The YBCO grain boundaries consist of (100), (010) and (110) segments (*continues overleaf*).

of the surface in the vicinity of the photoresist edge which is shadowed during milling. The width of the area depends on the thickness of the photoresist and the milling angle. This shadowed region can, after removal of the photoresist, induce a 45° rotation of the YBCO along the [001] YBCO axis (see Fig. 14.17). Two YBCO grain boundaries are thus introduced, see Figs. 14.17(b) and (c). They are separated by a distance corresponding to the width of the

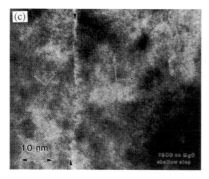

Fig. 14.17 (*cont.*).

shadowed area. It should be noted that the grain boundaries are meandering. One reason for the wavy morphology is the uneven profile of the photoresist edge. The boundaries consist of {100} and {110} segments similar to the boundaries on bi-crystal substrates. Any deviation of the step-edge direction from the ⟨100⟩ MgO or the ⟨110⟩ MgO would thus result in a meandering boundary.

14.6 Patterned and modified surfaces

It should be noted that the fabrication of devices requires patterning which involves, for example, wet etching and/or ion beam milling. The structure of the thin films and the conditions for epitaxial growth of following layers are changed by the patterning procedure [14.69, 14.70]. The patterning may introduce damage, thus changing the physical properties of the superconductor. The effects that have to be considered are surface damage due to the patterning procedure and also the depletion of oxygen [14.71]. The superconducting characteristics of YBCO depend strongly on the oxygen content. The mobility of oxygen is appreciable along the $(001)_{YBCO}$ plane at relatively low temperatures [14.72]. YBCO is also sensitive to humidity that may induce a decomposition of the YBCO resulting in the formation of $BaCuO_3$. It should be noted that in a c-axis geometry of the patterned film the diffusion is considerably more rapid along the in-plane direction than in the perpendicular direction. A method for avoiding the time-dependent degradation is to deposit a cap layer which prevents the long term diffusion.

The high T_c thin films are also incorporated into multilayer structures. The different layers are patterned and now the change in surface morphology with the growth conditions for subsequent layers has to be considered in addition to the effect of structural damage during patterning. The patterning not only

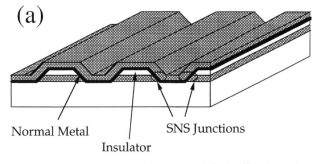

(a)

Normal Metal   SNS Junctions

Insulator

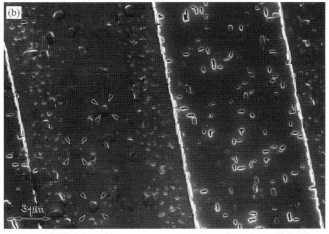

(b)

$3\mu m$

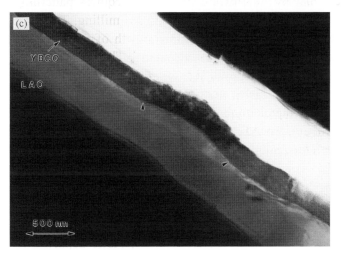

(c)

YBCO

LAO

$500\ nm$

Fig. 14.18. (a) A schematic representation of a series of step-edge super-conductor–normal metal–superconductor Josephson junctions. (b) SEM micrograph of a surface of a series of junctions illustrated in (a). Regions where the growth of the YBCO is affected by substrate surface features introduced by the ion milling are arrowed. (c) Cross-section TEM micrograph showing an area corresponding to the arrowed regions in (b).

changes the morphology of planar surfaces, see Fig. 14.18, but there are unavoidable abrupt changes in height where individual layers are to be separated by intermediate layers. The roughening of planar surfaces due to ion-beam thinning can give rise to changes in the epitaxial orientation relation-ships, e.g. *a*-axis-oriented YBCO particles, or cause local voids in the films, see Fig. 14.18.

It is thus important to optimize the patterning parameters in order to reduce the detrimental effect of the surface roughness. Abrupt changes in local surface slopes constitute particularly potent surface features. A subsequent anneal or a thin buffer layer can smooth such undesired features [14.70]. The method of improving the surface morphology depends on the specific application and material combination.

The steps illustrated in Fig. 14.18 can give rise to the introduction of grain boundaries following the mechanisms described in Section 14.5. This is not desired in structures where a high critical current density of the YBCO is required. The grain boundary formation can be avoided by carefully controlling the angle of the step slope. In the case of a $SrTiO_3$ step the slope of the step should not exceed 45° [14.61]. Another method which is common for all types of steps is to reduce the probability for nucleation on the step by, for example, diminishing the area of the step [14.73]. The YBCO will then nucleate on the planar surfaces on either side of the step. The orientation of the YBCO on the step will be determined by these two areas.

As unexpected defects appear due to the patterning procedure, it is important to investigate the structure of the thin films and to locate the origin of the disturbed structure. Low temperature scanning electron microscopy is one technique to locate the weak areas [14.74]. Particular attention can subse-quently be paid to those areas in the transmission electron microscope. Once the origin is identified the data presented in the above paragraphs can be used to improve the structure by altering the material combination, the film deposi-tion parameters, the patterning conditions and/or the surface treatments.

14.7 Multilayer structures

Integrated circuits based on YBCO require the use of multilayer structures where the individual YBCO films are separated by either metals, semiconduc-tors or insulators depending on the specific application. It is necessary that all layers are epitaxial in order to obtain high quality YBCO films at all levels of the multilayer. The layers should also be devoid of particles of either secondary phases or other orientations since particles may disturb the epitaxial orientation relationships. In addition, protruding particles can cause microshorts between

(a)

(b)

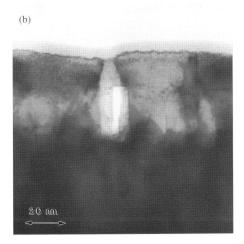

Fig. 14.19. TEM micrographs of YBCO/SrTiO$_3$ (STO)/PrGaO$_3$ (PGO) multilayers. (a) An interface between a bottom YBCO film and a 200 nm PGO film. The PGO film develops voids and pinholes which provide microshorts between the top and the bottom YBCO films. (b) The pinholes develop facets. (c) The formation of pinholes can be avoided by limiting the thickness of the PGO, in this case to 40 nm. A STO layer is inserted before another PGO layer is grown. In addition, the STO layer separates the YBCO from the PGO in order to prevent the chemical reaction. (d) A comparison between a single PGO layer which is 200 nm thick and a STO/PGO multilayer consisting of four STO and three PGO layers (*continues overleaf*).

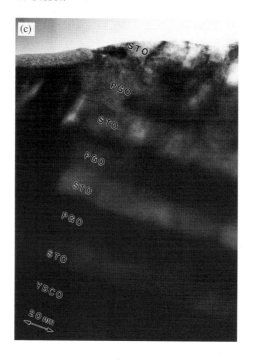

Fig. 14.19 (*cont.*).

the layers. An optimum multilayer structure thus consists of epitaxial layers with atomically flat interfaces without particles.

Most films develop surface roughness as the film thickness increases [14.75]. Once the roughness is initiated it can be pronounced by surface mobility of the film species. The redistribution of the film material can be driven by a mechanism which minimizes the surface energy and favors surfaces other than that parallel to the substrate interface, see Fig. 14.19. There are several parameters that affect the degree of roughness and examples are surface mobility, deposition temperature and angle of incidence for the impinging film species.

The formation of pinholes due to surface roughening can be avoided by optimization of the film deposition parameters. Experience shows, however, that the processing parameter window can be narrower than desired for a reproducible fabrication procedure. The use of a non-superconducting multilayer interspersing YBCO layers is a solution to the problem of irreproducibility. The use of $SrTiO_3/PrGaO_3$ (STO/PGO) intermediate multilayers is an example of a successful combination which significantly opens the processing window [14.76].

An initial 10 nm STO layer provides a continuous coverage of the bottom YBCO film and prevents interdiffusion between the PGO and the YBCO. Subsequent PGO films grow epitaxially with high quality up to a thickness of 40 nm whereafter voids form in the films. The voids deteriorate the quality of the subsequent films and the quality decreases with increasing thickness of the intermediate multilayer. Despite the presence of voids the top film of YBCO grows epitaxially and the total intermediate non-superconducting multilayer is continuous without any pinholes. The dielectric constant of the multilayer can be varied by varying the thicknesses of the individual STO and PGO layers [14.77]. The introduction of voids is also a tool to vary the physical properties of the multilayer.

The technique of interspersing a layer with thin layers that eliminate pinhole formation can be applied to other systems as well. The method is, for example, successful in $STO/PrBa_2Cu_3O_y$ multilayers [14.70]. A layer-by-layer growth mode favors high quality multilayers with large critical thicknesses. The multilayer technique also opens possibilities to continuously vary the physical properties of the intermediate layers by combining two materials with different behaviors.

14.8 SNS Josephson junctions

Multilayers are also used in the fabrication of superconducting–normal–super-

(a)

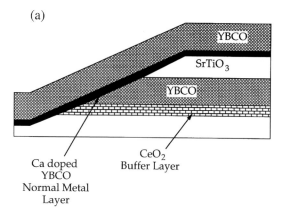

Ca doped
YBCO
Normal Metal
Layer

CeO$_2$
Buffer Layer

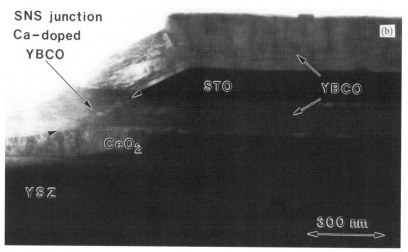

Fig. 14.20. (a) A schematic illustration of a step-edge SNS junction. The geometry allows the use of the high critical current density of YBCO and the long coherence length in the (001)$_{YBCO}$ planes; (b) Cross-section TEM micrograph of a step-edge SNS junction consisting of YBCO/Ca doped region/YBCO.

conducting (SNS) type devices . A variety of junction geometries are evaluated where the edge junctions constitute a potential geometry [14.78–14.90], see Fig. 14.20. Both noble metals and metallic oxides are candidate barrier materials, but the metallic oxides have a number of advantages in the high T_c oxide based junctions. The oxides offer the possibility of fabricating junctions with well defined interfaces while the noble metals have the disadvantage of relatively easily diffusing into the adjacent YBCO layers at their high deposition temperatures [14.91] (a property which could be utilized when making low resistance Ag contacts). It is also necessary to maintain the epitaxy through all layers due to the anisotropy of YBCO and the pronounced effect of grain boundaries on the properties of YBCO. Among the metallic oxides that have been incorporated in edge junction geometries are $CaRuO_3$ [14.82], $La_{0.5}Sr_{0.5}CoO_3$ [14.88], $Y_{0.7}Ca_{0.3}Ba_2Cu_3O_{7-x}$ [14.85], $YBa_2Cu_{2.79}Co_{0.21}O_{7-x}$ [14.83, 14.87, 14.84], $La_{1.4}Sr_{0.6}CuO_4$ [14.88], normal YBCO [14.80], $Y_{1-x}Pr_xBa_2Cu_3O_7$ [14.78, 14.81, 14.84, 14.86] and Nb doped $SrTiO_3$ [14.79].

Reproducible junction characteristics depend on the ability to control the structure and the thickness of the barrier layers. In addition, the interfaces between the YBCO and the barrier layer may have a much stronger influence on the junction properties than the barrier layer itself [14.82]. Interfacial dislocations and oxygen depletion of the YBCO adjacent to the barrier layer are structural details that need to be considered.

The morphology of the YBCO edge is instrumental in determining the structure and characteristics of the YBCO edge junctions. In general, it is therefore not possible to draw conclusions concerning the behavior of edge junctions from corresponding planar multilayers [14.82]. Both the slope of the edges and the local edge morphology affects the orientation of the YBCO as well as the nucleation of secondary phases, such as Y_2O_3. The secondary phases can have completely different electrical characteristics thus changing the behavior of the junction. It is therefore necessary to carefully control the YBCO edge morphology.

14.9 Other orientations

One reason for using the edge geometry in a SNS junction is that it allows the use of the direction of high current density in the $(001)_{YBCO}$ planes for c-axis-oriented films. Another solution is to make planar junctions with a-axis-oriented films [14.92–14.94], see Fig. 14.21. The a-axis films can often be obtained by lowering the deposition temperature. However, the resulting films often suffer from inferior superconducting properties compared with the c-axis-oriented films. In addition, the films exhibit domain structures due to growth twins [14.95, 14.96], see Fig. 14.21. The lattice match between the substrate and the film on cubic (001) substrates is identical for two perpendicular orientation relationships. The domain size is determined by the density of nuclei and film growth rate. The c-axis is in the plane of the film and rotates $90°$ when passing a domain boundary.

The domain effect can be avoided by using substrates devoid of the four-fold symmetry in the plane of the substrate interface. $NdGaO_3$ is a substrate with an orthorhombic structure ($a = 0.5431$ nm, $b = 0.5499$ nm and $c = 0.7710$ nm) on which high quality c-axis films can be grown. There are indeed reports of a-

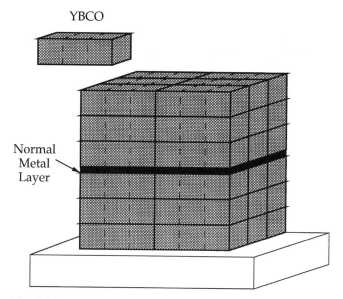

YBCO

Normal
Metal
Layer

Fig. 14.21. A schematic showing the geometry of a planar SNS junction. *a*-Axis-oriented films most often exhibit domain structures where the $[001]_{YBCO}$ rotates 90° when passing a domain boundary, see Fig. 14.4.

axis oriented YBCO films devoid of growth twins on NdGaO$_3$ substrates [14.97]. The *a*-axis orientation is obtained by lowering the deposition temperature.

The use of template layers is an alternative method for obtaining *a*-axis-oriented YBCO films [14.92, 14.98]. The principle is to deposit a layer on which the YBCO will grow epitaxially with the *a*-axis orientation at the optimum deposition temperature. PrBa$_2$Cu$_3$O$_z$ (PBCO) is a suitable crystal structure since it is similar to that of YBCO but with slightly larger lattice parameters. The larger lattice parameters further promote the *a*-axis orientation of the PBCO at lower deposition temperatures. Once the *a*-axis orientation is established in the PBCO, the YBCO layer inherits the *a*-axis orientation. However, a domain structure due to the growth twins is still present on substrate surfaces with four-fold symmetry.

The four-fold symmetry of (001) cubic substrates can be avoided by using (110) substrates instead [14.17, 14.98–14.101]. One of the ⟨100⟩ axes in the (001) substrate surface is then replaced by a ⟨110⟩ axis, see Fig. 14.9(a). The intention is that the *c*-axis of the YBCO should align with the remaining ⟨100⟩ substrate axis. However, a mixture of {110} and {103} is usually obtained, see Fig. 14.22. This effect in combination with the lower deposition temperature

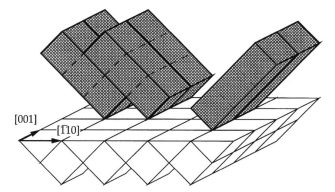

Fig. 14.22. Schematic illustration of {103}YBCO on (110)SrTiO$_3$ substrates where the four-fold symmetry of a (001)SrTiO$_3$ substrate interface is avoided. In general, the (110)YBCO orientation, see Fig. 14.9(a), is the most desired type of film since {103} YBCO exhibits domain structures.

rendering films with inferior superconducting characteristics, makes it favorable to use PBCO template layers for the {110} films as well. It should be noted that the thermal expansion coefficient along the c-axis is relatively large in comparison with the most common substrates used for YBCO films, see Fig. 14.10. The large mismatch in thermal expansion coefficient limits the film thickness of (110) oriented YBCO films [14.17]. Cracks form during cooling from the deposition temperature for a film thickness greater than a critical value, as described in Section 14.4. The phenomenon is similar to the case of c-axis-oriented films on Si substrates [14.16, 14.19].

14.10 Other layered high T_c superconductors

The information about the c-axis-oriented YBCO films presented above can be used to deduce the likely behavior of c-axis oriented thin films of other Cu—O based high T_c superconductors. The epitaxial orientation relationships will mainly be determined by lattice match between the oxygen sublattice of the films and the substrates [14.102]. Observations in, for example, epitaxial Bi$_2$Sr$_2$CaCu$_2$O$_8$ [14.52], Tl$_2$Ba$_2$CaCu$_2$O$_8$ [14.103, 14.104] and HgBa$_2$CaCu$_2$O$_{6+\delta}$ [14.105, 14.106] confirm that this is true. There is an important difference when comparing YBCO and Bi-based compounds to Tl- and Hg-based compounds. The former superconductors can be made *in situ*, while the latter require more complicated fabrication procedures with post-deposition annealing treatments in high partial pressures of Tl and Hg respectively.

The same methods to introduce grain boundaries, i.e. bi-crystal substrates, step-edges and bi-epitaxial techniques, also apply to the other superconductors [14.52, 14.107–14.109]. The details of the interfacial interactions between the substrates and the thin films need to be investigated in order to find the optimum conditions and geometries for each individual system. The chemical interactions must also be considered when integrating the superconductors into multilayer structures. Intermediate non-superconducting layers that are inert to the YBCO may react chemically with the other superconductors. It is also necessary to carefully consider the *ex situ* processes of the Tl- and Hg-based superconductors producing multilayer structures.

14.11 Conclusions

Transmission electron microscopy, atomic force microscopy and scanning tunneling microscopy constitute powerful techniques for the characterization of microstructural evolution of thin films and the effect of surface perturbations on film growth. The physical behavior of the high T_c thin films is governed by their microstructure. The influence of grain boundaries, defects and doping elements demands a stringent control of the thin-film microstructure. The knowledge of how individual defects affect the characteristics can in turn be used to control the local behavior of the superconductor by introducing specific defects in predetermined positions.

As the dimensions of devices decrease, the importance of individual interfaces and defects increases. The microstructure must simultaneously be controlled down to an atomic scale and the growth of epitaxial single crystal films on patterned and perturbed surfaces must be mastered. It is thus important to study the growth mechanisms and the interactions between the substrate and the film in order to manipulate the fine scale microstructure. Important structural characteristics are established already in the initial stages of the thin-film growth, and the interactions between the film and the substrate and also between the layers in a multilayer structure play a vital role for the thin-film evolution. The examples given in this chapter provide information of what aspects to consider when developing a new device based on high T_c epitaxial thin films. The data presented here are intended to rationalize this work.

Acknowledgments

Stimulating interactions, collaborations and discussions with colleagues at Chalmers University of Technology/Göteborg University, Conductus Inc. in Sunnyvale, IBM T.J. Watson Research Center in Yorktown Heights, the

Institute of Crystallography in Moscow, the Ioffe Physical-Technical Institute in St. Petersburg and Stanford University have been a constant source of inspiration. Financial support from the Swedish Materials Consortium Superconducting Materials and TFR (Swedish Research Council for Engineering Sciences) is gratefully acknowledged. The crystal structures were drawn using the program Ideal Microscope 1.0.

References

[14.1] P. Chaudhari *et al.*, *Phys. Rev. Lett.* **60**, 1653 (1988).

[14.2] D. Dimos *et al.*, *Phys. Rev. Lett.* **61**, 219 (1988).

[14.3] J. Mannhart *et al.*, *Phys. Rev. Lett.* **61**, 2476 (1988).

[14.4] R. Gross *et al.*, *Phys. Rev. B* **42** 10735 (1990).

[14.5] S. E. Russek *et al.*, *Appl. Phys. Lett.* **57** 1155 (1990).

[14.6] D. K. Lathrop *et al.*, *Appl. Phys. Lett.* **58** , 1095 (1991).

[14.7] Z. G. Ivanov *et al.*, *Appl. Phys. Lett.* **59**, 3030 (1991).

[14.8] D. A. Smith, M. F. Chisholm & J. Clabes, *Appl. Phys. Lett.* **53**, 2344 (1988).

[14.9] Yu. Boikov *et al.*, *J. Appl. Phys.* **72**, 199 (1992).

[14.10] C. G. Tretiatchenko, *Physica C* **198**, 7 (1992).

[14.11] E. Olsson & S. L. Shinde, Chapter 4 in *Interfaces in superconducting systems*, S. L. Shinde & D. Rudman (Springer Verlag, New York 1992).

[14.12] J. MacManus-Driscoll, T. H. Geballe & J. C. Bravman, *J. Appl. Phys.* **75**, 412 (1994).

[14.13] M. G. Norton, S. R. Sommerfelt & C. B. Carter, *Appl. Phys. Lett.* **56**, 2246 (1990).

[14.14] G. Brorsson *et al.*, *J. Appl. Phys.* **75**, 7958 (1994).

[14.15] A. Mogro-Campero, *Supercond. Sci. Technol.* **3**, 155 (1990).

[14.16] D. K. Fork *et al.*, *Appl. Phys. Lett.* **57**, 1161 (1990).

[14.17] E. Olsson *et al.*, *Appl. Phys. Lett.* **58**, 1682 (1991).

[14.18] D. B. Fenner *et al.*, *J. Appl. Phys.* **69**, 2176 (1991).

[14.19] A. L. Vasiliev *et al.*, *Physica C* **244** 373.

[14.20] A. L. Vasiliev *et al.*, *Supercond. Sci. Technol.* **3**, 155 (1990).

[14.21] A. L. Vasiliev *et al.*, *Physica C.* **253**, 297 (1995).

[14.22] A. Bardal *et al.*, *J. Mater. Res.* **8**, 2112 (1993).

[14.23] A. Bardal, Th. Matthée, J. Wecker & K. Samwer, *J. Appl. Phys.* **75**, 2902 (1994).

[14.24] D. K. Fork *et al.*, *Appl. Phys. Lett.* **58**, 2432 (1991).

[14.25] C. A. Copetti *et al.*, *Appl. Phys. Lett.* **63**, 1429 (1993).

[14.26] Yu. A. Boikov *et al.*, *Supercond. Sci. Technol.* **9**, A178 (1996).

[14.27] J. A. Alarco *et al.*, *Appl. Phys. Lett.* **61**, 723 (1992).

[14.28] J. A. Alarco *et al.*, *J. Appl. Phys.* **75**, 3202 (1994).

[14.29] Q. X. Jia, S. Y. Lee, W. A. Anderson & D. T. Shaw, *Appl. Phys. Lett.* **59**, 1120 (1991).

[14.30] D. K. Fork, K. Nashimoto & T. H. Geballe, *Appl. Phys. Lett.* **60**, 1621 (1992).

[14.31] L. D. Chang, M. Z. Tseng, E. L. Hu & D. K. Fork, *Appl. Phys. Lett.* **60**, 1753 (1992).

[14.32] J. H. van der Merwe, *J. Appl. Phys.* **34**, 123 (1963).

[14.33] J. W. Matthews & A. E. Blakeslee, *J. Cryst. Growth* **29**, 118 (1974); J. W. Matthews & A. E. Blakeslee, *J. Cryst. Growth* **27**, 273 (1975).

[14.34] W. D. Nix, *Metall. Trans.* A **20A**, 2217 (1989).

[14.35] A. Gupta *et al.*, *Phys. Rev. Lett.* **64**, 3191 (1990).

[14.36] S. K. Streiffer *et al.*, *Phys. Rev.* B **43**, 13007 (1991).

[14.37] A. Catana *et al.*, *Appl. Phys. Lett.* **60**, 1016 (1992).

[14.38] L. B. Freund, *J. Appl. Mech.* **54**, 553 (1987).

[14.39] H. Ledbetter & M. Lei, *J. Mater. Res.* **6**, 2253 (1991).

[14.40] D. J. Eaglesham & M. Cerullo, *Phys. Rev. Lett.* **64**, 1943 (1990).

[14.41] D. E. Jesson, S. J. Pennycook, J. Z. Tischler & J. D. Budai, *Phys. Rev. Lett.* **70**, 2293 (1993).

[14.42] M. D. Thouless, E. Olsson & A. Gupta, *Acta Metall. Mater.* **40**, 1287 (1992).

[14.43] R. F. Cook, T. R. Dinger & D. R. Clarke, *Appl. Phys. Lett.* **51**, 454 (1987).

[14.44] M. F. Chisholm & S. J. Pennycook, *Nature* **351**, 47 (1991); N. D. Browning *et al.*, *Physica C* **212**, 185 (1993).

[14.45] J. A. Alarco & E. Olsson, *Phys. Rev.* B **52**, 13625 (1995).

[14.46] Y. Zhu, Y. L. Corcoran & M. Suenaga, *Interface Science* **1**, 361 (1993).

[14.47] A. H. King & A. Singh, *J. Appl. Phys.* **74**, 4627 (1993).

[14.48] S. E. Babcock, X. Y. Cai, D. L. Kaiser & D. C. Larbalestier, *Nature* **347**, 167 (1990).

[14.49] D. J. Lew *et al.*, *Appl. Phys. Lett.* **65**, 1584 (1994).

[14.50] J. A. Alarco *et al.*, *Ultramicroscopy* **51**, 239 (1993).

[14.51] P. G. Quincey, *Appl. Phys. Lett.* **64**, 517 (1994).

[14.52] B. Kabius *et al.*, *Physica C* **231** 123 (1994).

[14.53] C. Træholt *et al.*, Physica C **230**, 425 (1994).

[14.54] J. W. Seo *et al.*, *Physica C* **245**, 25 (1995).

[14.55] K. Lee & I. Iguschi, *Appl. Phys. Lett.* **66**, 769 (1995).

[14.56] D. J. Miller *et al.*, *Appl. Phys. Lett.* **66**, 2561 (1995).

[14.57] K. Char *et al.*, *Appl. Phys. Lett.* **59**, 733 (1991).

[14.58] K. Char, M. S. Colclough, L. P. Lee & G. Zaharchuk, *Appl. Phys. Lett.* **59**, 2177 (1991).

[14.59] S. J. Rosner, K. Char & G. Zaharchuk, *Appl. Phys. Lett.* **60**, 1010 (1992).

[14.60] J. A. Dewards *et al.*, *Appl. Phys. Lett.* **60**, 2433 (1992).

[14.61] C. L. Jia *et al.*, *Physica C* **175**, 545 (1991).

[14.62] C. L. Jia *et al.*, *Physica C* **196**, 211 (1992).

[14.63] J. Ramos, Z. G. Ivanov, E. Olsson, & T. Claeson, *Superconductivity (BHTSC'92)*, May 25–29, 1992 Beijing, China.

[14.64] H. R. Yi *et al.*, *Appl. Phys. Lett.* **65** 1177 (1994); M. Gustafsson, E. Olsson, H. R. Yi, D. Winkler & T. Claeson *MRS 1995 Fall Meeting*, Boston.

[14.65] B. Vuhic *et al.*, *J. Appl. Phys.* **77**, 2591 (1995).

[14.66] N. G. Chew *et al.*, *Appl. Phys. Lett.* **60**, 1516 (1992).

[14.67] J. Ramos *et al.*, *Appl. Phys. Lett.* **63**, 2141 (1993).

[14.68] J. A. Alarco *et al.*, *Physica C* **247**, 263 (1995).

[14.69] S. Gevorgian, E. Carlsoon & E. Olsson, *Appl. Phys. Letts.* **67**, 1615 (1995).

[14.70] M. Gustafsson *et al.*, *Proceedings, SCANDEM Forty-seventh annual meeting*, eds T. Beisvåg, T.-H. Iversen & J. K. Solberg, p. 72 (1995).

[14.71] S. K. Tolpygo *et al.*, *Physica C* **209**, 211 (1993).

[14.72] S. J. Rothman, J. L. Routbort, U. Welp & J. E. Baker, *Phys. Rev.* B **44**, 2326 (1991).

[14.73] E. Olsson & K. Char, *Interface Science* **1**, 371 (1993); E. Olsson & K. Char, *Proceedings, SCANDEM Forty-seventh annual meeting*, eds T. Beisvåg, T.-H. Iversen & J. K. Solberg, p. 113 (1995).

[14.74] J. Mannhart *et al.*, *Science* **245**, 839 (1989); R. Gerdeman *et al.*, *J. Appl. Phys.* **76**, 8005 (1994).
[14.75] G. S. Bales *et al.*, *Science* **249**, 264 (1990).
[14.76] E. Olsson, G. Brorsson, P. Å. Nilsson & T. Claeson, *Appl. Phys. Lett.* **63**, 1567 (1993).
[14.77] G. Brorsson, E. Olsson, P. Å. Nilsson & T. Claeson, *J. Appl. Phys.* **75**, 827 (1994).
[14.78] J. Gao, W. A. M. Arnink, G. J. Gerritsma & H. Rogalla, *Physica C* **171**, 126 (1990).
[14.79] D. K. Chin & T. Van Duzer, *Appl. Phys. Lett.* **58**, 753 (1991).
[14.80] B. D. Hunt, M. C. Foote & L. J. Bajuk, *Appl. Phys. Lett.* **59**, 982 (1991).
[14.81] E. Polturak *et al.*, *Phys. Rev. Lett.* **67**, 3038 (1991).
[14.82] E. Olsson & K. Char, *Appl. Phys. Lett.* **64**, 1292 (1994).
[14.83] K. Char, L. Antognazza & T. H. Geballe, *Appl. Phys. Lett.* **65**, 904 (1994).
[14.84] B. D. Hunt *et al.*, *Physica C* **230**, 141 (1994).
[14.85] Q. X. Jia *et al.*, *Physica C* **228**, 160 (1994).
[14.86] B. Ghyselen *et al.*, *Physica C* **230**, 327 (1994).
[14.87] G. Koren & E. Polturak, *Physica C* **230**, 340 (1994).
[14.88] K. Char, L. Antognazza & T. H. Geballe, *Appl. Phys. Lett.* **63**, 2420 (1993).
[14.89] R. H. Ono *et al.*, *Appl. Phys. Lett.* **59**, 1126 (1991).
[14.90] M. S. Dilorio *et al.*, *Appl. Phys. Lett.* **58**, 2552 (1991).
[14.91] Z. H. Gong *et al.*, *Appl. Phys. Lett.* **63**, 836 (1993).
[14.92] A. Inam *et al.*, *Appl. Phys. Lett.* **57**, 2484 (1990).
[14.93] T. Hashimoto *et al.*, *Appl. Phys. Lett.* **60**, 1756 (1992).
[14.94] I. Takeuchi *et al.*, *Appl. Phys. Lett.* **66**, 1824 (1995).
[14.95] C. B. Eom *et al.*, *Science* **249**, 1549 (1990).
[14.96] J. F. Hamet *et al.*, *Physica C* **198**, 293 (1992).
[14.97] K. H. Young & J. Z. Sun, *Appl. Phys. Lett.* **59**, 2448 (1991).
[14.98] R. Ramesh, A. Inam, D. L. Hart & C. T. Rogers, *Physica C* **170**, 325 (1990).
[14.99] Y. Enomoto, T. Murakami, M. Suzuki & K. Moriwaki, *Jpn. J. Appl. Phys.* **26**, L1248 (1987).
[14.100] M. Matsumoto, H. Akoh & S. Takada, *J. Appl. Phys.* **66**, 3907 (1989).
[14.101] G. Linker *et al.*, *Sol. State Comm.* **69**, 249 (1989).
[14.102] R. Guo, A. S. Bhalla, L. E. Cross & R. Roy, *J. Mater. Res.* **9**, 1644 (1994).
[14.103] L. G. Johansson *et al.*, *Proceedings from MOS Eugene*, July 27–31, 1993.
[14.104] A. P. Bramley, S. M. Morley, C. R. M. Grosvenor & B. Pecz, *Appl. Phys. Lett.* **66**, 517 (1995).
[14.105] C. C. Tsuei, A. Gupta, G. Trafas & D. Mitzi, *Science* **263**, 1259 (1994).
[14.106] L. Kruzin-Elbaum, C. C. Tsuei & A. Gupta, *Nature* **373**, 679 (1995).
[14.107] E. Sarnelli, P. Chaudhari, W. Y. Lee & E. Esposito, *Appl. Phys. Lett.* **65**, 362 (1994).
[14.108] T. Nabatame *et al.*, *Appl. Phys. Lett.* **65**, 776 (1994).
[14.109] Y. F. Chen *et al.*, *Inst. Phys. Conf. Ser.* **148**, 1335 (1995).